Studies on the Taxonomy of Crustaceans

Crustaceana Monographs constitutes a series of books on carcinology in its widest sense. Contributions are handled by the Series Editor and may be submitted through the office of Koninklijke Brill NV, P.O. Box 9000, 2300 PA Leiden, The Netherlands.

Published in this series:

CRM 001 - *Stephan G. Bullard*	Larvae of anomuran and brachyuran crabs of North Carolina
CRM 002 - *Spyros Sfenthourakis et al. (eds.)*	The biology of terrestrial isopods, V
CRM 003 - *Tomislav Karanovic*	Subterranean Copepoda from arid Western Australia
CRM 004 - *Katsushi Sakai*	Callianassoidea of the world (Decapoda, Thalassinidea)
CRM 005 - *Kim Larsen*	Deep-sea Tanaidacea from the Gulf of Mexico
CRM 006 - *Katsushi Sakai*	Upogebiidae of the world (Decapoda, Thalassinidea)
CRM 007 - *Ivana Karanovic*	Candoninae (Ostracoda) from the Pilbara region in Western Australia
CRM 008 - *Frank D. Ferrari & Hans-Uwe Dahms*	Post-embryonic development of the Copepoda
CRM 009 - *Tomislav Karanovic*	Marine interstitial Poecilostomatoida and Cyclopoida (Copepoda) of Australia
CRM 010 - *Carrie E. Schweitzer et al.*	Systematic list of fossil decapod crustacean species
CRM 011 - *Peter Castro et al. (eds.)*	Studies on Brachyura: a homage to Danièle Guinot
CRM 012 - *Patricio R. De los Ríos-Escalante*	Crustacean zooplankton communities in Chilean inland waters
CRM 013 - *Katsushi Sakai*	Axioidea of the world and a reconsideration of the Callianassoidea (Decapoda, Thalassinidea, Callianassida)
CRM 014 - *Charles H.J.M. Fransen et al. (eds.)*	Studies on Malacostraca: Lipke Bijdeley Holthuis Memorial Volume
CRM 015 - *Akira Asakura et al. (eds.)*	New Frontiers in Crustacean Biology: Proceedings of the TCS Summer Meeting, Tokyo, 20-24 September 2009
CRM 016 - *Danielle Defaye et al. (eds.)*	Studies on Freshwater Copepoda: a Volume in Honour of Bernard Dussart
CRM 017 - *Hironori Komatsu et al. (eds.)*	Studies on Eumalacostraca: a homage to Masatsune Takeda
CRM 018 - *Masahiro Dojiri & Ju-Shey Ho*	Systematics of the Caligidae, copepods parasitic on fishes
CRM 019 - *Darren Yeo et al. (eds.)*	Advances in Freshwater Decapod systematics and biology
CRM 020 - *Finn Viehberg et al. (eds.)*	The Recent and Fossil meet Kempf Database Ostracoda Festschrift Eugen Karl Kempf — Proceedings of the 15th International German Ostracodologists' Meeting
CRM 021 - *L.A.M. Neethling & A. Avenant-Oldewage*	Branchiura — A compendium of the geographical distribution and a summary of their biology
CRM 022 - *Célio Magalhães et al. (eds.)*	Studies on Decapoda and Copepoda in Memory of Michael Türkay
CRM 023 - *Natalia Sukhikh et al. (eds.)*	Studies on *Eurytemora*: Proceedings of the *Eurytemora* Conference, St. Petersburg, 2019

Studies on the Taxonomy of Crustaceans

Prof. Ruiyu Liu Memorial Volume

By
Xinzheng Li, Wenliang Liu and Wei Jiang
(Editors)

CRUSTACEANA MONOGRAPHS, 24

BRILL

LEIDEN | BOSTON

This book was originally published as *Crustaceana* 93, 11-12 (2020), 1233-1546.

Cover: Professor Ruiyu Liu. Photo provided by Xiaoqing Ji.

Editors' addresses:
Xinzheng Li, Institute of Oceanology, Chinese Academy of Sciences, Qingdao 266071, P.R. China; e-mail: lixzh@qdio.ac.cn
Wenliang Liu, Yangtze River Delta Estuary Wetland Station, Shanghai Key Lab for Urban Ecological Processes and Eco-Restoration, School of Ecological and Environmental Science, East China Normal University, Shanghai, 200241, P.R. China; e-mail: wlliu@sklec.ecnu.edu.cn
Wei Jiang, Institute of Oceanology, Chinese Academy of Sciences, Qingdao 266071, P.R. China; e-mail: jiangwei@qdio.ac.cn

Library of Congress Cataloging-in-Publication Data

Names: Li, Xinzheng, editor.
Title: Studies on the taxonomy of crustaceans : prof. Ruiyu Liu memorial volume / [edited] by Xinzheng Li, Wenliang Liu, and Wei Jiang.
Description: Leiden ; Boston, MA : BRILL, [2021] | Series: Crustaceana monographs, 1570-7024 ; 24 | This book was originally published as Crustaceana 93, 11-12 (2020), 1233-1546. | Includes bibliographical references and index.
Identifiers: LCCN 2021022759 | ISBN 9789004464339 (hardback) | ISBN 9789004464353 (ebook)
Subjects: LCSH: Crustacea. | Crustacea–Classification.
Classification: LCC QL435 .S894 2021 | DDC 595.3–dc23
LC record available at https://lccn.loc.gov/2021022759

ISBN 978 90 04 46433 9 (hardback)
ISBN 978 90 04 46435 3 (e-book)

This book is printed on acid-free paper and produced in a sustainable manner.

CONTENTS

First published as a Special Issue of *Crustaceana*. The bracketed page numbers in this book refer to the page numbers in *Crustaceana* 93, 11-12 (2020). When citing from this book, refer to these bracketed page numbers.

NOTE ON ORIGINAL PUBLICATION

The chapters in this book were originally published in the journal *Crustaceana*, Vol. 93, Issues 11-12 (2020), pp. 1233-1546.

Xinzheng Li, Wenliang Liu and Wei Jiang, "Preface," *Crustaceana* 93(11-12): 1233-1234. Online Publication Date: 25 Nov. 2020. DOI: https://doi.org/10.1163/15685403-00004060.

Wenliang Liu, Wei Jiang and Xinzheng Li, "In memoriam Professor Ruiyu Liu (J. Y. Liu) (4 November 1922-16 July 2012)," *Crustaceana* 93(11-12): 1235-1265. Online Publication Date: 25 Nov. 2020. DOI: https://doi.org/10.1163/15685403-00004061.

Yi-Jia Shih, Ping-Ho Ho and Bella S. Galil, "Contributions to the knowledge of Leucosiidae VII. *Liusius* gen. nov. (Crustacea, Brachyura)," *Crustaceana* 93(11-12): 1269-1276. Online Publication Date: 25 Nov. 2020. DOI: https://doi.org/10.1163/15685403-bja10082.

Yifan Wang, Shuxin Xu, Chunchao Zhu, Xinnan Jia, Xianmin Zhou and Jiexin Zou, "The complete mitochondrial genome of the freshwater crab *Longpotamon kenliense* (Decapoda, Brachyura, Potamidae) with phylogenetic consideration," *Crustaceana* 93(11-12): 1277-1293. Online Publication Date: 25 Nov. 2020. DOI: https://doi.org/10.1163/15685403-bja10085.

Siying Mao and Chao Huang, "Descriptions of a new species of *Minpotamon* Dai & Türkay, 1997, and a monotypic new genus of aquatic freshwater crab (Brachyura, Potamidae) from eastern Guangdong, China," *Crustaceana* 93(11-12): 1295-1313. Online Publication Date: 25 Nov. 2020. DOI: https://doi.org/10.1163/15685403-bja10060.

Pengfei Wang, Zewei Zhang and Hongying Sun, "*Ruiyupotamon*, a new genus and four new species of freshwater crabs from northwestern Yunnan, China (Decapoda,

Brachyura, Potamidae)," *Crustaceana* 93(11-12): 1315-1341. Online Publication Date: 25 Nov. 2020. DOI: https://doi.org/10.1163/15685403-bja10059.

Bee Yan Lee and Peter K. L. Ng, "The identity of *Hyastenus pleione* (Herbst, 1803) and description of a new species from China (Decapoda, Brachyura, Majoidea, Epialtidae)," *Crustaceana* 93(11-12): 1343-1360. Online Publication Date: 25 Nov. 2020. DOI: https://doi.org/10.1163/15685403-bja10064.

Chao Huang, Hsi-Te Shih and Shane T. Ahyong, "The freshwater crab genus *Lacunipotamon* Dai, Song, He, Cao, Xu & Zhong, 1975 (Decapoda, Brachyura, Potamidae), with descriptions of two new species from southwestern China," *Crustaceana* 93(11-12): 1361-1379. Online Publication Date: 25 Nov. 2020. DOI: https://doi.org/10.1163/15685403-bja10067.

Tin-Yam Chan, "*Sicyonella liui* sp. nov., a new sergestid shrimp (Decapoda, Dendrobranchiata) discovered from Madagascar," *Crustaceana* 93(11-12): 1383-1390. Online Publication Date: 25 Nov. 2020. DOI: https://doi.org/10.1163/15685403-bja10008.

Yanrong Wang, Chaodong Zhu and Zhongli Sha, "Description of a new species of the genus *Lebbeus* White, 1847 (Decapoda, Thoridae) from the Bohai Sea, China," *Crustaceana* 93(11-12): 1393-1403. Online Publication Date: 25 Nov. 2020. DOI: https://doi.org/10.1163/15685403-00003938.

Yixiong Cai, "Species of the *Caridina nilotica* group in China, with description of one new species (Crustacea, Decapoda, Atyidae)," *Crustaceana* 93(11-12): 1405-1422. Online Publication Date: 25 Nov. 2020. DOI: https://doi.org/10.1163/15685403-bja10086.

Charles H. J. M. Fransen, "New records of the sponge-symbiotic shrimp genus *Anchistioides* from the Indo-West Pacific (Decapoda, Caridea)," *Crustaceana* 93(11-12): 1423-1447. Online Publication Date: 25 Nov. 2020. DOI: https://doi.org/10.1163/15685403-00003964.

Wenliang Liu and Gary C. B. Poore, "A new species of the genus *Eucalliaxiopsis* Sakai, 2011 from the South China Sea (Decapoda, Axiidea, Eucalliacidae)," *Crustaceana* 93(11-12): 1451-1458. Online Publication Date: 25 Nov. 2020. DOI: https://doi.org/10.1163/15685403-bja10029.

Chien-Hui Yang and Tin-Yam Chan, "A new slipper lobster of the genus *Galearctus* Holthuis, 2002 (Decapoda, Scyllaridae) from Madagascar," *Crustaceana*

93(11-12): 1461-1470. Online Publication Date: 25 Nov. 2020. DOI: https://doi.org/10.1163/15685403-bja10010.

Shane T. Ahyong and Chia-Wei Lin, "The mantis shrimp superfamily Eurysquilloidea confirmed from Taiwan: *Liusquilla formosa* gen. et sp. nov.," *Crustaceana* 93(11-12): 1473-1482. Online Publication Date: 25 Nov. 2020. DOI: https://doi.org/10.1163/15685403-00003941.

Yami Zheng, Zhonge Hou and Shuqiang Li, "*Gammarus liuruiyui* sp. nov. (Amphipoda, Gammaridae), a new blind species from Tian Shan, China," *Crustaceana* 93(11-12): 1485-1501. Online Publication Date: 25 Nov. 2020. DOI: https://doi.org/10.1163/15685403-00003939.

Jianmei An, Qianqian Xi and Gustav Paulay, "A new species of the genus *Parioninella* (Epicaridea, Bopyridae, Pseudioninae) from Australia," *Crustaceana* 93(11-12): 1503-1511. Online Publication Date: 25 Nov. 2020. DOI: https://doi.org/10.1163/15685403-00003940.

Meng-Chen Yu, Takefumi Yorisue, Gregory A. Kolbasov and Benny K. K. Chan, "A new species of the sponge-inhabiting barnacle *Neocasta* (Archaeobalanidae, Acastinae) in Kyushu, Japan," *Crustaceana* 93(11-12): 1515-1532. Online Publication Date: 25 Nov. 2020. DOI: https://doi.org/10.1163/15685403-bja10087.

Lin Ma, Minxiao Wang and Xinzheng Li, "A new species of *Stygiopontius* (Copepoda, Siphonostomatoida, Dirivultidae) from a deep-sea cold seep in the South China Sea," *Crustaceana* 93(11-12): 1535-1546. Online Publication Date: 25 Nov. 2020. DOI: https://doi.org/10.1163/15685403-bja10023.

[When citing this volume, please refer to *Crustaceana* 93 (2020), 1233-1546]

PREFACE

On the 16th of July 2012, the eminent Chinese carcinologist Prof. Ruiyu Liu (known as J. Y. Liu or Jui-yu Liu to his carcinological colleagues and friends around the world) passed on. No longer would his friends, colleagues and students be able to rely on his invaluable knowledge of taxonomy, systematics, phylogeny, macrobenthos and marine culture for their work. China and the carcinological world, will miss his selfless enthusiasm and wisdom.

Prof. Liu was a polymath! As his students, we were always surprised he knew the Latin names of so many plant and microbe species! And Latin is not the easiest language for Chinese scientists. We have also been deeply touched by his meticulousness, patience and kindness to colleagues and students alike — and not just those in his disciplines. Even today, seven years after he left us, we still often talk about and recall the deeds of Prof. Liu with the greatest of respect.

Prof. Liu's research fields covered a variety of disciplines in marine biology. He was one of the founders of carcinology, marine macrobenthic ecology, and the founder of marine crustacean culture in China. Since the 1940s, he and his students have carried out taxonomic studies on many crustacean groups such as Decapoda, Cirripedia, Amphipoda, Stomatopoda and Mysida; mastering their taxonomy, species compositions in habitats, distributions and overall diversity. His first monograph, "The Economic Shrimps from Northern China", published in 1955, is a classic crustacean reference, still widely cited in China. Subsequently, a series of monographs on the crustacean fauna of China, including three volumes of "Fauna Sinica", and almost 100 papers on crustacean taxonomy, were published by him and his collaborators. These works, mostly on crustacean diversity in China, substantially expanded our knowledge of faunal compositions in Chinese waters, their zoogeographic characteristics as well as the phylogenetic relationships among many groups. In the 1960s, Prof. Liu divided the macrobenthic communities of China seas and defined several key habitats. For example, he found for the first time that the cold-water community in the deep-water area of the central Yellow Sea was protected by the cold waters of the Yellow Sea Cold Water Mass. He also highlighted that this community was part of the Fauna of the North Pacific Temperate Zone, distinct from the warm water fauna of the Indo-West Pacific as well as the East and South China Seas. In the domain of aquaculture, Prof. Liu

 | DOI:10.1163/9789004464353_002

initiated marine shrimp culture in China and was the pioneer for the mass culture of the Chinese prawn (*Penaeus chinensis*) and white prawn (*Penaeus vannamei*).

As one of the most experienced and senior marine scientists in China, Prof. Liu successively presided over many national and international comprehensive surveys, for example: "The comprehensive resource investigation of Chinese coast and shallow waters" (1958-1960), "The national comprehensive survey of resources of coastal and intertidal zones" (1981-1987), and "The Sino-German joint survey of marine biodiversity on Hainan Island" (1990-1992). He was also a leader in academic societies — he was the president or vice president of the Chinese Society for Oceanology and Limnology, the Chinese Society for Oceanography, the China Society of Fisheries, and the Chinese Crustacean Society. The "Checklist of Marine Biota of China Seas", edited by Prof. Liu has been hailed as a landmark monograph by many international colleagues. Because of his outstanding contributions to scientific research in China, Prof. Liu was elected as member of the Chinese Academy of Sciences in 1997. In 2007, The Crustacean Society presented him with the "Outstanding Research Contribution Award" in recognition of his numerous contributions to crustacean research.

A total of 42 colleagues contributed 18 manuscripts to this issue, they are not only Prof. Liu's students or his students' students, but also close colleagues and friends for many years. During his lifetime, Prof. Liu published 97 papers and monographs on crustacean taxonomy, more than 70 papers and monographs on macrobenthos, and 43 papers on aquaculture. The contributions in this issue, however, are mainly on the taxonomy of crabs, shrimps and other crustacean groups, a field for which he is best known for by the international community.

As guest editors of this issue, we would like to thank Drs. Charles H. J. M. Fransen and Peter K. L. Ng for their very kindly support of the publication and continually urging us to complete the work. We are also deeply appreciative of the chief editor of *Crustaceana*, Dr. J. C. von Vaupel Klein, for his support of this special issue. We are also mostly grateful to the anonymous reviewers and all the authors — friends of Prof. Liu — who have contributed to the success of this issue.

September 2020, *Guest Editors for the present issue*

XINZHENG LI
WENLIANG LIU
WEI JIANG

[When citing this volume, please refer to *Crustaceana* 93 (2020), 1233-1546]

Prof. Ruiyu Liu Memorial Issue

IN MEMORIAM PROFESSOR RUIYU LIU (J. Y. LIU) (4 NOVEMBER 1922-16 JULY 2012)

BY

WENLIANG LIU[1,3]), WEI JIANG[2,4]) and XINZHENG LI[2,5])

[1]) Yangtze River Delta Estuary Wetland Station, Shanghai Key Lab for Urban Ecological Processes and Eco-Restoration, School of Ecological and Environmental Science, East China Normal University, Shanghai, 200241, P.R. China

[2]) Institute of Oceanology, Chinese Academy of Sciences, Qingdao 266071, P.R. China

Fig. 1. Ruiyu Liu in the laboratory of the Institute of Oceanology, Chinese Academy of Sciences, Qingdao, 26 May 2004.

[3]) Corresponding author; e-mail: wlliu@sklec.ecnu.edu.cn

[4]) e-mail: jiangwei@qdio.ac.cn

[5]) e-mail: lixzh@qdio.ac.cn

 | DOI:10.1163/9789004464353_003

EARLY YEARS, SCHOOLING, FAMILY

Prof. Ruiyu Liu (J. Y. Liu) was born in Laoting county, a small town at the coast of Bohai Sea of Hebei Province, China on the 4th of November 1922. His father, Guanying Liu, was the general manager of Laoting Changchun Yifa Co., Ltd. After the founding of the People's Republic of China in 1949, he was the manager of Tianjin Xinguang Knitting Factory while his mother managed the family. During his childhood, a vast variety of sea life became his playmates, which might have heralded his destiny as a marine biologist. He graduated from Laoting County First Primary School in 1936, and entered the famous Baoding Yude Middle School in 1937, and was transferred to the Changli Huiwen Middle School the following year. He studied in Peiping Provincial Senior High School from 1938 to 1941. At that time, the school's teaching level was very high, especially the lectures and experiments of the late Mr. Xiwu Chen, his biology teacher. Through the microscope, he saw the splendid world of life and developed an immense interest in biology. In 1941, he was admitted to the department of Fujen University to do a major in biology. He graduated for his Bachelor of Science degree in 1945. In the same year he was introduced by the late Prof. Liangqing Li (the pioneer of algology in China) to the School of Medicine at Peking University, as an assistant professor of the Department of Pharmacology. One year later, he entered the Institute of Zoology, National Peiping Academy of Sciences, as research assistant with the help of the late Mr. Fuqing Yang (who served as Deputy Minister of Fisheries of China) and became a student of the late Prof. Jiarui Shen (C. J. Shen), the father of carcinology in China, whose subsequent three students were all taxonomists: the late Profs. Aiyun Dai (crabs), Chongzhou Zhang (myriapods) and Daxiang Song (spiders). Ruiyu then started 67 years of marine biology research until his passing away on July 16, 2012.

Behind every great man or woman there is usually a great life-partner, and Ruiyu Liu's wife Zhongpu Kang was of invaluable support to him throughout his career. Zhongpu also graduated from Fujen University, and became an educator after graduation. The couple met at the university campus, married in 1948, and Zhongpu was pleased to play the supporting family role in the relationship. Two sons were born to the happy couple, Yuankang Liu, following Ruiyu's family name, Liu, in 1950 and Youliu Kang following Zhongpu's family name, Kang, in 1952; 'Yuan' and 'You' are different words and euphemistic expressions for love in Chinese, the morals of the two sons' names symbolizes their sincere love. Ruiyu was in deep sorrow when Zhongpu passed away of asthma in spring 2008. He realized he would never watch the Beijing Opera together with her again and chat with her.

He still kept working, powerfully supported by his scientific belief and the greatest desire, educating the succeeding taxonomists of every marine taxa until the end of his life. Apart from working, he was good at brush calligraphy and painting. He liked traditional Chinese food, such as Tianjin Stuffed bun, Shredded Fish with Flower Pepper Oil (a Qingdao dish facing with the risk of lost), some sweet taste dishes (Squirrel-Shaped Mandarin Fish and Soy Sauce-seasoned Grass Carp Brisket), especially two Beijing dishes, Beijing Quanjude Roast Duck and Donglaishun Mutton Hotpot. He also liked sweets, especially ice cream and chocolate. Whenever his international counterparts visited Qingdao, he was sure to treat them very well and share his local favorites and the famous Qingdao Beer with them. Although he didn't like to drink much, he insisted on having his beer served cold. To save time, he invented a simple but creative way to eat instant noodles for lunch: he brought a dish with cooked instant noodles and heated it in the microwave. It was efficient, nutritious and delicious (fig. 2).

RESEARCH FIELD

Prof. Liu is one of the most eminent carcinologists and leading scientists in marine biology and ecology, and founder of marine benthic ecology study in China. He organized and participated in several major national and international oceanographic research projects and studies on marine biological and living resources, as well as the study of crustacean taxonomy and biology.

Prof. Liu began his carcinology research on taxonomy of stomatopods and the life history of palaemonid prawn, under the guidance of Prof. Jiarui Shen, at the Zoological Institute, National Peiping Academy, in Beijing starting from 1946. After the founding of the People's Republic of China and the establishment of the Chinese Academy of Sciences in 1949, the first marine research institute, the Qingdao Marine Biological Laboratory of the Chinese Academy of Sciences was established in Qingdao, Shandong province in 1950. Prof. Liu followed Profs. Xi Zhang (conchologist, the director of the Zoological Institute) and Zhongyan Qi (conchologist, the founder of the Chinese Society of Malacology) in moving to Qingdao in 1954, and was from then on engaged in marine biological research for fifty-eight years. The Marine Biological Laboratory was reconstructed into the Institute of Marine Biology in 1957 and was further reconstructed into the Institute of Oceanology, Chinese Academy of Sciences in 1959. Prof. Liu served successively as assistant, assistant professor (1952), associate professor (1956), professor (1980), research director of invertebrate biological laboratory (1966-1978), vice-director of the institute (1978-1984), director of the institute (1984-1987), and was elected Member of the Chinese Academy of Sciences in 1997.

Prof. Liu promoted the advancement of marine biotaxonomy study in China. Before the establishment of the People's Republic of China, only the decapod

Fig. 2. Daily life. A, Youth in 1941; B, Ruiyu and his first son in 1950s; C, Ruiyu, Zhongpu and their great-granddaughter in 2008; D, 89th birthday in 2011.

groups of crabs and shrimps were extensively studied by Jiarui Shen (crabs) and Zhaoqi Yu (S. C. Yu) (shrimps). Since the early 1950s, Prof. Liu started a comprehensive investigation on the Chinese marine crustacean fauna and resources. He published his first monograph, 'Economic Shrimps of North China' in 1955, which has been proved to be a very useful reference for students in carcinology in both China and neighboring countries. He also published another monograph, 'Penaeoid shrimps of the South China Sea' (Liu & Zhong, 1986). He and his group have been working for several decades on systematic studies of the decapod, stomatopod, mysidacean, cirripedian, and cumacean fauna. During their investigations of the crustacean fauna of the China seas, they completed five volumes in the series 'Fauna Sinica', namely Mysidacea (Liu & Wang, 2000), Thoracic Cirripedia (Liu

& Ren, 2006), Decapoda: Palaemonoidea (Liu et al., 2007), Stomatopoda (in press) and Decapoda: Dendrobranchiata (in press). They summarized the species composition, biodiversity and zoogeography characteristics of the four major groups (shrimps, barnacles, mysids and mantis shrimps) of the Chinese crustacean fauna.

With regard to reproductive biology and resource enhancement research of shrimps, Prof. Liu and his assistants devoted their research to the reproductive biology of *Penaeus chinensis* (Osbeck, 1765) ever since 1952. They first cleared its breeding habits and life history and subsequently recognized that stock enhancement by releasing artificially reared fries or spats of marine species is one of the effective measures for increasing living resources in the natural environment. Prof. Liu therefore initiated and carried out a key project 'Shrimp resource enhancement experiment' by releasing artificially reared shrimp juveniles which was carried out in Jiaozhou Bay, Yellow sea in the period of 1981–1993. Based on results of several years of multidisciplinary studies on physico-chemical and biological environmental conditions, with regard to living resources, Liu and his group suggested appropriate body size, number of juvenile shrimps and site for releasing, and predicted the possible stock size of young shrimps and production of recaptured shrimp in the autumn fishing season. The results showed that restocked shrimp population could be an effective supplement to local shrimp stock. Compared to the catches before 1987 and in 1993 when no shrimp juveniles were released in the Bay, catches in 1984–1986 and 1988–1992 increased significantly during the research. The success of his experiment provided valuable references to similar work and promoted the progress of the country's shrimp resource enhancement practice. He published papers and monographs such as, 'Ecology and living resources of Jiaozhou Bay' (Liu, 1992). He also initiated the marine shrimp culture in China and was the pioneer of Chinese prawn (*Penaeus chinensis* (Osbeck, 1765)) and white prawn (*Penaeus vannamei* Boone, 1931) culture introducing these to China.

Prof. Liu is the founder of marine benthic ecology studies in China. The first multipurpose oceanographic research vessel of China, the "Venus", was constructed in 1957 and enabled Chinese marine scientists to carry out oceanographic research on their own in the vast seas along the China coasts. He was appointed to pioneer the benthic ecology and resources study, a key field of marine biology connected to ecosystem food chain analyses and biological productivity measurements. He found and recorded for the first time cold water benthic communities dominating the Yellow Sea Cold Water Mass, and studied the structure and ecology of benthic communities in the China seas. He found an ecological barrier existing between the Yellow Sea cold water fauna (dominated by temperate faunal elements of the North Pacific region) and the East China Sea warm water fauna (dominated by the Indo-West Pacific faunal components), tentatively stretching from the Yangtze estuary northeastward to Jeju Island of Korea, and further to

Tsushima Island between Korea and Japan, while some eurythermal warm water species, usually with high abundance, expanded their distribution range northward through a shallow "corridor" to the northern Yellow Sea shallow area and the Bohai Sea. The results of many biodiversity studies over a long period of time in the China seas were consolidated in the 'The People's Republic of China National Natural Geography Atlas' edited by him and his group (Liu et al., 1965).

Prof. Liu was fully involved with marine biodiversity and conservation research ever since 2000. In order to strengthen the conservation of biodiversity and natural resources, a series of four monographs under the title 'China Red List of Endangered Species (of vertebrate groups)' was published in the 1990s. A new Red List entitled 'China Species Red List' of animal and plant species (including terrestrial, freshwater and marine mammals, amphibians, reptiles, birds, fishes, invertebrates, and higher plants) has just been published by the IUCN Biodiversity Group, with the threatened category of species assessed on the basis of the newly adopted 'IUCN Red List Endangered Categories and Criteria' (2000). He was responsible for composing the part of marine invertebrates in 'China Species Red List Volume I' (Wang & Xie, 2004) and Volume III (Wang & Xie, 2005). He assessed many marine species to be "endangered" in this Red List: the Chinese penaeid shrimp, *Penaeus chinensis*, an endemic species of large size and high economic value, and with high production of more than 40,000 tons in 1979 in China, became rare in its natural habitat in North China waters because of overfishing and deterioration of spawning and nursing ground in the Bohai and Yellow Sea was assessed endangered (EN); the horse shoe crab, *Tachypleus tridentatus* (Leach, 1819) and two other species of Xiphosura were very abundant in the South China Sea, particularly in the Beibu Gulf, but now assessed as endangered (EN) because of over exploitation; among the 256 species of scleractinian corals, 26 species were assessed to be endangered (EN), and all the others vulnerable (VU); 23 species of Mollusca were assessed vulnerable (VU), 22 critically endangered (CR), and 12 extinct (EX); 56 species of decapod Crustacea were assessed endangered (EN); 270 species of fish were assessed endangered (EN), 19 critically endangered (CR) and 4 extinct (EX), also one of the most important fishery resources, the large yellow croaker *Larimichthys crocea* (Richardson, 1846), was assessed to be endangered (EN); the most serious deterioration was among the 150 species of holothurid echinoderms of which 53 were assessed endangered (EN) due to overexploitation.

Prof. Liu joined two international cooperative projects, 'Census of Marine Life (CoML)' and the 'Species 2000 China Node' in the early 21st century. In order to provide the most recent fundamental data on Chinese marine biodiversity, he led almost 50 taxonomists to study different taxa to be published in a comprehensive book entitled 'Checklist of marine biota of China seas' (Liu, 2008). The checklist recorded 22 629 marine species within 46 phyla. The authority of carcinology, Prof. Patsy A. McLaughlin (Western Washington University) praised the book as "a landmark book".

He received a honorary Ph.D. degree from Inha University of Korea in 1995 for his contribution to Yellow Sea research in oceanography and marine biology. Prof. Liu received the first prize in the National Science and Technology Progress Award in 1992 and the Award for Research Excellence of the Crustacean Society in 2007 acknowledging his contributions to the development of marine biology and carcinology in China.

His passion for science and his responsibility for the future of marine science research supported his powerful work until his health seriously deteriorated in 2012. Still, he and his students published 11 new species and 16 papers including a famous review paper entitled 'Status of Marine Biodiversity of the China Seas' (Liu, 2012) between 2012 and 2014.

TEACHING

Prof. Liu was a knowledgeable and strict teacher. He was still active in education even in the last moments of his life. He attended his three student's PhD dissertation defenses and helped them to get their doctoral degrees, even though facing serious illness in May 2012. One month later he was hospitalized, and after about another month he was gone forever. During his short stay in the hospital, he donated a million yuan Renminbi to establish a scholarship to reward excellent students studying marine science.

In total, Prof. Liu supervised four master theses, 18 doctoral theses, and 6 postdoctoral work reports. Under his supervision, seven taxonomists graduated with doctoral degrees: Wei Jiang (Brachyura), Wenliang Liu (Astacidea, Axiidea, Gebiidea and Isopoda), Ping Ning (fishes), Junlong Zhang (Mollusca), Ning Xiao (Echinodermata), Xuwen Wu (Polychaeta), and Yang Li (Cnidaria). Many successful Chinese scientific careers were born under Prof. Liu's inspiration and supervision, including Profs. Jianhai Xiang, Linsheng Song, and Xinzheng Li.

He was always genuine about passing on his knowledge, and inspired coming generations to conduct scientific research. He was a naturally inquisitive and very determined man, he had an exceptionally broad biological knowledge, from taxonomy and systematics to functional morphology, and from evolutionary biology to ecology, zoogeography and aquaculture. Students would secretly refer to him as a walking encyclopedia, and had the deepest respect for him (fig. 3).

It was also a great concern to Prof. Liu that the values of a traditional zoological education were slowly being eroded and degraded by new genetic approaches

Fig. 3. Ruiyu Liu and the students. A, On board of the R.V. 'Xiangyanghong VIII', talking with pupils, 1st June 1986; B, examining specimens with Prof. Xianqiu Ren, July 1990; C, Jiang Wei's PhD defense, 27th May 2009; D, Wenliang Liu's PhD defense, 24th May 2010; E, Junlong Zhang, Ning Xiao and Ping Ning's PhD defenses, 20th May 2012.

that appeared to offer quick and easy answers, whilst ignoring whole organisms and their ecological role. He disagreed with the idea that biology students could become focused on laboratory and computer processes only. He demanded from students that they would explain the real nature instead of becoming a "paper maker" or even a "data game player". He respected every student's scientific interests and treated everyone as future scientist. He once asked his student, "Are you interested in crustacean or all invertebrate taxonomy? And if the latter, then I'll teach you to be an invertebrate zoologist." He was always glad to chat with any student who was devoted to taxonomy, tell them funny stories about natural history, his own life experiences, delicious food, and share his favorite Godiva chocolates with them. He was always keeping a natural curiosity in his students. One day a student showed him a rare alpheid shrimp specimen collected during a Yellow Sea survey in 2009, he immediately began to check it and illustrated the specimen in an hour. He was so happy at that time, just like a child getting his favorite toy. His passion for nature infected his students forever.

In addition to his professional duties, he was tireless in trying to spread knowledge of marine biology to the general public. He guided his students to produce numbers of guidebooks, general scientific and pictorial books, newspaper columns, popular magazines, as well as help out various television programmes and so on.

FIELD EXPEDITIONS AND INTERNATIONAL COMMUNICATION

Prof. Liu loved field work, he was interested in virtually all habitats, and wherever he went he would collect. His field work group was always jokingly called "fishing village", and he naturally would be the village chief.

He led or joined hundreds of field expeditions and research cruises, some comprehensive and important surveys are: Sino-Soviet Marine Biological Survey to Yellow Sea and Hainan Island (1957–1960); National Comprehensive Oceanographic Survey carried out in the continental shelf of the China Sea (1958–1960); Sino-Vietnamese Comprehensive Oceanographic Survey to Beibu Gulf, North South China Sea (1959–1960, 1962); Chinese Academy of Sciences Xisha Islands Marine Biological Investigation (1975–1986); National Coastal Zone Resources Comprehensive Survey (1981–1987); Shandong Province Coastal Zone Resources Comprehensive Survey (1981–1988); Sino-Korean Joint Research on Yellow Sea Oceanography (1991–1999).

The research vessel "Kexue III" was put into service in 2007 when Prof. Liu was 85 years old. He arranged the Yellow Sea Cold Water Mass Survey that year, but he could not participate the whole survey because of his poor health. He guided his "fishing villagers" to complete the first station and was very happy to be on board of the research vessel for just one day. When the exploration team returned he wanted to know everything about the catch and treated the team with a delicious dinner (fig. 4).

International communication and cooperation were very important for him. He attended the plenary session of the Intergovernmental Oceanographic Commission, UNESCO in 1982 and 1985, and was engaged in the China-United States cooperation in Marine and Fishery Science and Technology project from 1983 to 1985. He organized the "Blue Agriculture Topic" during the International Conference on Agricultural Science and Technology (Chinese Ministry of Science and

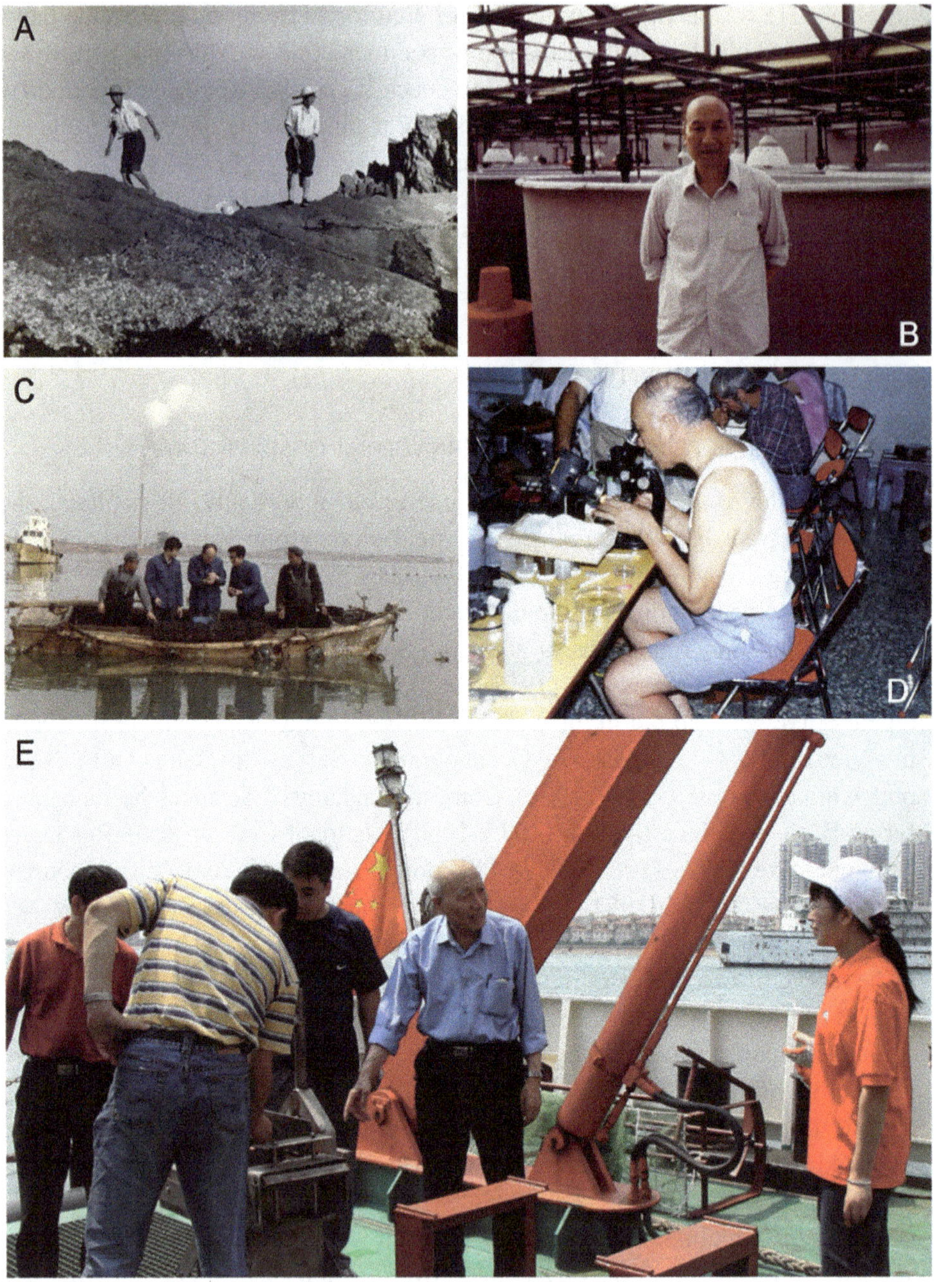

Fig. 4. Ruiyu Liu on field expedition. A, Collecting specimens, 1957; B, at prawn farm of Tianjin, June 1980; C, introducing *Argopecten irradians* into Shandong, 1981; D, Hainan island, preparing specimens, November 1990; E, on board of the R.V. 'Kexue III', 2007.

Technology) in Beijing in 2001 and guided the organization of the Seventh International Crustacean Congress in Qingdao in 2010. He has visited more than 20 countries and regions where he made friends with many international counterparts (fig. 5).

SCIENTIFIC SOCIETIES AND EDITORSHIP

Prof. Liu also passionately contributed to academic activities and journals. He served as president (1988–1992), president emeritus (1992–2012) of the Chinese Society for Oceanology and Limnology; vice-president of Chinese Society of Oceanography (1982–1987) and China Fisheries Society (1987–1996); Council Member (1990–2012) and Standing Council Member (1994–2004) of the China Zoological Society; president (1981–2002), president emeritus (2002–2012) of the Chinese Crustacean Society; member of the International Crustacean Council (1998–2001); corresponding member of the Senckenberg Natural History Society (1992–2012); scientific steering committee member of the Land-Ocean Interactions in the Coastal Zone project of the International Geosphere and Biosphere Programme (1992–1994); member of the presidium (1994–1999) and honorary president (1999–2012) of the International Society of Yellow Sea Research; and served as deputy editor-in-chief (1987–2003) and advisor (2003–2012) of 'Acta Oceanologica Sinica'; vice editor-in-chief (1985–1990), editor-in-chief (1990–1994) and advisor (2002–2012) of 'Oceanologia et Limnologia Sinica'; chief editor (1984–1988) and advisor (2002–2012) of 'Studia Marina Sinica'; member of the editorial board of 'Scientia Sinica' (1985–1996), 'China Science Bulletin' (1985–1996), 'Acta Zoologica Sinica'(1980–2012), 'Acta Zootaxonomica Sinica' (1985–2012), 'Chinese Journal of Oceanology and Limnology' (1982–2000), 'Journal of Fisheries of China' (1980–1996), 'Journal of Fishery Science of China' (1998–2012), 'Resources and Environment in the Yangtze River Basin' (1992–2012), 'Fauna Sinica'(1973–2012), 'Estuarine, Coastal and Shelf Science' (UK) (1985–2000), 'Aquaculture' (Netherlands) (1983–1988); and co-editor-in-chief of 'The Yellow Sea' (Korea) (1995–1999).

What worried him most was the continuation, inheritance and development of traditional marine biotaxonomy. While he won the highest award of science and technology in Qingdao in 2009, his first thought at the time was to donate the bonus of RMB five hundred thousand yuan to establish an academic journal specialized in crustacean taxonomy or even marine invertebrate taxonomy and to encourage young researchers to devote their life to this field of research.

Fig. 5. Ruiyu Liu on international communication. A, Paris, July 1977; B, USA, April 1978; C, Aberdeen University, UK, June 1979; D, Thailand, November 1983; E, Leiden, with Dr. Lipke Bijdeley Holthuis, November 1992; F, Fukuoka, with Dr. Sadayoshi Miyake, 26th March 1996.

PUBLICATIONS AND NEW TAXA

Prof. Liu started his publication career in 1949 with papers on taxonomy of stomatopods and the life history of palaemonid shrimps. Since than he produced a total of more than 210 papers and monographs on taxonomy, systematics, ecology, zoogeography and aquaculture. He described 2 new genera, 52 new species and 1 new subspecies, including not only crustaceans but also cnidarians, polychaetes and mollusks; meanwhile, 1 genus, 24 species (excluding what is done in this volume, see appendix I) have so far been named after him, these taxa also included cnidarians, polychaetes, molluscs, echinoderms, and even sea weed besides the crustaceans. It is therefore our great pleasure that a further 3 new genera and 11 new species are named after Prof. Liu in this volume.

ACKNOWLEDGEMENTS

This account of Prof Ruiyu Liu's life is based not only on the fond memories we had with him, but has also relied on the information and materials provided by his colleagues, assistants and students. In particular those of Profs. Yongliang Wang, Xixing Liu, Fengshan Xu, Xianqiu Ren and Suping Zhang; Drs. Zhongli Sha, Junlong Zhang, Ning Xiao, Xuwen Wu and Yang Li; Ms. Xiaohong Li and Ms. Xiaoqing Ji. We are grateful to his family for providing some personal information that helped us to fill in the story of his life. We especially want to thank Dr. Peter K. L. Ng for giving us the idea of a special book for Prof. Liu even in 2010 during the Seventh International Crustacean Congress in Qingdao, a "homage" volume for his 90th birthday. Sadly, the "homage" volume should be changed to "memorial" volume two years later, and he still encouraged us to heal from sorrow and continue the volume. We also thank him and Dr. Charles H. J. M. Fransen for the language editing of the manuscript.

APPENDIX I: TAXA NAMED IN HONOUR OF RUIYU LIU

Genera:

- *Liuimysis* Wang, 1998 (Mysida: Mysidae)
- *Liusquilla* Ahyong & Lin, 2020 (Stomatopoda: Eurysquillidae)
- *Liusius* Shih, Ho & Galil, 2020 (Brachyura: Leucosiidae)
- *Ruiyupotamon* Wang, Zhang & Sun, 2020 (Brachyura: Potamidae)

Species:

- *Gracilaria tenuistipitata* var. *liui* Zhang & Xia, 1988 (Rhodophyta: Gracilariaceae)
- *Sternaspis liui* Wu, Salazar-Vallejo & Xu, 2015 (Polychaeta: Sternaspidae)
- *Antillophos liui* Zhang & Zhang, 2014 (Mollusca: Nassariidae)
- *Cadella liui* Xu & Zhang, 2018 (Mollusca: Tellinidae)
- *Stygiopontius liui* Ma, Wang & Li, 2020 (Copepoda: Dirivultidae)

Arcoscalpellum liui Sha & Ren, 2015 (Thoracica: Scalpellidae)
Neoacasta liui Yu, Yorisue, Kolbasov & Chan, 2020 (Thoracica: Archaeobalanoidea)
Bathyconchoecia liui Yin, Chen & Li, 2014 (Halocyprida: Halocyprididae)
Petalophthalmus liui Wang, 1998 (Mysida: Mysidae)
Gammaropsis liuruiyui Ren, 1992 (Amphipoda: Photidae)
Gammarus liuruiyui Zheng, Hou & Li, 2020 (Amphipoda: Gammaridae)
Eriopisa liuzhangi Ren & Sha, 2016 (Amphipoda: Eriopisidae)
Minicopenaeon liuruiyui An, Boyko & Li, 2013 (Isopoda: Bopyridae)
Procepon liuruiyui An, Williams & Yu, 2009 (Isopoda: Bopyridae)
Parioninella liuruiyui An, Xi & Paulay, 2020 (Isopoda: Bopyridae)
Metapenaeopsis liui Crosnier, 1987 (Dendrobranchiata: Penaeidae)
Sicyonella liui Chan, 2020 (Dendrobranchiata: Sergestidae)
Caridina sheni Cai, 2020 (Caridea: Atyidae)
Paranchistus liui Li, Bruce & Manning, 2004 (Caridea: Palaemonidae)
Synalpheus liui Wang & Sha, 2015 (Caridea: Alpheidae)
Systellaspis liui Sha & Wang, 2015 (Caridea: Oplophoridae)
Urocaridella liui Wang, Chan & Sha, 2015 (Caridea: Palaemonidae)
Manuscaris liui Wang & Sha, 2016 (Caridea: Alvinocarididae)
Lebbeus liui Wang, Zhu & Sha, 2020 (Caridea: Thoridae)
Austinogebia liui Liu & Sha, 2015 (Gebiidea: Upogebiidae)
Upogebia ruiyui Liu, 2015 (Gebiidea: Upogebiidae)
Eucalliaxiopsis liui Liu & Poore, 2020 (Axiidea: Eucalliacidae)
Ciliopagurus liui Forest, 1995 (Anomura: Diogenidae)
Pagurus liui Han, An & Sha, 2016 (Anomura: Paguridae)
Galearctus yuanshi Yang & Chan, 2020 (Achelata: Scyllaridae)
Cymonomus liui Ahyong & Ng, 2011 (Brachyura: Cymonomidae)
Hyastenus ducator Lee & Ng, 2020 (Brachyura, Epialtidae)
Teoswamon scolasticum Mao & Huang, 2020 (Brachyura: Potamidae)
Lacunipotamon yuanshi Huang, Shih & Ahyong, 2020 (Brachyura: Potamidae)
Amphiura (*Fellaria*) *liui* Liao, 2004 (Echinodermata: Amphiuridae)

APPENDIX II: TAXA DESCRIBED OR ERECTED BY OR WITH RUIYU LIU

CNIDARIA
- ACTINIARIA
  - Species
    - *Aulactinia sinensis* Li & Liu, 2012
    - *Phytocoetes sinensis* Li, Liu & Xu, 2013

POLYCHAETA
- EUNICIDAE
  - Species
    - *Eunice uschakovi* Wu, Sun & Liu, 2013
    - *Eunice hainanensis* Wu, Sun, Liu & Xu, 2013
    - *Eunice carrerai* Wu, Sun, Liu & Xu, 2013

MOLLUSCA
- MYIDAE
  - Species
    - *Sphenia elongate* Zhang, Xu & Liu, 2012

CRUSTACEA
BALANOMORPHA
Species
Astroclita longicostata Ren & Liu, 1979
Balanus rhizophorae Ren & Liu in Ren, 1989
Conopea sinensis (Ren & Liu, 1978) (as *Conopea*)
Tesseropora alba Ren & Liu, 1979
SCALPELLOMORPHA
Species
Teloscalpellum album Liu & Ren, 1985
STOMATOPODA
Genus
Sinosquilla Liu & Wang, 1978
Species
Anchisquilla fasciaticauda Liu & Wang, 1998
Harpiosquilla sinensis Liu & Wang, 1998
Sinosquilla hispida Liu & Wang, 1978
Sinosquilla sinica Liu & Wang, 1978
CUMACEA
Species
Cyclaspis linguiloba Liu & Liu, 1990
Hemileucon bidentatus Liu & Liu, 1990
Lamprops hexaspinula Liu & Liu, 1990
MYSIDA
Genus
Paraleptomysis Liu & Wang, 1983
Species
Anisomysis (Anisomysis) minuta Liu & Wang, 1983 (as *Acanthomysis*)
Notacanthomysis laticauda (Liu & Wang, 1980) (as *Acanthomysis*)
Orientomysis crassispinosa (Liu & Wang, 1980) (as *Acanthomysis*)
Orientomysis leptura (Liu & Wang, 1980) (as *Acanthomysis*)
Orientomysis meridionalis (Liu & Wang, 1983) (as *Acanthomysis*)
Orientomysis rotundicauda (Liu & Wang, 1980) (as *Acanthomysis*)
Orientomysis serrata (Liu & Wang, 1980) (as *Acanthomysis*)
Orientomysis tenella (Liu & Wang, 1983) (as *Acanthomysis*)
Paraleptomysis sinensis Liu & Wang, 1983
Pleurerythrops monospinosus Liu & Wang, 1986
DECAPODA
Dendrobranchiata
Species
Hadropenaeus spinicaudatus Liu & Zhong, 1983
Kishinouyepenaeopsis incisa (Liu & Wang, 1986) (as *Parapenaeopsis*)
Metapenaeopsis sinica Liu & Zhong, 1988
Metapenaeopsis tenella Liu & Zhong, 1988
CARIDEA
Species
Alpheus xishaensis Liu & Lan, 1980
Palaemon guangdongensis Liu, Liang & Yan, 1990
Palaemon tenuidactylus Liu, Liang & Yan, 1990
Subspecies

Coutierella tonkinensis guangdongensis (Liu, Liang & Yan, 1990) (as *Palaemonetes*)
AXIIDEA
Species
Michaelcallianassa sinica Liu & Liu, 2009
Gourretia sinica Liu & Liu, 2010
Paragourretia laevidactyla (Liu & Liu, 2010) (as *Gourretia*)
Michelea takeda Liu & Liu, 2012
Neotrypaea hainanensis (Liu & Liu, 2014) (as *Nihonotrypaea*)
GEBIIDEA
Species
Axianassa sinica Liu & Liu, 2010
Neogebicula holthuisi Liu & Liu, 2010
Paragebicula bijdeleyi Liu & Liu, 2010
Austinogebia monospina Liu & Liu, 2012
Acutigebia serrifera Liu & Liu, 2013
Acutigebia laticauda Liu & Liu, 2013
Mantisgebia multispinosa Liu & Liu, 2013
BRACHYURA
Species
Paragoneplax chenae Jiang & Liu, 2011
Sakaina glabra Jiang & Liu, 2011
Sakaina granulata Jiang & Liu, 2011

APPENDIX III: PROF. RUIYU LIU BIBLIOGRAPHY

1949

LIU, R. Y., 1949a. On some species of *Squilla* (Crustacea Stomatopoda) from China coasts. Contr. Inst. Zool. Nat. Acad. Peiping, **5**(1): 27-47, pls. 1-6, figs. 1-17.

LIU, R. Y., 1949b. On a fresh-water prawn, *Leander modestus* Heller, and its larval development. Contr. Inst. Zool., Nat. Acad. Peiping, **5**(5): 173-189, pls. 18-21.

1951

XIA, W. P., Z. Y. QI, R. Y. LIU & X. T. MA, 1951. The aquatic animals in Baiyangdian Lake and adjacent waters. Oceanol. Limnol. Sinica, **1**(1): 72-76.

1955

LIU, R. Y., 1955a. Economic shrimps and prawns of northern China: i-iii, 1-73, pls. 1-24. (Science Press, Beijing).

LIU, R. Y., 1955b. On *Penaeus orientalis* Kishinouye. Bull. Biol. Sci., **5**: 17-20.

1956

LIU, R. Y., 1956. Notes on two species of *Acetes* of the family Sergestidae (Crustacea Decapoda) from the coasts of north China. Acta Zool. Sinica, **8**(1): 29-40, pls. 1-4. [In Chinese, with English abstract.]

SHEN, J. R. & R. Y. LIU, 1956. Shrimps and crabs of China: 1-99. (Science Popularization Press, Beijing).

1957

LIU, R. Y., 1957. Prawns of the genus *Palaemon* and *Macrobrachium*. Bull. Biol. Sci., **6**: 14-23. [In Chinese.]

GURJANOVA, E. F., J. Y. LIU, O. A. SCARLATO, P. V. USCHAKOV, B. L. WU & C. Y. TSI, 1958. A short report on the intertidal zone of the Shantung Peninsula (Yellow Sea). Bulletin of the Institute of Marine Biology of the Chinese Academy of Sciences, **1**(2): 1-43. [In Russian and Chinese, with English abstract.]

1959

LIU, R. Y., 1959. Notes on the economic macrurous crustacean fauna of the Yellow Sea and the East China Sea. Oceanol. Limnol. Sinica, **2**(1): 35-42. [In Chinese, with English abstract.]

LIU, R. Y., S. C. WU & N. TSAI, 1959. Studies on the life history of *Penaeus orientalis* Kishinouye. Transactions of the symposium on embryology: 15-17.

1960

LIU, R. Y. (ed.), 1960. The fishing atlas of Bohai Sea, Yellow Sea and East China Sea, 1958-1959. II. The oceanographic atlas: charts 369-465. (Institute of Oceanology, Chinese Academy of Sciences, Science Press, Beijing). [In Chinese.]

1961

LIU, R. Y., ET AL., 1961. The data record, national comprehensive oceanographic survey of China seas. Vol. 5. Zooplankton and benthos in Bohai Sea, Yellow Sea and East China Sea. Oct. 1958-Dec. 1959. (Office of the Oceanographic Comprehensive Survey, Section of Marine Science, State Commission of Science and Technology).

1962

LIU, R. Y., ET AL., 1962. On the biomass of benthos in South China Sea. Proc. sympos. animal ecol. & taxonomy: 87-88. [In Chinese.]

TCHANG, S. I., R. Y. LIU, ET AL., 1962. The characteristics and economic significance of invertebrate fauna of China seas. Proceedings of the fifth meeting of the west pacific fisheries research commission: 13-14. [In Russian and Chinese.]

1963

LIU, R. Y. & F. S. XU, 1963. Preliminary studies on the benthic fauna of the Yellow Sea and the East China Sea. Oceanologia et Limnologia Sinica, **5**(4): 306-321. [In Chinese, with English abstract.]

LIU, R. Y., 1963. Zoogeographical studies on the macrurous crustacean fauna of the Yellow Sea and the East China Sea. Oceanologia et Limnologia Sinica, **5**(3): 230-244. [In Chinese, with English abstract.]

LIU, R. Y., ET AL., 1963a. The fishing atlas of coastal waters of Zhejiang province. 3. Benthos, figs. 113-191. (Aquatic Resources Commission of Zhejiang Province).

LIU, R. Y., ET AL., 1963b. The atlas, comprehensive oceanographic survey of China seas. Vol.7. Benthos of Bohai Sea, Huanghai Sea and East China Sea. Oct. 1958-Dec. 1959, pls. 1-244. (Office of the Oceanographic Comprehensive Survey, Section of Marine Sciences, States Commission of Science and Technology).

LIU, R. Y., ET AL., 1963c. The atlas, comprehensive oceanographic survey of China seas. Vol. 14. Benthos of South China Sea. Jan. 1959-Dec. 1960, pls. 1-263. (Office of the Oceanographic Comprehensive Survey, Section of Marine Sciences, States Commission of Science and Technology).

SHEN, J. R. & R. Y. LIU, 1963. Preliminary studies on the characteristics of the crab fauna of China seas. Oceanologia et Limnologia Sinica, **5**(2): 139-153. [In Chinese, with English abstract.]

1964

LIU, R. Y. & Y. L. WANG, 1964. Studies on Stomatopoda of China Sea, I. Genus *Squilla*. Abstracts, academic annual meeting of Chinese society of oceanology and limnology: 97-98.

LIU, R. Y., ET AL., 1964a. Original data, national comprehensive oceanographic survey, 1959-60. Vol. 9, distributions of biomass and main species of zoo-benthos in South China Sea, Jan. 1959-Nov. 1960: 1-952. (Office of the Oceanographic Comprehensive Survey, Section of Marine Sciences, States Commission of Science and Technology).

LIU, R. Y., ET AL., 1964b. Ecology of the benthic fauna in coastal sea off Zhejiang province. Reports of aquatic living resources of coastal waters of Zhejiang province. Marine biology: 267-301. (Zhejiang Provincial Commission of Fisheries Resources).

1965

LIU, R. Y. & Z. C. TANG, 1965. Chart 101, marine benthic fauna and explanation text. In: The physical geography atlas of the People's Republic of China: 32-35.

LIU, R. Y. (ed.), 1965a. Science reports, China-Vietnam joint oceanographic comprehensive survey of Beibu (Tonkin) Gulf. Volumes 1-7. (States Commission of Science and Technology). [In Chinese and Vietnamese.]

LIU, R. Y., ET AL., 1965b. Chapter 6. Marine benthos. In: Science reports of China-Vietnam joint oceanographic comprehensive survey of Tonkin Gulf: 1-98, pls. 1-86. (States Commission of Science and Technology). [In Chinese and Vietnamese.]

LIU, R. Y., ET AL., 1965a. The quantitative distribution and the ecological characteristics of benthos of Bohai Sea and north-western Yellow Sea. Transactions of the 30th annual scientific sympysium of the China zoological society: 17.

LIU, R. Y., 1965a. Studies on *Acetes* from Chinese coasts. Abstracts of the 30th anniversary symposium of the China zoological society: 131.

SHEN, J. R. & R. Y. LIU, 1965. Shrimps and crabs of China: 1-138. (The Youth Press, Beijing).

1975

LIU, R. Y., 1975. On a collection of Stomatopod Crustacea from the Xisha Islands, Guangdong province, China. Studia Marina Sinica, **10**: 183-198, figs. 1-6, pl. 1. [In Chinese, with English abstract.]

1976

SHEN, J. R. & R. Y. LIU, 1976. Shrimps and crabs of China: 1-145. (Science Press, Beijing).

1977

LIU, R. Y. & D. G. CAO, 1977. Spawning and larval rearing of *Penaeus kerathurus* in Algeria. National commision of science & marine research center, Algeria: 1-6. [In Chinese and French.]

1978

LIU, R. Y. & Y. L. WAN, 1978. Description of a new genus and two new species of stomatopod Crustacea from the South China Sea. Oceanologia et Limnologia Sinica, **9**(1): 89-94, figs. 1-2, pl. 1. [In Chinese, with English abstract.]

REN, X. Q. & R. Y. LIU, 1978. Studies on Chinese Cirripedia (Crustacea) I. Genus *Balanus*. Studia Marina Sinica, **13**: 119-196, pls. 1-11. [In Chinese, with English abstract.]

1979

LIU, R. Y. & Z. C. TANG, 1979. Chapter 5. Marine biology, 3. Benthic fauna. In: The physical geography of China. Geography of China seas: 184-200. (Science Press, Beijing).

REN, X. Q. & R. Y. LIU, 1979. Studies on Chinese Cirripedia (Crustacea) II. Family Tetraclitidae. Oceanologia et Limnologia Sinica, **10**(4): 338-353, pls. 1-4. [In Chinese, with English abstract.]

ZENG, C. K. & R. Y. LIU, 1979. Chapter 5. Marine biology. In: The physical geography of China. Geography of China seas: 159-160, 211-213. (Science Press, Beijing).

1980

LIU, R. Y. & S. W. WANG, 1980. Five new species of the genus *Acanthomysis* (Crustacea Mysidacea) from the South China Sea. Oceanologia et Limnologia Sinica, **11**(4): 320-334, pl. 4. [In Chinese, with English abstract.]

LIU, R. Y. & J. Y. LAN, 1980. On a collection of the genus *Alpheus* (Crustacea Decapoda) from the Xisha Islands, Guangdong province, China. Studia Marina Sinica, **17**: 77-115. [In Chinese, with English abstract.]

1981

LIU, R. Y. & D. G. CAO, 1981. Problems in the development of shrimp culture of China. Status of fisheries production, **37**: 1-5. (States Fisheries Administration, People's Republic of China).

1983

LIU, R. Y. & S. W. WANG, 1983a. On a new genus of Mysidacea, *Paraleptomysis* gen. nov. from the South China Sea. Oceanologia et Limnologia Sinica, **14**(3): 203-210. [In Chinese, with English abstract.]

LIU, R. Y. & S. W. WANG, 1983b. On three new species of Mysidacea (Crustacea) from the coastal waters of Guangdong, China. Oceanologia et Limnologia Sinica, **14**(6): 522-530. [In Chinese, with English abstract.]

LIU, R. Y. & Z. R. ZHONG, 1983. On a new genus and two new species of Solenocerid shrimps (Crustacea, Penaeoidea) from South China Sea. Oceanologia et Limnologia Sinica, **1**(2): 171-176. [In Chinese, with English abstract.]

LIU, R. Y., 1983a. Thirty years of marine ecology studies in China. Proceedings of the second Chinese oceanological and limnological science conference: 303-313. [In Chinese, with English abstract.]

LIU, R. Y., ET AL., 1983. Ecology of macrobenthos of the East China Sea and adjacent waters. Proc. int. symposium sediment. cont. shelf, spec. ref. East China Sea: 795-818.

LIU, R. Y., 1983b. Shrimp mariculture studies in China. Proc. 1st. intern. conf. warm water aquaculture, Crustacea, Honolulu, 1983: 82-90.

1984

LIU, R. Y., 1984a. Perspectives to marine biology. Abstract volume, the 4th national congress of Chinese society of oceanology and limnology: 5-6.

LIU, R. Y., 1984b. Advances in shrimp culture in China. Proc. 1st. int. conf. penaeid prawns/shrimps. Manila, 1984: 163.

LIU, R. Y., ET AL., 1984. Ecological analysis of tagged released Chinese shrimp, *Penaeus chinensis* in resources enhancement experiment. Abstracts, presentations at the 4th congress of the Chinese society of oceanology and limnology, Qingdao, 1984: 644.

1985

LIU, R. Y. & X. Q. REN, 1985. Studies of Chinese Cirripedia (Crustacea) VI. Suborder Lepadomorpha. Studia Marina Sinica, **25**: 179-281, pls. 1-12. [In Chinese, with English abstract).]

LIU, R. Y., 1985a. Strategy for the development of marine fisheries of Shandong province. Qilu (Shandong) Fisheries, **1985**(1): 2-6.

LIU, R. Y., 1985b. The exploitation of living resources in Jiaozhou Bay. Proc. sympos. compreh. exploit. Jiaozhou Bay, Yellow Sea, Qingdao, 1985: 22-26.

LIU, R. Y., ET AL., 1985. On the development strategy of fisheries in Jiaozhou Bay. Proc. sympos. compreh. exploit. Jiaozhou Bay, Yellow Sea, Qingdao, 1985: 27-29.

LIU, R. Y., 1985c. Conservation and enhancement of coastal zone living resources. Natural resources research. Transactions of the first symposium of the China Research Society of Natural Resources: 193-197.

1986

LIU, R. Y., 1986. Marine benthic crustacean studies in China. Transactions of the Chinese Crustacean Society, **1**: 7-12. [In Chinese, with English abstract.]

LIU, R. Y. & S. W. WANG, 1986. Studies on Mysinae (Crustacea Mysidacea) of the northern South China Sea. Studia Marina Sinica, **26**: 159-202. [In Chinese, with English abstract.]

LIU, R. Y. & W. Q. ZANG, 1986. On the species name of *Penaeus orientalis* Kishinouye. Trans. Chinese Crust. Soc., **1**: 212-213.

LIU, R. Y. & Y. L. WANG, 1986. Studies on *Parapenaeopsis* (Crustacea, Decapoda, Penaeidae) of Chinese waters. Trans. Chinese Crust. Soc., **1**: 214-215.

LIU, R. Y. & Z. R. ZHONG (eds.), 1986. Penaeoid shrimps of the South China Sea: 1-278, pls. 1-5. (Agricultural Publishing House, Beijing). [In Chinese, with English abstract.]

LIU, R. Y., ET AL., 1986a. Ecological characteristics of macrobenthos of the Yellow Sea and East China Sea. Studia Marina Sinica, **27**: 153-173. [In Chinese, with English abstract.]

LIU, R. Y. & R. J. LIN, 1986. Achievements in crustacean mariculture studies in China. Trans. Chinese Crust. Soc., **1**: 37-45. [In Chinese, with English abstract.]

LIU, R. Y., Y. H. CUI & F. S. XU, 1986. Biological characteristics of penaeid shrimps in Jiaozhou Bay, Yellow Sea. Trans. Chinese Crust. Soc., **1**: 245-246. [In Chinese.]

LIU, R. Y., ET AL., 1986b. Protocol (concise rules), national coastal zone and beach resources comprehensive survey. Part 8, biological survey: 1-67. (China Ocean Press, Beijing).

1987

LIU, R. Y. & Y. L. WANG, 1987. Studies on Chinese species of the genus *Parapenaeopsis* (Decapoda, Crustacea). Oceanologia et Limnologia Sinica, **18**(6): 523-539. [In Chinese, with English abstract.]

LIU, R. Y., B. Z. LUO, ET AL., 1987. Effects of the Three Gorge Project on estuarine biology and fishery resources. In: Treatise on impact of the Three Gorge Project on ecology and environment and the study of possible countermeasures: 403-446. (Science Press, Beijing).

LIU, R. Y. & B. Z. LUO, 1987a. Demonstration report of impact of the Three Gorge Project on ecology and environment. Section 10. Effect on ecology and environment of the estuary and adjacent seas: 69-75.

LIU, R. Y. & B. Z. LUO, 1987b. Three Gorges Project affects the ecology and fishery around estuary of Yangtze River. Bulletin of Soil and Water Conservation, **7**(4): 37-40.

LIU, R. Y., 1987a. Marine crustaceans. Encyclopedia sinica, volume atmospheric, oceanographic and hydographic sciences: 351-353. (China Encyclopedia Press, Beijing).

LIU, R. Y., 1987b. Marine zoobenthos. Encyclopedia sinica, volume atmospheric, oceanographic and hydographic sciences: 312-314. (China Encyclopedia Press, Beijing).

LIU, R. Y. & Y. H. CUI, 1987a. Deep sea ecology. Encyclopedia sinica, volume atmospheric, oceanographic and hydographic sciences: 668-669. (China Encyclopedia Press, Beijing).

LIU, R. Y. & Y. H. CUI, 1987b. Marine ecology. Encyclopedia sinica, volume atmospheric, oceanographic and hydographic sciences: 376-378. (China Encyclopedia Press, Beijing).

LIU, R. Y. & G. Z. XU, 1987a. Marine living resources. Encyclopedia sinica, volume atmospheric, oceanographic and hydographic sciences: 388-389. (China Encyclopedia Press, Beijing).

LIU, R. Y. & G. Z. XU, 1987b. Marine sound producing animals. Encyclopedia sinica, volume atmospheric, oceanographic and hydographic sciences: 326-328. (China Encyclopedia Press, Beijing).

WANG, S. W. & R. Y. LIU, 1987. Preliminary study of the subfamily Gastrosaccinae (Crustacea Mysidacea) of the South China Sea. Studia Marina Sinica, **28**: 205-231. [In Chinese, with English abstract.]

ZENG, C. K., R. Y. LIU & G. Z. XU, 1987. Marine biology. Encyclopedia sinica, volume atmospheric, oceanographic and hydographic sciences: 384-387. (China Encyclopedia Press, Beijing).

1988

LIU, R. Y. & B. Z. LUO, 1988. Chapter 6. Effect of Three Gorge Project on ecology and environment of Changjiang River estuary and adjacent seas. In: Studies on impact of the Three Gorge Project on ecology and environment and possible countermeasures: 116-140. (Science Press, Beijing).

XIAO, Y. S., R. Y. LIU & Y. H. CUI, 1988. Recruitment of postlarval prawn, *Penaeus orientalis*, to a tidal estuary of Jiaozhou Bay, China. Chin. J. Oceanol. Limnol., **6**(4): 290-299.

1989

LIU, R. Y., ET AL., 1989. Summary report, Shandong province coastal zone and beach resources comprehensive investigation: 1-57, tbls. 1-5, charts 1-2. (Academic Press, Beijing).

LIU, R. Y., 1989. Marine biological resources enhancement studies. Proc. 2nd general assembly, division of earth sciences., Chin. acad. sci.: 31-38.

LIU, R. Y., X. Q. LIANG & S. L. YAN, 1990. A study of the Palaemoninae (Crustacea Decapoda) from China I. *Macrobrachium*, *Leander* and *Leandrites*. Trans. Chinese Crust. Soc., **2**: 102-134. [In Chinese, with English abstract.]

SHEN, J. R., R. Y. LIU & S. W. WANG, 1989. Mysidacea in waters off the north China coasts. Studia Marina Sinica, **30**: 189-227. [In Chinese, with English abstract.]

1990

LIU, R. Y., X. Q. LIANG & S. L. YAN, 1990. A study of the Palaemoninae (Crustacea Decapoda) from China II. *Palaemon*, *Exopalaemon*, *Palaemonetes* and *Leptocarpus*. Studia Marina Sinica, **31**: 229-265, pl. 1. [In Chinese, with English abstract.]

LIU, H. & R. Y. LIU, 1990. Study on Cumacea (Crustacea Malacostraca) of the offshore waters of north China. Studia Marina Sinica, **31**: 195-228. [In Chinese, with English abstract.]

LIU, R. Y., 1990a. In memory of Professor Shen Jiarui (C. J. Shen, 1902-1975). The founder of China's carcinology. Trans. Chinese Crust. Soc., **2**: 1-7. [In Chinese and English.]

LIU, R. Y. & J. S. CHEN (eds.), 1990. Comprehensive report, Shandong province coastal zone and beach resources comprehensive investigation: 1-793, charts 1-4. (China Science and Technology Publishing House, Beijing).

LIU, R. Y., 1990b. Studies on marine ecology of Yellow Sea in China. The first proc. intern. symp. mar. sci. Yellow Sea, Inchon, Korea, 1989. Yellow Sea Research, **3**: 51-79. [In Korean and Chinese.]

LIU, R. Y., 1990c. Mariculture studies in north China. The first proc. intern. symp. mar. sci. Yellow Sea, Inchon, Korea, 1989. Yellow Sea Research, **3**: 121-141. [In Korean and Chinese.]

LIU, R. Y., 1990d. Present status and future prospects for shrimp mariculture in China. The culture of cold-tolerant shrimp: proc. Asian-U. S. workshop on shrimp culture: 16-28.

LIU, R. Y., 1990e. Resource enhancement of Chinese shrimp, *Penaeus orientalis*. Bull. Mar. Sci., **47**(1): 124-133.

LIU, R. Y., 1990f. Mariculture studies in north China. The first proc. intern. symp. mar. sci. Yellow Sea, Inchon, Korea, 1989. Yellow Sea Research, **3**: 121-141. [In Korean and Chinese, with English abstract.]

LIU, R. Y., 1990g. The Chinese shrimp: 205-206; shrimps and prawns: 1317-1319. Encyclopedia sinica, volume on agriculture & fishery production. (China Encyclopedia Press).

1991

LIN, Q. W., R. Y. LIU & J. H. XIANG, 1991. Studies of the morphological structure and physiological and biochemical functions of the spermatozoa in Chinese shrimp *Penaeus chinensis* I. Ultrastructure of the spermatozoa. Oceanologia et Limnologia Sinica, **22**(5): 397-401, pls.1, 2. [In Chinese, with English abstract.]

LIU, H. & R. Y. LIU, 1991. The comparative morphology of larval stages of the main penaeid shrimps, *Penaeus chinensis*, *P. penicillatus* and *P. merguiensis* in China Seas. Proc. 2nd IOC. WESTPAC Symposium: 287-292.

LIU, R. Y., 1991a. Subphylum Crustacea. Encyclopedia sinica, volume biology: 696-699. (China Encyclopedia Press, Beijing).

LIU, R. Y., 1991b. Class Cephalocarida: 1676-1677; class Mystacocarida: 1955; class Malacostraca: 1234-1235; subclass Phyllocarida: 2037-2038; subclass Eumalacostraca: 2187 (class Malacostraca). Encyclopedia sinica, volume biology. (China Encyclopedia Press, Beijing).

LIU, R. Y., 1991c. Super order Syncarida: 519-520; order Anaspidacea:1280; superorder Pancarida: 1179. encyclopedia sinica, volume biology. (China Encyclopedia Press, Beijing).

LIU, R. Y., 1991d. Superorder Peracarida: 1017; order Spelaeogriphacea: 1960; order Mysidacea: 799-800; order Cumacea: 863-864; order Amphipoda: 292; order Tanaidacea: 2139. Encyclopedia sinica, volume biology. (China Encyclopedia Press, Beijing).

LIU, R. Y., 1991e. Superorder Eucarida: 2190. Encyclopedia sinica, volume biology. (China Encyclopedia Press, Beijing).

LIU, R. Y., 1991f. Order Decapoda: 1505-1507; genus *Penaeus*: 293-294; genus *Acetes*: 958; genus *Macrobrachium*: 2169-2170; genus *Exopalaemon*: 26; genus *Crangon*: 544; genus *Spongicola*: 859. Encyclopedia sinica, volume biology. (China Encyclopedia Press, Beijing).

LIU, R. Y., 1991g. Genus *Homarus*: 13; genus *Procambarus*: 13; Thalassinidea: 483; genus *Panulirus*: 899-900. Encyclopedia sinica, volume biology. (China Encyclopedia Press, Beijing).

LIU, R. Y., 1991h. Merostomata: 2197; horseshoe crab: 562. Encyclopedia sinica, volume biology. (China Encyclopedia Press, Beijing).

LIU, R. Y., 1991i. In memory of Dr. Shao-Wen Ling (1907-1990). Trans. Chin. Crustacean Soc., **3**: 1-5. [In Chinese and English.]

LIU, R. Y., Y. H. CUI & F. S. XU, 1991. Recapture rate of released shrimp *Penaeus chinensis* in resource enhancement experiment in Jiaozhou Bay, Yellow Sea. Proc. 2nd intern. symp. mar. sci. Yellow Sea, July 27-29, 1991, Qingdao, China: 240-242.

LIU, R. Y., Y. H. CUI, ET AL., 1991. Part I, Chapter 9, marine biology: 216-268. Part II, Chapter 5. Marine living resources: 385-410. In: Science report of the national comprehensive survey of coastal zone and beach resources of China. (China Ocean Press).

XIANG, J. H., R. Y. LIU, L. H. ZHOU & H. F. WANG, 1991. A preliminary study on the chromosome in giant tiger prawn *Penaeus monodon* Fabricius. Ann. Res. Rept., Experim. Mar. Biol. Lab., Inst. Oceanol., Academia Sinica, **1991**: 115-121, pl. 1. [In Chinese, with English abstract.]

1992

HE, H. Q., A. L. LAWRENCE & R. Y. LIU, 1992. Evaluation of dietary essentiality of fat-soluble vitamins, A, D, E and K for penaeid shrimp (*Penaeus vannamei*). Aquaculture, **103**: 177-185.

LIU, R. Y. & B. Z. LUO, 1992. Impacts of the Three Gorges Project on the ecology and environment of the Changjiang River estuary and adjacent waters. Studia Marina Sinica, **33**: 1-13.

LIU, R. Y. (ed.), 1992. Ecology and living resources of Jiaozhou Bay: 1-460, pls. 1-10. (Science Press, Beijing). [In Chinese with English abstract.]

LIU, R. Y., 1992a. Chapter 1.7. Study of marine crustacean. In: C. K. TSENG, ET AL. (eds.), Marine science study and its prospect in China. I. Development of marine biology in China: 27-33 [in Chinese], 431-437 [in English]. (Qingdao Publishing House). [In Chinese and English.]

LIU, R. Y., 1992b. Perspectives to shrimp mariculture research in China. In: J.-K. WANG & P. V. DEHADRAI (eds.), Aquaculture research needs for 2000 A. D.: 103-113.

LIU, R. Y., F. S. XU, D. Y. SUN, Y. H. CUI & H. F. WANG, 1992. Benthic fauna in the Changjiang River estuary and prediction of influence of the Three Gorges Project on it. Studia Marina Sinica, **33**: 237-247. [In Chinese, with English abstract.]

LIU, R. Y., J. H. XIANG, Y. Q. WU & B. L. ZHANG, 1992. Prediction of the impacts of the Three Gorges Project on the invertebrate resources in the Changjiang River estuary. Studia Marina Sinica, **33**: 249-264. [In Chinese, with English abstract.]

LIU, R. Y., Y. H. CUI & S. H. CHEN, 1992. On the development trend of shrimp culture. Trans. Chin. Crustacean Soc., **3**: 145-150. [In Chinese, with English abstract.]

TSENG, C. K. & R. Y. LIU, 1992. Advances and perspectives of China's marine biology. Abstracts of presentations at the 6th congress of division members of the Chinese academy of sciences: 196-199. [In Chinese.]

WANG, S. W. & R. Y. LIU, 1992. Crustacea Mysidacea from the seas off north and west coast of Antarctic Peninsula. Trans. Chin. Crustacean Soc., **3**: 318-334. [In Chinese, with English abstract.]

XIANG, J. H., L. H. ZHOU & R. Y. LIU, 1992. Induction of the tetraploids of the Chinese shrimp, *Penaeus chinensis*. Marine Sciences, **4**: 55-61. [In Chinese, with English abstract.]

ZHOU, L. H., J. H. XIANG, R. Y. LIU & D. G. YU, 1992. Meiotic and mitotic chromosome complements from the testis of three species of shrimps in the genus *Penaeus*. Trans. Chin. Crustacean Soc., **3**: 190-195. [In Chinese, with English abstract.]

1993

LIU, R. Y., 1993a. Zoogeographical characteristics of the shrimp fauna of the Yellow Sea. Abstract Volume, International Senckenberg Symposium Crustacea Decapoda. Frankfurt a. M. Oct. 18-22, 1993: 36-37.

LIU, R. Y., Y. H. CUI & F. S. XU, 1993a. Recapture rate of released shrimp *Penaeus chinensis* in resource enhancement experiment in Jiaozhou Bay, Yellow Sea. Oceanologia et Limnologia Sinica, **24**(2): 137-142. [In Chinese, with English abstract.]

LIU, R. Y., Y. H. CUI & F. S. XU, 1993b. On the recapture rate of tagged Chinese shrimp released in Jiaozhou Bay, Yellow Sea. Marine Sciences, **6**: 39-42. [In Chinese, with English abstract.]

LIU, R. Y., 1993b. Problems and countermeasures in the exploitation of China's marine biological resources. In: Division of Earth Sciences, Chinese Academy of Sciences (ed.), Resources in China, potentials, trend and strategies: 304-310. (Beijing Press).

1994

LIU, H. & R. Y. LIU, 1994a. Comparative study on postlarval development of penaeid shrimp *Penaeus chinensis*, *P. penicillatus* and *P. merguiensis*. Studia Marina Sinica, **35**: 179-238. [In Chinese, with English abstract.]

LIU, H. & R. Y. LIU, 1994b. Comparative studies on the larval development of the penaeid shrimps, *Penaeus chinensis*, *P. merguiensis* and *P. penicillatus*. Chin. J. Oceanol. Limnol., **12**(4): 295-307.

LIU, R. Y. & D. G. CAO, 1994. Mariculture of penaeid shrimp in China. In: D. ZHOU, ET AL. (eds.), Oceanology of China sea, **1**: 331-344. (Kluwer Academic Publishers, Dordrecht).

LIU, R. Y. & Z. R. ZHONG, 1994. Crustacea Decapoda: shrimps and lobsters. In: Z. HUANG (ed.), Marine species and their distributions in China's seas: 545-568. (China Ocean Press, Beijing).

TSENG, C. K. & R. Y. LIU, 1994. Chapter 52. Marine sciences. In: China today. Chinese academy of sciences. Book III: 533-568.

WANG, S. W. & R. Y. LIU, 1994. A faunal study of the Mysidacea (Crustacea) from Nansha Islands and it's adjacent waters. In: Study on marine biota and biogeography of Nansha Islands and its adjacent South China Sea. I: 61-110. [In Chinese, with English abstract.]

1996

HUANG, B. & R. Y. LIU, 1996. The applications of system analysis methods in fishery resources study and ecological environmental engineering. Marine Sciences, **1996**: 18-20.

LI, X. Z. & R. Y. LIU, 1996. Ten new record species of Pontoniinae from China (Decapoda: Palaemonidae). Acta Zootaxonomica Sinica, **21**(4): 504, 497. [In Chinese.]

LIU, R. Y. & Y. H. CUI (eds.), 1996. Marine biota of coastal zone of China seas — report of the national comprehensive survey of coastal zone and beach resources of China: 1-445. (China Ocean Press, Beijing).

LIU, R. Y., D. G. CAO & C. Q. HU, 1996. Advances in the study of penaeid shrimp explosive epidemic diseases in China. Proceedings of the 2nd symposium on diseases of cultured penaeid shrimp: 183-186. [In Chinese, with English abstract.]

LIU, R. Y. & X. C. WENG, ET AL., 1996. New results of the China/Korea joint study on oceanography of the Yellow Sea. Oceanologia et Limnologia Sinica, **27**(6): 657-666. [In Chinese, with English abstract.]

WANG, Z. H. & R. Y. LIU (eds.), 1996. Proceedings of the 2nd symposium on diseases of cultured penaeid shrimp. Department of society activities, China association for science and technology: 1-228. (Qingdao Ocean University Press, Qingdao). [In Chinese, with English abstracts.]

1997

LI, X. Z. & R. Y. LIU, 1997. A preliminary study on the phylogeny of Pontoniinae (Decapoda: Palaemonidae). Oceanologia et Limnologia Sinica, **28**(4): 383-393. [In Chinese, with English abstract.]

LIU, R. Y. & D. X. HU, 1997. Essentials and goals of the science plan of land /ocean interactions in coastal zone (LOICZ). Frontier of Geo-Science, **4**(1-2): 194, 220.

LIU, R. Y., 1997. Impact of human activities on marine biodiversity. Symposium on coastal ocean resources and environment (Hong Kong, 11-12 August 1997): 39-46.

WANG, S. W. & R. Y. LIU, 1997. Mysidacea fauna of the East China Sea. Studia Marina Sinica, **38**: 191-222. [In Chinese, with English abstract.]

1998

HUANG, B. & R. Y. LIU, 1998a. The statistical models of abiotic environmental factors in the Jiaozhou Bay enhancement and aquaculture ecosystem. Journal of East China Normal University Natural Science, Special Issue of Zoology: 131-134.

HUANG, B. & R. Y. LIU, 1998b. The relationship among subsystems in the enhancement and aquaculture ecosystem: the typical bay. Journal of East China Normal University (Natural Science): 121-125. [In Chinese, with English abstract.]

HUANG, B., Q. YU & R. Y. LIU, 1998. Assessment of the contents of several subsystems biomass in enhancement and aquaculture ecosystem: Jiaozhou Bay. Journal of East China Normal University (Natural Science): 126-130. [In Chinese, with English abstract.]

LEE, Y. C., Y. S. QIN & R. Y. LIU (eds.), 1998. Yellow Sea atlas: 1-524. (Ho Yong Publishing Co., Korea).

LI, T. W., J. H. XIANG & R. Y. LIU, 1998. Construction of cDNA library of shrimp *Penaeus chinensis* (Crustacea, Decapoda). Acta Zoologica Sinica, **44**(2): 237-238. [In Chinese, with English abstract.]

LI, T. W., M. J. DING, J. ZHANG, J. H. XIANG & R. Y. LIU, 1998. Studies on the pustule disease of abalone (*Haliotis discus hannai* Ino) on the Dalian coast. Journal of Shellfish Research, **17**: 707-711.

LIU, R. Y. & B. Z. LUO, 1998. Chapter 6. Effect of the Three Gorges Project on the environment and ecology of the Changjiang Estuary and adjacent sea. In: Z. SHE (ed.), The impacts of the Three Gorges Project on ecology and environment with countermeasures: 124-147. (Science Press, Beijing).

LIU, R. Y. & Y. L. WANG, 1998. On two new species of the Squillidae and Harpiosquillidae (Crustacea Stomatopoda) from the South China Sea. Oceanologia et Limnologia Sinica, **29**(6): 588-596. [In Chinese, with English abstract.]

LIU, R. Y., 1998a. Scientific problems in the sustainable development of marine biological resources. In: G. ZHOU (ed.), Progress of science and technology and development of disciplines: 101-105. [In Chinese, with English abstract.]

LIU, R. Y., 1998b. On the exploitation of South China Sea resources. In: Q. HE (ed.), Proceedings, strategies of resources exploitation and sustainable development of the South China Sea: 25-35.

LIU, R. Y., 1998c. Scientific problems on sustainable development of fishery resources in the Yangtze River estuary and adjacent sea. Challenges to science and technology in the new centenary. — Selected presentations at the 1998 Shanghai science and technology forum: 144-145.

LIU, R. Y., X. Z. LI & J. LIU, 1998. Present status and prospect of taxonomic study on marine animals of China. In: D. NIU (ed.), Review and perspectives of biological systematics of China: 98-102.

SONG, L. S., J. H. XIAN, C. X. LI, L. H. ZHOU & R. Y. LIU, 1998. Studies on genetic variation and phylogenetic relationships among six species of *Penaeus* genus by REPD markers. Acta Zoologica Sinica, **44**(3): 353-359. [In Chinese, with English abstract.]

WANG, Y. L. & R. Y. LIU, 1998. Stomatopod fauna of the northern South China Sea and the Nansha Islands. Studies on marine fauna and flora and biogeography of the Nansha Islands and neighbouring South China Sea, III: 131-142.

1999

HUANG, B. & R. Y. LIU, 1999. Construction of Jiaozhou Bay marine ecosystem models, analysis of their characteristics and control measures. Studia Marina Sinica, **41**: 227-233.

HUANG, B., R. Y. LIU & X. H. LI, 1999. Analysis on the sex ratio of male and female number of *Metapenaeus joyneri* population in Jiaozhou Bay. Journal of Fisheries of China, **23**(4): 429-431.

LIU, R. Y. & S. W. WANG, 1999. Studies on Mysidacea fauna of the China seas. In: Zoological studies in China: 18-30. (China Forestry Press). [In Chinese, with English abstract.]

LIU, R. Y. & X. C. WENG (eds.), 1999. Chart set of oceanography. In: L. KE (ed.), National atlas of the People's Republic of China: 1-283. (China Map Press).

LIU, R. Y. & Y. L. WANG, 1999. The stomatopod fauna of the China seas. In: F. R. SCHRAM & J. C. VON VAUPEL KLEIN (eds.), Crustaceans and the biodiversity crisis, Proceedings of the Fourth International Crustacean Congress, Amsterdam, The Netherlands July 1998, **1**: 569-582.

LIU, R. Y., 1999a. Marine biodiversity conservation in the South China Sea. Proc. European-Asian workshop on investigation and management of mediterranean and South China Sea coastal zone, November 1998, Hong Kong: 351-360.

LIU, R. Y., 1999b. Strictly managing, intensively studying, keep sustainable development of fisheries of Shandong province. Science and Management, **1999**(1): 8-11.

WU, C. G., L. H. ZHOU, J. H. XIANG & R. Y. LIU, 1999. Morphology and structure of thelycum in *Trachypenaeus curvirostris* and *Penaeus chinensis*. Journal of Fisheries of China, **23**(3): 223-228. [In Chinese, with English abstract.]

2000

LIU, R. Y. & S. W. WANG, 2000. Fauna sinica, volume Mysidacea (Arthropoda Crustacea Malacostraca): 1-326. (Science Press, Beijing). [In Chinese, with English abstract.]

LIU, R. Y., H. Y. YU & X. Z. LI, 2000. The Chinese names of isopod families Cirolanidae and Cymothoida. Chinese Journal of Zoology, **35**(5): 54-55.

LIU, R. Y., 2000. Marine resources and their reasonable and sustainable utilization. Lectures and talks of academicians and scientists in Hainan Province — towards sciences and new centenary: 123-147. (Hainan Provincial Association for Science and Technology).

2001

LIU, R. Y., ET AL., 2001. Prediction of long term change of abundance of large invertebrates in the Jiaozhou Bay, Yellow Sea. Oceanologia et Limnologia Sinica, **32**(3): 274-279.

LIU, R. Y., 2001a. Marine biological production and resources sustainable exploitation. In: D.-J. YU (ed.), Offshore resources protection and sustainable development. 2000 international forum on marine science & technology and economic development: 54-62. (Qingdao, China).

LIU, R. Y., 2001b. Sustainable development of mariculture production in China. Essays from academicians on marine science & technology and economic development, Qingdao, China, July 15-16, 2001: 32-38.

LIU, R. Y., 2001c. Resources of shrimps, lobsters, crabs and their sustainable exploitation. In: Z. ZHOU & R. HU (eds.), Ocean exploitation strategies of China in the 21 centuary — proceedings of the symposium on the exploitation of South China Sea marine resource: 543-571. (The Ocean Press, Beijing).

2002

LIU, R. Y., 2002. Sustainable development of China's mariculture. In: B. ZHANG (ed.), Studies on shrimp culture: 3-7. (The Ocean Press, Beijing).

2003

LI, X. Z. & J. Y. LIU, 2003. Report on Pontoniinae shrimps (Crustacea: Decapoda) collected by joint Chinese-German marine biology expeditions to Hainan Island, South China Sea II. *Harpiliopsis, Ischnopontonia, Jocaste, Palaemonella, Periclimenaeus, Periclimenella*. Chinese Journal of Oceanology and Limnology, **21**(2): 154-165.

LI, X. Z., R. Y. LIU & X. Q. LIANG, 2003. Study on the zoogeography of Chinese palaemonoid fauna. Transactions of the Chinese Crustacean Society, **4**: 125-147. [In Chinese, with English abstract.]

LIU, R. Y. & X. Q. REN, 2003. Advances in the study of thoracic cirripedian phylogeny and systematics. Transactions of the Chinese Crustacean Society, **4**: 89-105. [In Chinese, with English abstract.]

LIU, R. Y., 2003. Introduction of an updated classification of the recent crustaceans. Transactions of the Chinese Crustacean Society, **4**: 78-88. [In Chinese, with English abstract.]

LIU, R. Y., 2003b. On the unification of the scientific name of Penaeid shrimp (Crustacea: Decapoda). Transactions of the Chinese Crustacean Society, **4**: 106-124. [In Chinese, with English abstract.]

LIU, R. Y., 2003c. My research works on crustaceology and marine ecology studies. Transactions of the Chinese Crustacean Society, **4**: 20-34.

LIU, R. Y., 2003d. Present status of mariculture in China. In: C.-S. LEE & A. VENTURA (eds.), Status of aquaculture in China: 1-11.

SONG, D. X. & R. Y. LIU, 2003. Advances in the study of crustacean phylogeny. Transactions of the Chinese Crustacean Society, **4**: 61-77. [In Chinese, with English abstract.]

XIAN, W. W., R. Y. LIU & X. H. ZHU, 2003. Carbon budget of bastard halibut *Paralichthys olivaceus* in relation to body weight and temperature. Chinese Journal of Oceanology and Limnology, **21**: 134-140.

2004

LIU, R. Y., 2004a. Crustacea Decapoda, Merostomata Xiphosura. In: S. WANG & Y. XIE (eds.), China species red list, vol. 1. Red list: 88-91, 99. (Higher Education Press, Beijing). [In Chinese and English.]

LIU, R. Y., 2004b. "China species red list" and the diversity of China's marine biota. Daziran (China Nature), **2004**(3): 15-17.

LIU, R. Y., 2004c. Zoogeography of the hippolytid fauna (Crustacea: Decapoda) of China seas. Abstracts book, the 19th international zoological congress, August, 2004, Beijing: 473.

LIU, R. Y., 2004d. The China species red list and the present status of China's marine biodiversity. Daziran (China Nature), **2004**(6): 15-17.

LIU, R. Y., 2004e. On sustainable exploitation of marine biological resources in China. Science and Technology Review, **2004**(11): 28-31. [In Chinese, with English abstract.]

XIAN, W. W., R. Y. LIU & B. Z. LUO, 2004. Environment of the Changiang estuary before the sluice construction in the Three Gorges Reservoir. Resources and Environment in the Yangtze Basin, **13**: 119-123.

2005

LIU, R. Y., 2005. Crustacea Decapoda, Merostomata Xiphosura. In: S. WANG & Y. XIE (eds.), China species red list, vol. 3. Text of red list: 424-459, 503-505. (Higher Education Press, Beijing). [In Chinese and English.]

XIAN, W. W., B. KANG & R. Y. LIU, 2005. Jellyfish blooms in the Yangtze estuary. Science, **307**: 41.

2006

LIU, R. Y. & X. Q. REN (eds.), 2006. Fauna sinica. Invertebrata. Volume Crustacea: Thecostraca: Thoracica: 1-611. (Science Press, Beijing). [In Chinese, with English abstract.]

LIU, R. Y., 2006. Marine biodiversity study and biotechnology exploitation in China. Journal of Marine Science and biotechnology, **1**(1): 40-47.

ZHANG, Z. & R. Y. LIU, 2006. Studies on the subfamily Scyllarinae (Crustacea: Decapoda: Palinura) from China coast. Marine Sciences, **5**: 18-27. [In Chinese with English abstract.]

2007

JIANG, W., H. L. CHEN & R. Y. LIU, 2007. New record of two species of genus *Asthenognathus* (Crustacea: Decapoda: Pinnotheridae) from China seas. Oceanologia et Limnologia Sinica, **38**(1): 77-83. [In Chinese with English abstract.]

LI, X. Z., R. Y. LIU, X. Q. LIANG & G. X. CHEN, 2007. Fauna sinica. Volume Crustacea: Decapoda: Palaemonoidea: 1-369. (Science Press, Beijing).

SHA, Z. L. & R. Y. LIU, 2007. Study on *Athanas* Leach of Alpheidae (Crustacea: Decapoda) from China seas. Acta Zootaxonomica Sinica, **32**(4): 749-755.

2008

LIU, R. Y. (ed.), 2008. Checklist of marine biota of China seas: i-xiv, 1-1267. (Science Press, Beijing). [In Chinese and English.]

SHA, Z. L. & R. Y. LIU, 2008. Studies on genus *Arete* Stimpson of Alpheidae (Crustacea: Decapoda) from China seas. Acta Zootaxonomica Sinica, **33**(1): 7-9.

2009

LIU, W. L. & R. Y. LIU, 2009. *Michaelcallianassa sinica* sp. nov. (Crusacea, Decapoda, Thalassinidea, Callianassidae) from the South China Sea. Zootaxa, **2294**: 39-46.

2010

LIU, W. L. & R. Y. LIU, 2010a. Two new species of the axiidean genus *Gourretia* De Saint Laurent, 1973 (Decapoda, Ctenochelidae) from South China Sea. Journal of Crustacean Biology, **30**(4): 745-756.

LIU, W. L. & R. Y. LIU, 2010b. The subfamily Neogebiculinae (Decapoda, Thalassinidea, Upogebiidae) from South China Sea. Crustaceana Monographs, **14**: 389-401.

LIU, W. L. & R. Y. LIU, 2010c. First record of the gebiidean genus *Axianassa* Schmitt, 1924 (Decapoda, Laomediidae) in the west Pacific, with descriptions of a new species from the South China Sea. The Raffles Bulletin of Zoology, **58**(2): 193-197.

2011

JIANG, W. & R. Y. LIU, 2011a. On a new species of *Paragoneplax* Castro (Crustacea, Brachyura, Goneplacidae), *P. chenae* sp. nov. from South China Sea. Chinese Journal of Oceanology and Limnology, **29**(1): 213-216.

JIANG, W. & R. Y. LIU, 2011b. New species and new records of pinnotherid crabs (Crustacea: Decapoda: Brachyura) from the Yellow Sea. Zoologischer Anzeiger, **250**(4): 488-496.

JIANG, W. & R. Y. LIU, 2011c. A new record of deep-water crab, *Hephthopelta cribrorum* Rathbun, 1932 (Crustacea: Decapoda: Chasmocarcinidae) from the East China Sea. Oceanologia et Limnologia Sinica, **42**(6): 880-883. [In Chinese with English abstract.]

LIU, R. Y., 2011. Progress of marine biodiversity studies in China seas. Biodiversity Science, **19**: 614-626.

XIAO, N., Y. L. LIAO & R. Y. LIU, 2011. Records of the genus *Henricia* Gray, 1840 (Echinodermata: Asteroidea: Echinasteridae) from Chinese waters. Zootaxa, **3115**: 1-20.

2012

LI, Y. & R. Y. LIU, 2012. *Aulactinia sinensis*, a new species of sea anemone (Cnidaria: Anthozoa: Actiniaria) from Yellow Sea. Zootaxa, **3476**: 62-68.

LIN, F. J., Y. LIU, Z. L. SHA, L. M. TSANG, K. H. CHU, T. Y. CHAN, R. Y. LUI & Z. X. CUI, 2012. Evolution and phylogeny of the mud shrimps (Crustacea: Decapoda) revealed from complete mitochondrial genomes. BMC Genomics, **13**: 631.

LIU, W. L. & R. Y. LIU, 2012a. *Michelea takeda* sp. nov. (Crustacea, Decapoda, Axiidea, Micheleidae) from the South China Sea. Crustaceana Monographs, **17**: 175-181.

LIU, W. L. & R. Y. LIU, 2012b. A new species of the genus *Austinogebia* Ngoc-Ho, 2001 (Crustacea, Decapoda, Gebiidea, Upogebiidae) from northern China. Zootaxa, **3243**: 59-64.

SHI, H. F., R. Y. LIU, Z. L. SHA & J. P. MA, 2012. Complete mitochondrial DNA sequence of *Stenopus hispidus* (Crustacea: Decapoda: Stenopodidea) and a novel tRNA gene cluster. Marine Genomics, **6**: 7-15.

ZHANG, J. L., F. S. XU & R. Y. LIU, 2012a. The Myidae (Mollusca, Bivalvia) from Chinese waters with description of a new species. Zootaxa, **3383**: 39-60.

ZHANG, J. L., F. S. XU & R. Y. LIU, 2012b. Community structure changes of macrobenthos in the South Yellow Sea. Chinese Journal of Oceanology and Limnology, **30**(2): 248-255.

2013

LIU, R. Y., 2013. Status of marine biodiversity of the China seas. PLoS ONE, **8**(1): 1-24.

LIU, W. L. & R. Y. LIU, 2013a. Two new species of *Acutigebia* (Crustacea: Decapoda: Gebiidea: Upogebiidae) from the South China Sea. The Raffles Bulletin of Zoology, **67**(2): 571-577.

LIU, W. L. & R. Y. LIU, 2013b. A new species of the genus *Mantisgebia* Sakai, 2006 (Crustacea, Decapoda, Gebiidea, Upogebiidae) from the South China Sea. Zootaxa, **3637**: 592-596.

WU, X. W., R. P. SUN & R. Y. LIU, 2013. A new species of *Eunice* (Polychaeta: Eunicidae) from Hainan Island, South China Sea. Chinese Journal of Oceanology and Limnology, **31**(1): 134-139.

2014

LI, Y., R. Y. LIU & K. D. XU, 2013. *Phytocoetes sinensis* n. sp. and *Telmatactis clavata* (Stimpson, 1855), two poorly known species of Metridioidea (Cnidaria: Anthozoa: Actiniaria) from Chinese waters. Zootaxa, **3637**: 113-122.

LIU, W. L. & R. Y. LIU, 2014. A new species of the genus *Nihonotrypaea* Manning and Tamaki, 1998 (Crustacea, Decapoda, Axiidea, Callianassidae) from the South China Sea. Zookeys, **457**: 35-44.

WU, X. W., R. P. SUN, R. Y. LIU & K. D. XU, 2013. Two new species of *Eunice* Cuvier, 1817 (Polychaeta, Eunicidae) from the coral reefs of Hainan Island with a key to 16 species of *Eunice* from China seas. Zootaxa, **3652**: 249-264.

XIAO, N., R. Y. LIU, S. YUAN & Z. L. SHA, 2013. A preliminary phylogenetic analysis of *Luidia* (Paxillosida: Luidiidae) from Chinese waters with cytochrome oxidase subunit I (COI) sequences. Journal of Ocean University of China, **12**: 459-468.

2015

CHENG, J., Z. L. SHA & R. Y. LIU, 2015. DNA barcoding of genus *Metapenaeopsis* (Decapoda: Penaeidae) and molecular phylogeny inferred from mitochondrial and nuclear DNA sequences. Biochemical Systematics and Ecology, **61**: 376-384.

YANG, C. H., Z. L. SHA, T. Y. CHAN & R. Y. LIU, 2015. Molecular phylogeny of the deep-sea penaeid shrimp genus *Parapenaeus* (Crustacea: Decapoda: Dendrobranchiata). Zoologica Scripta, **44**: 312-323.

First received 23 April 2020.
Final version accepted 28 July 2020.

DECAPODA, BRACHYURA

[When citing this volume, please refer to *Crustaceana* 93 (2020), 1233-1546]

Prof. Ruiyu Liu Memorial Issue

CONTRIBUTIONS TO THE KNOWLEDGE OF LEUCOSIIDAE VII. *LIUSIUS* GEN. NOV. (CRUSTACEA, BRACHYURA)

BY

YI-JIA SHIH[1,2]), PING-HO HO[2]) and BELLA S. GALIL[3,4])

[1]) Fisheries College, Jimei University, No. 43, Yindou Road, Jimei District, Xiamen City 361010, P.R. China

[2]) Department of Environment Biology and Fishery Science, National Taiwan Ocean University, 2 Pei-Ning Road, Keelung, Taiwan 201, R.O.C.

[3]) The Steinhardt Museum of Natural History, Tel-Aviv University, Tel-Aviv 69978, Israel

ABSTRACT

A new genus, *Liusius*, is established for the East Asian leucosiid crab *Leucosia longimaculata* Chen & Fang, 1991. The new genus differs from *Leucosia* s. str. and allied genera in having the shaft of the male first gonopod coiled eight times on its axis, bearing a distal setose lobe topped by a digitate process, and prominently swollen female vulvae opening distad. The molecular phylogenetic analysis of the 16S rDNA supports the morphological observations. The species is redescribed and illustrated.

Key words. — Decapoda, Leucosiidae, *Leucosia*, *Liusius*, new genus

RÉSUMÉ

Un nouveau genre, *Liusius*, est établi pour le crabe Leucosiidae *Leucosia longimaculata* Chen & Fang, 1991, d'Asie de l'Est. Le nouveau genre diffère de *Leucosia* s. str. et genres apparentés en ayant le canal du premier gonopode mâle enroulé huit fois sur son axe, portant un lobe distal pourvu de soies et surmonté d'un processus digitiforme, et l'ouverture des vulves femelles notablement enflées distalement. L'analyse phylogénétique moléculaire du 16S rDNA corrobore les observations morphologiques. L'espèce est redécrite et illustrée.

Mots clés. — Decapoda, Leucosiidae, *Leucosia*, *Liusius*, genre nouveau

INTRODUCTION

Since the millennium the definition of the leucosiid genus *Leucosia* Weber, 1795 has changed substantially, and several new genera have been established

[4]) Corresponding author; e-mail: bgalil@tauex.tau.ac.il

 | DOI:10.1163/9789004464353_004

(Galil, 2003a, b; 2005a, b; 2006a, b; Galil & Ng, 2010). Yet, the generic status of a number of species remain unresolved (see Ng et al., 2008). Herein, a new genus is established for *Leucosia longimaculata* Chen & Fang, 1991, described from specimens collected in the Taiwan Straits (Chen & Fang, 1991). The new genus differs from others formerly included within *Leucosia* in having the shaft of the male first gonopod coiled eight times on its axis, bearing a distal setose lobe topped by a digitate process, and possessing prominently swollen female vulvae facing distad.

The molecular analysis of the 16S ribosomal RNA (rDNA) supports the morphological observations. The species is redescribed and illustrated. Measurements provided are carapace width (including lateral teeth of carapace) × carapace length (including frontal lobes), in millimetres. Abbreviations: G1, male first gonopod; juv., juvenile; ovi., ovigerous.

The material examined is deposited in the following institutions: Academia Sinica (ASIZ), Taipei, Taiwan; National Taiwan Ocean University (NTOU), Keelung, Taiwan; National Kaohsiung University of Science and Technology (NKUST), Kaohsiung, Taiwan; National Museum of Marine Biology and Aquarium (NMMB), Pingtung, Taiwan; and Zoological Reference Collection (ZRC) of the Lee Kong Chian Natural History Museum, National University of Singapore, Singapore.

MATERIAL AND METHODS

Molecular analysis

DNA extraction and PCR amplification.— For the phylogenetic analysis, DNA was extracted from specimens of *Coleusia anatum* (Herbst, 1783), and *C. urania* (Herbst, 1801); *Euclosiana crosnieri* (Chen, 1989), *E. scitula* Galil, 2003, *E. unidentata* (De Haan, 1841); *Leucosia craniolaris* (Linnaeus, 1758); *Soceulia alainia* Galil, 2006, *S. major* (Chen & Ng, 2003); and *Liusius longimaculata* (table I). The DNA samples were obtained from the walking leg muscle or eggs. The PUREENE D-7000A DNA Isolation Kit (QBIOgene, Carlsbad, CA) was used to extraction total genomic DNA by the manual. Total DNA was storage at −20°C refrigerator for the next step. 16Sar-L: 5′-CGCCTGTTTATCAAAAACAT-3′ and 16Sbr-H: 5′-CCGGTCTGAACTCAGATCACGT-3′ was the primer used (Palumbi et al., 1991). Volumes of each reaction was performed in 25 μl including DNA template with 1 μl, mixed with 12.5 μl Master Mix buffer (2.5 unit/25 μl), 0.5 μl for each forward and reverse primer (1 μl, 0.4 μM) and 10.5 μl distillation-distillation H_2O (ddH_2O). Thermal cycling conditions for PCR included an initial denaturing temperature of 94°C for 2 min, followed by 40 cycles of 94°C for

TABLE I

List of species, their catalogue numbers and GenBank accession numbers for gene sequences, used in this study

No.	Species	Catalogue number	GenBank accession number of 16S rDNA
Subfamily Leucosiinae Miers, 1886			
Genus *Coleusia* Galil, 2006			
1	*C. anatum* (Herbst, 1783)	NTU-L20140815	MH673437
2	*C. urania* (Herbst, 1801)	NTOU-L19889114	MH673436
Genus *Euclosiana* Galil, 2003			
3	*E. crosnieri* (Chen, 1989)	NTOU-L20140518	MH673425
4	*E. scitula* Galil, 2003	ZRC 2009.0856	MH673427
5	*E. unidentata* (de Haan, 1841)	ZRC 2009.0283	MH673426
Genus *Leucosia* Weber, 1795			
6	*L. craniolaris* (Linnaeus, 1758)	ZRC 2014.0177	MH673433
Genus *Liusius* gen. nov.			
7	*L. longimaculata* (Chen & Fang, 1991)	ASIZ-C133	MH673434
Genus *Soceulia* Galil, 2006			
8	*S. alainia* Galil, 2006	ZRC 2009.0554	MH673431
9	*S. major* (Chen & Ng, 2003)	ZRC 2003.0192	MH673432

35 s, 50°C~55°C for 35 s, and 72°C for 60 s, and a final extension step at 72°C for 10 min. Following PCR assay, 5 μl of each PCR product was confirmed by electrophoresis on a 1% agarose gel. The gel was strained by 0.5 μg/ml ethidium bromide staining, and imaged by ultraviolet transillumination system. All positive PCR amplicons were cut off from gel, purified and sequenced by Mission Biotech Ltd. (Taipei, Taiwan).

Sequence analyses.— Nine 16S rDNA sequences were obtained (table I). The sequences from the individual specimens were aligned using CLUSTAL-W program by BioEdit version 7.2.3 with default gap penalties and modified. After alignment, the 16S rDNA sequence was 410 bp. The best-fitting model for sequence evolution was determined by jModeltest (v0.1.1) (Posada, 2008), selected by the Akaike information criterion (AIC) and Bayesian information criterion (BIC). The best model both was obtained to GTR+I+G+invgamma in this study. Both maximum likelihood (ML) and Bayesian (BI) analyses were performed. The ML analysis was carried out calculating by using the analyses in RAxMLv7.0.3 (Stamatakis, 2006) and BI analysis in MrBayes v3.2.6 program (Ronquist & Huelsenbeck, 2003). ML inferences were set with the best-known tree and the node support values inferred out of 200 trees and 1000 bootstraps. BI analysis and All Markov chain Monte Carlo (MCMC) algorithms ran for 1 000 000 generations, sampling one tree every 1,000 generations. Both Bayesian posterior probability (PP) and ML bootstrap values (BP) for clades were only displayed when they were greater than 0.70 and 50%. Genetic diversities between species

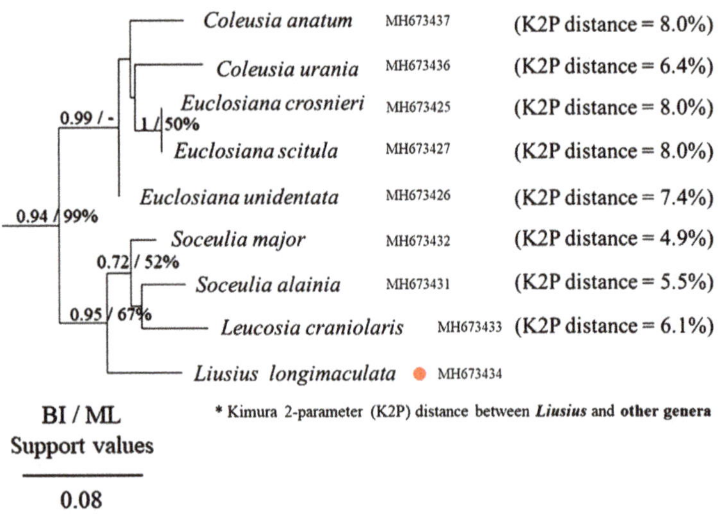

Fig. 1. Phylogenetic analysis of leucosiid crabs based on 16S rDNA by Bayesian inference (BI) tree. Probability values and bootstrap values at the nodes represent support values for BI and maximum likelihood (ML). The values above 0.70 and 50% are shown, respectively.

depended on pairwise estimates of Kimura 2-parameter (K2P) distance (Kimura, 1980) by MEGA v.5.2.2 (Tamura et al., 2011).

The topology of the phylogenetic tree (fig. 1) was obtained through BI and ML analysis, performed by Bayesian inference. The result supports the establishment of *Liusius* gen. nov. for *Leucosia longimaculata* (BI Pp/ML BP = 0.95/67%), as well as revealing that it is more closely related to *Leucosia* and *Soceulia*, than to *Coleusia* and *Euclosiana*.

SYSTEMATICS

Family LEUCOSIIDAE Samouelle, 1819
Subfamily LEUCOSIINAE Samouelle, 1819
Liusius gen. nov.

Type species.— *Leucosia longimaculata* Chen & Fang, 1991; gender: feminine.

Description.— Carapace subpentagonal, globose; hepatic region slightly swollen; regions of carapace indistinct. Dorsal surface of carapace smooth, minutely punctate. Frontal region produced, upcurved, laterally concave. Male abdominal sulcus deep, nearly reaching buccal cavity; its lateral margin bearing a distinct ridge fitting into suture between abdominal segments. Male abdomen with 2nd segment minute; segments 3-5 fused, proximally slightly swollen; 6th segment large, trapezoid, as long as fused segment; telson triangular; shaft of G1 distally

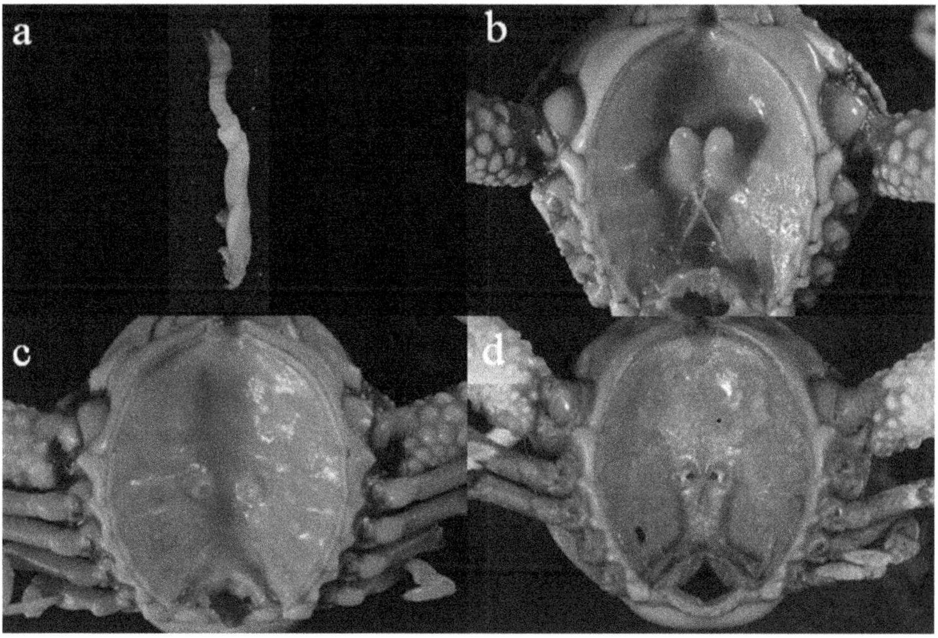

Fig. 2. a, Gonopod of *Liusius longimaculata* (Chen & Fang, 1991) (holotype, male, 28.0 × 30.8 mm, No. S54-B5); b, vulvae of *L. longimaculata* (female, 28.6 × 30.0 mm, NTOU-L20010924); c, vulvae of *Euclosiana crosnieri* (Chen, 1989) (female, 25.2 × 27.3 mm, NTOU-L0015); d, vulvae of *Coleusia urania* (Herbst, 1801) (female, 37.3 × 40.0 mm, NTOU-L19880114).

coiled eight times on its axis (figs. 2a, 3a, b), terminating in setose lobe topped by slender process. G2 short, slender, distally leaf-like. Female abdomen with segments 3-6 fused, greatly enlarged, shield-like; telson triangular; vulvae swollen, hemispherical, facing distad (fig. 2b, 3c, d).

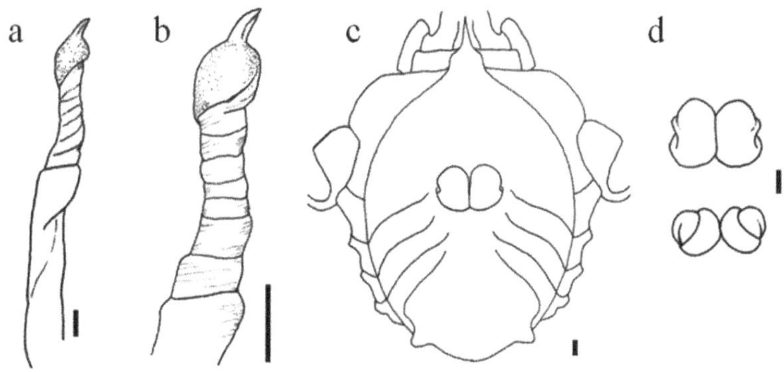

Fig. 3. *Liusius longimaculata* (Chen & Fang, 1991), a-b, male first pleopod (G1) of right side; c, vulvae in abdominal cavity; d, views of female vulvae (above: facing viewer; below: anterior facing). Scales = 1.0 mm.

Remarks.— *Liusius* gen. nov. differs greatly from *Coleusia* Galil, 2006, *Soceulia* Galil, 2006, and *Euclosiana* Galil & Ng, 2010, in lacking a denticle medially on the sixth abdominal segment of male, and the shape of G1 and vulvae. G1 is coiled on its axis thrice in *Euclosiana* and *Coleusia*, once in *Soceulia*, but eight times in *Liusius*, and the vulvae are greatly swollen, hemispherical, rather than only slightly swollen, as in *Euclosiana* and *Coleusia* (fig. 2c, d).

The divergence of pairwise distance is generally between 4.9% to 8% for intergeneric taxa. The 16S rDNA marker is commonly used for identification of brachyurans. However, the threshold of 16S rDNA divergence of intergenic K2P distance are not fully agreed upon. We regarded the value as high enough to warrant establishing a new genus.

Etymology.— The genus is named in honour of Dr. Liu Ruiyu, a pioneer of crustaceology in China, for his outstanding contributions to marine biology and marine benthic ecology. Gender masculine.

Liusius longimaculata (Chen & Fang, 1991)
(figs. 2-4)

Leucosia longimaculata Chen & Fang, 1991: 183, fig.1; Chen & Sun, 2002: 426-427, fig. 192; Ng et al., 2008: 95.
Leucosia anatum (Herbst, 1783) — Jeng et al., 1998: 121.

Material examined.— Taiwan strait: holotype, one male, 28.0 × 30.8 mm, No. S54-B5, 5 Jul. 1988; allotype, one female, 20.0 × 22.5 mm, No. S19-B5, 29 Aug. 1987; paratypes, one male, 26.2 × 29.0 mm, one female (juv.), 12.0 × 14.0 mm, No. 57Z46-B4, 14 Jan. 1984. — Taiwan: YiLan: Daili, one male, 27.2 × 30.5 mm, NMNS-19961120, 20 Nov. 1996; Dashi, one male, 26.7 × 29.5 mm, ASIZ-C133, 18 Aug. 1985; Nanfangau, one female (ovi.), 28.6 × 30.0 mm, NTOU- L20010924, 24 Sep. 2001. Tainan: Anpeng, one female, 28.1 × 30.1 mm, NKUST, 16 Feb. 2014.

Description.— Dorsal surface of carapace shiny, sparsely punctate anteriorly. Frontal margin planar, medially with minute rounded denticle. Thoracic sinus with row of six granules above cheliped basis, median four largest. Thoracic sinus anteriorly defined by granulate, rounded, slightly overhanging margin of pterygostomian region. Propodus lenticular, its upper margin rounded, smooth, somewhat carinate; lower margin bearing two rows of conic granules diminishing in size distally. Thoracic sternites smooth, sparsely punctate. Sixth abdominal segment in male slightly concave lengthwise; distal margin arched; telson rounded. Shaft of G1 slightly sinuous, stout, tightly coiled eight times on its axis distally, distally bulbous, setose, topped by digitate process. G2 short, slender, distally leaf-like. Female vulvae swollen, hemispherical, facing distad.

Remarks.— Chen & Fang (1991: 185) described *Leucosia longimaculata*'s "First male pleopod stout, with six to eight spirals. . . .". However, their drawing (fig. 1f) reveals eight whorls. Examination of several specimens, including the

Fig. 4. Colour photo of *Liusius longimaculata* (Chen & Fang, 1991) (female, 28.1 × 30.1 mm, NKUST).

holotype and paratypes, listed above, accord with the drawing, though it is noted that the margins between whorls four and six may not be clearly discernible. The vulvae of the juvenile female differ from the adult, only slightly swollen and undeveloped.

Colour (in alcohol, fig. 4).— Carapace dorsally buff with pale areolate pattern laterally, marginal granules orange-red. Chelipeds darker buff, granules orange-red; fingers reddish-brown proximally, white distally. Pereiopods marked with brownish- orange bands.

ACKNOWLEDGEMENTS

We express our deepest respect and gratitude to Dr. Ming-Shiou Jeng (Biodiversity Research Center, Academia Sinica), Dr. Shih-Chang Chuang (Fisheries Research Institute, COA), Dr. Chia-Wei Lin (National Museum of Marine Biology and Aquarum, NMMBA), Mr. Chun-Yang Wei, Mr. Zhi-Wen Jeng, Mr. Yen-Feng Lee, and Mr. Ray-Ming Wang for the material from Taiwan. We are grateful to Prof. Peter K. L. Ng (Lee Kong Chian Natural History Museum (formerly the Raffles Museum of Biodiversity Research), National University of Singapore) for his valuable suggestions and comments.

REFERENCES

CHEN, H. L. & S. H. FANG, 1991. A new species of *Leucosia* (Crustacea: Brachyura: Leucosiidae) from the Taiwan straits. Raffles Bull. Zool., **39**(1): 183-185.

COSTA, F. O., J. R. DEWAARD, J. BOUTILLIER, S. RATNASINGHAM, R. T. DOOH, M. HAJIBABAEI & P. D. HEBERT, 2007. Biological identifications through DNA barcodes: the case of the Crustacea. Can. J. Fish. Aquat. Sci., **64**: 272-295.

GALIL, B. S., 2003a. Contribution to the knowledge of Leucosiidae I. The identity of *Leucosia craniolaris* (Linnaeus, 1758), and redefinition of the genus *Leucosia* Weber, 1795 (Crustacea: Brachyura). Zool. Med., **77**(20): 181-191.

GALIL, B. S., 2003b. Contribution to the knowledge of Leucosiidae II. *Euclosia* gen. nov. (Crustacea: Brachyura). Zool. Med., **77**(20): 331-347.

GALIL, B. S., 2005a. Contributions to the knowledge of Leucosiidae III. *Urnalana* gen. nov. (Crustacea: Brachyura). Zool. Med., **79**(2): 9-40.

GALIL, B. S., 2005b. Contributions to the knowledge of Leucosiidae IV. *Seulocia* gen. nov. (Crustacea: Brachyura). Zool. Med., **79**(2): 41-59.

GALIL, B. S., 2006a. Contributions to the knowledge of Leucosiidae V. *Coleusia* gen. nov. (Crustacea: Brachyura). Zool. Med., **80**(5): 55-69.

GALIL, B. S., 2006b. Contributions to the knowledge of Leucosiidae VI. *Soceulia* gen. nov. (Crustacea: Brachyura). Zool. Med., **80**(6): 71-79.

GALIL, B. S. & P. K. L. NG, 2010. On a collection of calappoid and leucosioid crabs (Decapoda, Brachyura) from Vanuatu, with description of a new species of Leucosiidae. Crustac. Monogr., **11**: 139-152.

JENG, M. S., K. T. SHAO, C. S. TZENG, F. L. FENG & S. H. WU, 1998. Survey and monitoring of natural ecological resources in the Northeast Coast National Scenic Area (IV). (Northeast Coast National Scenic Area Administration, Bureau of the Ministry of Transportation and Communications, Taipei).

KIMURA, M., 1980. A simple method for estimating evolutionary rates of base substitutions through comparative studies of nucleotide sequences. J. Mol. Evol., **16**: 111-120.

NG, P. K. L., D. GUINOT & P. J. F. DAVIE, 2008. Systema Brachyurorum: part I. An annotated checklist of extant brachyuran crabs of the world. Raffles Bull. Zool., (Suppl.) **17**: 1-286.

PALUMBI, S. R., A. MARTIN, S. ROMANO, W. O. MCMILLAN, L. STICE & G. GRABOWSKI, 1991. The simple fool's guide to PCR. (University of Hawaii Zoology Department, Honolulu).

POSADA, D., 2008. jModelTest: phylogenetic model averaging. Mol. Biol. Evol., **25**: 1253-1256.

RONQUIST, F. & J. P. HUELSENBECK, 2003. MrBayes 3: Bayesian phylogenetic inference under mixed models. Bioinformatics, **19**: 1572-1574.

STAMATAKIS, A., 2006. RAxML-VI-HPC: maximum likelihood-based phylogenetic analyses with thousands of taxa and mixed models. Bioinformatics, **22**: 2688-2690.

TAMURA, K., D. PETERSON, N. PETERSON, G. STECHER, M. NEI & S. KUMAR, 2011. MEGA5: molecular evolutionary genetics analysis using maximum likelihood, evolutionary distance, and maximum parsimony methods. Mol. Biol. Evol., **28**: 2731-2739.

VRIJENHOEK, R., 1994. DNA primers for amplification of mitochondrial cytochrome coxidase subunit I from diverse metazoan invertebrates. Mol. Mar. Biol Biotechnol., **3**(5): 294-299.

First received 12 September 2018.
Final version accepted 26 August 2020.

[When citing this volume, please refer to *Crustaceana* 93 (2020), 1233-1546]

Prof. Ruiyu Liu Memorial Issue

THE COMPLETE MITOCHONDRIAL GENOME OF THE FRESHWATER CRAB *LONGPOTAMON KENLIENSE* (DECAPODA, BRACHYURA, POTAMIDAE) WITH PHYLOGENETIC CONSIDERATION

BY

YIFAN WANG[1,2]), SHUXIN XU[1]), CHUNCHAO ZHU[1]), XINNAN JIA[1]), XIANMIN ZHOU[3]) and JIEXIN ZOU[1,3,4])

[1]) Research lab of Freshwater Crustacean Decapoda & Paragonimus, School of Basic Medical Sciences, Nanchang University, 461 Bayi Avenue, Nanchang City, Jiangxi Province 330006, People's Republic of China

[2]) Institute of Pathogen Biology, Jiangxi Academy of Medical Sciences, 461 Bayi Avenue, Nanchang City, Jiangxi Province 330006, People's Republic of China

[3]) Key Laboratory of Poyang Lake Environment and Resource Utilization, Ministry of Education, Nanchang University, 1299 Xuefu Avenue, Nanchang City, Jiangxi Province 330031, People's Republic of China

ABSTRACT

The authors herein report for the first time the complete sequence of the mitochondrial genome of *Longpotamon kenliense*. The results showed that the mitochondrial genome of *L. kenliense* is 18,499 bp in length, shares 37 genes and 1 control region with the typical metazoan mitochondrial genome, and has a strong A + T bias (74.48%), i.e., a characteristic of the metazoan mitochondrial genome. In addition, the authors used the 13 protein-coding genes from the mitochondrial genomes of *L. kenliense* and 67 other species of Brachyura available from the NCBI to estimate divergence times. The divergence time of *L. kenliense* was 14.39 Ma, which is close to the divergence times of *L. xiushuiense* and *L. yangtsekiense*, but later than that of *Sinopotamon yaanense*. The results indicate that the genera *Sinopotamon* and *Longpotamon* are closely related and that the genus *Sinopotamon* has a longer history, which may have been caused by geological events.

Key words. — Freshwater crab, *Longpotamon kenliense*, mitochondrial genome, phylogenetic relationship, divergence time estimation

RÉSUMÉ

Les auteurs rapportent ici pour la première fois la séquence complète du génome mitochondrial de *Longpotamon kenliense*. Les résultats montrent que le génome mitochondrial de *L. kenliense* a une longueur de 18 499 pb, partage 37 gènes et une région de contrôle avec le génome mitochondrial typique des métazoaires, et a un fort biais A + T (74,48%), i.e., une caractéristique du génome

[4]) Corresponding author; e-mail: jxzou@ncu.edu.cn

 | DOI:10.1163/9789004464353_005

mitochondrial des métazoaires. De plus les auteurs ont utilisé 13 gènes codant pour des protéines des génomes mitochondriaux de *L. kenliense* et de 67 autres espèces de Brachyoures disponibles à partir du NCBI pour estimer les temps de divergence. Le temps de divergence de *L. kenliense* a été de 14,39 Ma, ce qui est proche du temps de divergence de *L. xiushuiense* et *L. yangtsekiense*, mais plus tardif que celui de *Sinopotamon yaanense*. Les résultats indiquent que les genres *Sinopotamon* et *Longpotamon* sont étroitement liés et que le genre *Sinopotamon* est plus ancien, ce qui est peut être dû à des évènements géologiques.

Mots clés. — Crabe d'eau douce, *Longpotamon kenliense*, génome mitochondrial, relation phylogénétique, estimation des temps de divergence

INTRODUCTION

The genus *Longpotamon* (Crustacea: Malacostraca: Decapoda: Brachyura: Potamidae) is one of the most abundant species of true freshwater crabs in the world and is endemic to eastern Asia (Ng et al., 2008). In mainland China, the distribution of the genus *Longpotamon* covers a vast area from mountains and hills to plains. Due to geographical restrictions by various mountain ranges, river basins, and other natural regions (Dai, 1999), there is a high degree of species diversity. In mainland China, the genus is distributed in Guizhou Province and other provinces and mainly lives in mountain streams at altitudes of 200–500 meters above sea level or below (Dai, 1999). To understand the mechanism underlying the distribution of this species, it is important to know the genetic basis of its morphological variations. In this study, we first report the complete sequence of the mitochondrial genome of *Longpotamon kenliense* (Dai, 1997) and provide data (GenBank accession number MK584299) for species identification and future studies on phylogenetics and species conservation.

MATERIAL AND METHODS

Specimen collection and classification

In this study, *Longpotamon kenliense* specimens were collected by the laboratory team in Taipan village, Taipan Township, Taijiang County, Southeast Guizhou Autonomous Prefecture, Guizhou Province, China. Specimens were preserved in 95% ethanol prior to DNA extraction and stored at room temperature, and the fixing solution was replaced approximately every 24 hours and repeated 2 to 3 times to ensure that the sample was sufficiently fixed.

Morphological identification was based on the book "Fauna Sinica: Arthropoda Crustacea Malacostraca Decapod Parathelphusidae Potamidae" (Dai, 1999) and the holotype photo database of specimens deposited in the National Zoological Museum of the Institute of Zoology, Chinese Academy of Sciences, Beijing. The identification criteria included the features of the first gonopods (G1) and their natural position and the shapes of the anterolateral margin, carapace and chelipeds. The collected specimens were confirmed to be *Longpotamon kenliense*.

Sequencing and assembling the mitochondrial genome

In this study, the specimens of *Longpotamon kenliense* were collected in Taipan village, Taipan Township, Taijiang County, Southeast Guizhou Autonomous Prefecture, Guizhou Province, China (26°40′46.578″N 108°10′3.774″E, altitude: 668 m). The specimens were preserved in 95% ethanol prior to DNA extraction and stored at room temperature. An Aidlab Genomic DNA Extraction Kit (Aidlab, Beijing, P.R. China) was used according to manufacturer's instructions to extract all the DNA from the muscle tissue of an individual specimen. First, according to the general conserved region, primers were designed to amplify short segments of the 16S, 12S, COX1, ATP6, COX3, ND4, CYTB, and ND1 genes (Table I). Then, specific primers were designed to amplify the remaining sequences based on the conserved sequences that had been amplified; LA Taq polymerase was used for this series of long fragment amplifications. The PCR system consisted of 0.5 μl of TaKaRa LA Taq (5 U/μl), 5 μl of 10× LA Taq Buffer II (Mg^{2+} Plus), 8 μl of dNTP mixture, 2 μl of template DNA, 2 μl of the two primers, and enough ddH_2O to reach 50 μl. The PCR was carried out with LA Taq polymerase for 35 cycles at 94°C for 30 s and annealing at 50°C for 30 s, followed by extension at 72°C for 1 min per 1 kb. The PCR products were cloned into pMD18-T vector (Takara, Tokyo, Japan) for sequencing or directly sequenced using the dideoxynucleotide procedure with an ABI 3730 automatic DNA sequencer (Applied Biosystems, Foster City, CA, U.S.A.).

Mitochondrial genome annotation

Sequences were assembled using DNAStar software and manually adjusted to generate the complete sequence of mitochondrial DNA (Burland, 2000). Most tRNAs were annotated using the website server tRNA-scan SE 1.21 (Lowe & Chan, 2016). Protein-coding genes (PCGs) were annotated using SeqBuilder software (Burland, 2000). Other tRNAs and rRNAs were aligned to those of other, similar Brachyura species available from the NCBI, such as *Sinopotamon xiushuiense*, *Geothelphusa dehaani*, *Huananpotamon lichuanense* and *Potamiscus motuoensis*. Then BLAST was used to determine their locations. The mitochondrial genome map was constructed using the web server OGDraw v1.1 (http://ogdraw.mpimp-golm.mpg.de).

TABLE I
Primers for mitochondrial genome used in the present study

Fragment No.	Gene or region	Primer name	Sequence (5′ →3′)	Length (bp)
F1	*16S*	HNF1	AGGTCGAACAGACCTTCT	533
		HNR1	GGTTTCTTTGCCTGTTTATC	
F2	*16S-12S*	HNF2	GAGAGCTTTTCTTTTGTCCGACC	1444
		HNR2	GTAAGTTTAATTTAGGAAGGAGG	
F3	*12S*	HNF3	CAATTCTAGATACACTTTCCAG	618
		HNR3	GCCAGCATTCGCGGTTAAAC	
F4	*12S-ND2*	HNF4	CCTTAAAGCTTATTAGCTTCC	1077
		HNR4	GAAAGGGGTAGATAGAGAGG	
F5	*ND2*	HNF5	CTCCTCCTTAAACTAGGTAC	340
		HNR5	CTGAAGAGATAATAGAATAG	
F6	*ND2-COX1*	HNF6	GTCAGCAATAATCGGGGC	1022
		HNR6	GGAATGCTATATCAGGAGCTCC	
F7	*COX1*	HNF7	GAAGACTTATTGGTAATGATC	1164
		HNR7	ATCCAAGAAAATGTTGTGGG	
F8	*COX1-COX2*	HNF8	GGAGGAATTGCCCACTGATTCCC	874
		HNR8	GGGATTATAGTTCGGTTATCGAC	
F9	*COX2*	HNF9	GACAATTGGTCACCAATG	387
		HNR9	ATCGAAGAGGAAACTCAGT	
F10	*COX2-ATP6*	HNF10	GCTGACGCCGTACCAGGC	675
		HNR10	GTAGAAGCACCGAAGTGAG	
F11	*ATP6*	HNF11	GATTATCGACTTTGTTAGG	415
		HNR11	CGTACCCTTAAAGTCCCAGG	
F12	*ATP6-COX3*	HNF12	CACTTTCTTTATCTCTTCC	960
		HNR12	GAACCAAAAGCAGAGTCAGC	
F13	*COX3*	HNF13	CAGTAACATGAGCCCATCATG	401
		HNR13	TGATTGGAAATCAAATGTAC	
F14	*COX3-ND5*	HNF14	GTACTACATTCCTATTTATCTGC	1968
		HNR14	CGACTTTGAGTCAATTAGGGG	
F15	*ND5*	HNF15	CGAATATCTTGGTAATCTCC	332
		HNR15	GCGTTAGTACATTCTTCTAC	
F16	*ND5-ND4*	HNF16	TTAGCCCCTAATCCAGCTA	1652
		HNR16	GATAAGTAGGTTAAAAGCAGG	
F17	*ND4-CYTB*	HNF17	ACCTAATTCCAGTTAATTG	2019
		HNR17	CTCCTACTATTCAAGTATGG	
F18	*CYTB*	HNF18	GCATTTTCCAGAGTAGGTC	842
		HNR18	GGACGAGCACCGATTCAAG	
F19	*CYTB-ND1*	HNF19	TCGCTCTTGTCTTGTCTG	951
		HNR19	GGCTTCGAGCTGTAGCTC	
F20	*ND1*	HNF20	CAGCTAAACAAGACCCAAACC	540
		HNR20	GTGTTTTGGTAGGTGTAGC	
F21	*ND1-16S*	HNF21	GATACAACAGAAAAAGTAGGCTCC	581
		HNR21	GTGGGTAAGTAAAAGATCC	

Phylogenetic analysis

In the NCBI database, 66 brachyuran mitochondrial genomes have been published; only eight are from freshwater crabs: *Sinopotamon yaanense* (see Yuhui et al., 2017), *Longpotamon yangtsekiense* (see Yuhui et al., 2017), *Longpotamon xiushuiense* (see Wang et al., 2016), *Chinapotamon maolanense* (unpubl.), *Geothelphusa dehaani* (see Segawa & Aotsuka, 2005), *Huananpotamon lichuanense* (see Bai et al., 2018), *Potamiscus motuoensis* (unpubl.), and *Somanniathelphusa boyangensis* (see Jia et al., 2018). A phylogenetic analysis was performed using 13 PCG sequences for Brachyura from the NCBI together with the sequence data of the mitochondrial genome of *Longpotamon kenliense* (table II); *Clibanarius infraspinatus* (GenBank accession no. LN626968) and *Neopetrolisthes maculatus* (GenBank accession no. KC107816) were chosen as outgroups. In total, 69 species covering almost all Brachyura groups were analyzed. MAFFT software (Katoh et al., 2005) was used for sequences alignment. Gblocks Server (http://molevol.cmima.csic.es/castresana/Gblocks_server.html) was used to remove gaps. DAMBE software (Xia & Xie, 2001) was used to detect saturation of nucleotide substitutions for phylogenetic analysis. Based on the obtained nucleotide sequence, the best-fitting evolutionary model for the maximum likelihood (ML) method was selected as GTR + G + I using ModelGenerator (Vazquez et al., 2000). Then, the ML tree was established using MEGA 6.06 (Tamura et al., 2013), and 1000 bootstrap replicates were performed. For the Bayesian inference (BI) analyses, the best-fitting evolutionary model (GTR + G + I) was determined using MrModeltest 2.3. The BI analyses were performed with MrBayes 3.2.6 (Ronquist et al., 2012), which used Markov chain Monte Carlo (MCMC) with four chains run over 2 million generations. Sampling was performed every 1000 generations, and the initial 500 000 steps were discarded as burn-in. Tracer 1.6 was used to view the ESS values (each with an ESS value higher than 200). The consensus BI trees were visualized using FigTree 1.4.0 software.

Divergence time estimation

To further study the taxonomic status of *Longpotamon kenliense* among freshwater crabs, we estimated the divergence time of *L. kenliense*. The 66 mitochondrial genome sequences for Brachyura available from the NCBI were selected to estimate divergence times using 13 PCGs of the mitochondrial genome (table II). Five groups of fossil times were selected for divergence time corrections (table III) (Schweitzer & Feldmann, 2000; Karasawa et al., 2008; Karasawa et al., 2011; Van Bakel et al., 2012). The divergence time of *L. kenliense* was calculated based on a strict molecular clock using BEAST 1.8.2 software (Drummond & Rambaut, 2007) and normal prior distribution. The best-fitting evolutionary model was GTR + G +

TABLE II
Brachyura species included in the present phylogenetic analysis

Species	Family	Size (bp)	Accession number
Scylla olivacea	Portunidae	15 723	FJ827760
Scylla serrata	Portunidae	15 721	HM590866
Scylla paramamosain	Portunidae	15 824	JX457150
Scylla tranquebarica	Portunidae	15 833	FJ827759
Charybdis bimaculata	Portunidae	15 714	MG489891
Charybdis feriata	Portunidae	15 660	KF386147
Charybdis japonica	Portunidae	15 738	FJ460517
Charybdis natator	Portunidae	15 664	MF285241
Thalamita crenata	Portunidae	15 787	LK391945
Callinectes sapidus	Portunidae	16 263	AY363392
Portunus sanguinolentus	Portunidae	16 024	KT438509
Portunus pelagicus	Portunidae	16 157	KR153996
Portunus trituberculatus	Portunidae	16 026	AB093006
Monomia gladiator	Portunidae	15 878	MG770549
Helice latimera	Varunidae	16 246	KU589291
Helice tientsinensis	Varunidae	16 212	KR336555
Helicana wuana	Varunidae	16 359	KX344898
Cyclograpsus granulosus	Varunidae	16 300	LN624373
Eriocheir hepuensis	Varunidae	16 335	FJ455506
Eriocheir japonica	Varunidae	16 352	FJ455505
Eriocheir sinensis	Varunidae	16 353	KP126617
Varuna yui	Varunidae	15 915	MG756602
Hemigrapsus sanguineus	Varunidae	16 275	KX456205
Austinograea alayseae	Bythograeidae	15 611	KC851803
Austinograea rodriguezensis	Bythograeidae	15 611	JQ035658
Gandalfus puia	Bythograeidae	15 548	KR002727
Gandalfus yunohana	Bythograeidae	15 567	EU647222
Segonzacia mesatlantica	Bythograeidae	15 521	KY541839
Parasesarma tripectinis	Sesarmidae	15 612	KU343209
Clistocoeloma sinense	Sesarmidae	15 706	KU589292
Metopaulias depressus	Sesarmidae	15 765	KX118277
Sesarmops sinensis	Sesarmidae	15 905	KR336554
Sesarma neglectum	Sesarmidae	15 920	KX156954
Chinapotamon maolanense	Potamidae	17 130	MT134100
Potamiscus motuoensis	Potamidae	17 971	KY285013
Geothelphusa dehaani	Potamidae	18 197	AB187570
Longpotamon xiushuiense	Potamidae	18 460	KU042041
Longpotamon kenliense	Potamidae	18 499	MK584299
Longpotamon yangtsekiense	Potamidae	17 126	KY785880
Sinopotamon yaanense	Potamidae	17 885	KY785879
Huananpotamon lichuanse	Potamidae	15 380	KX639824
Ranina ranina	Raninidae	15 557	AB752308
Lyreidus brevifrons	Raninidae	16 112	KM983394
Umalia orientalis	Raninidae	15 466	KM365084
Homologenus malayensis	Homolidae	15 793	KJ612407

TABLE II
(Continued)

Species	Family	Size (bp)	Accession number
Moloha majora	Homolidae	15 903	KT182069
Maja crispata	Majidae	16 592	KY650651
Maja squinado	Majidae	16 598	KY650652
Grapsus tenuicrustatus	Grapsidae	15 858	KT878721
Pachygrapsus crassipes	Grapsidae	15 652	KC878511
Ocypode cordimanus	Ocypodidae	15 604	KT896743
Ocypode ceratophthalmus	Ocypodidae	15 564	LN611669
Xenograpsus testudinatus	Xenograpsidae	15 798	EU727203
Xenograpsus ngatama	Xenograpsidae	16 106	KY985236
Macrophthalmus japonicus	Macrophthalmidae	16 170	KU343211
Mictyris longicarpus	Mictyridae	15 548	LN611670
Ilyoplax deschampsi	Dotillidae	15 460	JF909979
Somanniathelphusa boyangensis	Parathelphusidae	17 032	KU042042
Damithrax spinosissimus	Mithracidae	15 817	KM405516
Pyrhila pisum	Leucosiidae	15 516	KU343210
Ashtoret lunaris	Matutidae	15 807	LK391941
Atergatis floridus	Xanthidae	16 180	MG792341
Atergatis integerrimus	Xanthidae	15 924	MG786939
Leptodius sanguineus	Xanthidae	15 480	KT896744
Myomenippe fornasinii	Menippidae	15 658	LK391943
Chaceon granulatus	Geryonidae	16 135	AB769383
Pseudocarcinus gigas	Eriphiidae	15 515	AY562127

I according to ModelGenerator (Vazquez et al., 2000). The Yule process (Gernhard, 2008) was used as the tree prior for the shared and random starting trees. In total, 500 000 000 generations were run, and the parameters were logged every 5000 generations, with the first 10% discarded as burn-in. Tracer 1.6 was used to determine the ESS value (each parameter ESS value was greater than 200). TreeAnnotator v1.8.2 software was used to generate the tree, and the divergence time was visualized using FigTree 1.4.0 software.

TABLE III
Fossil calibrations used in divergence time estimation

Classification	Species or genus with available mitochondrial genome data	Age
Family Portunidae	genus *Portunus*, genus *Callinectes*	33.9–56 Ma[a]
Family Portunidae	genus *Scylla*	33.9–37 Ma
Family Portunidae	genus *Charybdis*	33.9–35 Ma
Subsection Raninoida	*Umalia orientalis*, *Lyreidus brevifrons*, *Ranina ranina*	130.8–133.9 Ma
Family Homolidae	*Homologenus malayensis*, *Moloha majora*	145–152.1 Ma

[a] Ma, million years from the present.

TABLE IV
Nucleotide composition in 13 PCGs of the *Longpotamon kenliense* mitogenomes

Gene	Base composition (%)				A + T (%)	Length (bp)	Codon		Strand
	A	T	C	G			Start	Stop	
COX1	28.65	36.71	19.23	15.40	65.37	1539	ATG	TAA	H
COX2	31.83	37.65	19.62	10.90	69.48	688	ATG	T	H
COX3	27.02	40.53	18.56	13.89	67.55	792	ATG	TAA	H
CYTB	27.22	40.88	19.30	12.60	68.11	1135	ATG	T	H
ATP6	30.65	39.88	18.75	10.71	70.54	672	ATA	TAA	H
ATP8	28.93	50.94	15.72	4.40	79.87	159	ATG	TAG	H
ND1	438	26.35	19.79	8.23	71.98	960	ATT	TAA	L
ND2	29.35	44.58	18.41	7.66	73.93	1005	ATG	TAA	H
ND3	28.81	44.63	16.10	10.45	73.45	354	ATC	TAA	H
ND4	42.68	29.90	19.66	7.77	72.57	1338	ATG	TAA	L
ND4L	46.86	27.72	20.46	4.95	74.59	303	ATG	TAA	L
ND5	41.93	31.06	18.16	8.85	72.99	1729	ATG	T	L
ND6	26.55	48.30	19.16	5.99	74.85	501	ATA	TAA	H
Total	34.35	36.52	18.94	10.20	70.87	11 162	–	–	–

RESULTS

Genome structure and organization

Longpotamon kenliense had a full-length mitochondrial genome of 18 499 bp, which shared 37 genes (13 PCGs, 22 tRNA genes, and 2 rRNA genes) and 1 control region with the typical metazoan mitochondrial genome and showed strong A + T bias (74.48%). Among the genes, 23 genes were located in the H-strand, and 14 genes were located in the L-strand (table IV).

The length of protein-coding genes in the freshwater crab was approximately 14 800 bp, but the length of the noncoding regions was quite different (Ki et al., 2009). The total length of the protein-coding genes in *Longpotamon kenliense* was 14 739 bp, and the total length of the noncoding regions was 3760 bp. Overlapping and noncoding regions are frequently found in the mitochondrial genomes of decapods, which is also the case in freshwater crabs (Yamauchi et al., 2003; Sun et al., 2005; Ivey & Santos, 2007; Shen et al., 2007; Liu & Cui, 2010).

The mitochondrial genome of *Longpotamon kenliense* had 19 noncoding regions, which were between 1-1,664 bp in length, and the longest noncoding region was located between *12S rRNA* and *tRNA^{Ile}*. Eight overlapping regions 1–7 bp in length were observed, and the total length was 22 bp. The longest overlapping region was located between ND4 and ND4L. The mitochondrial genome map of *L. kenliense* is shown in fig. 1.

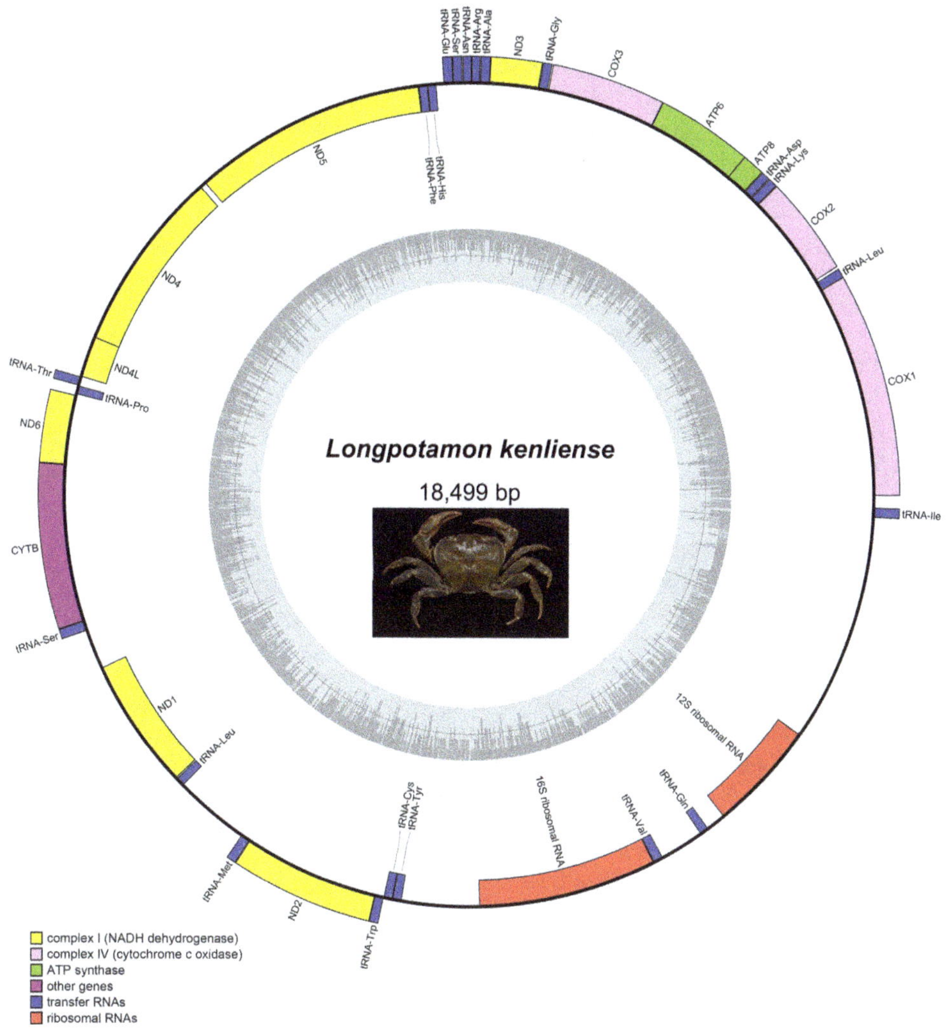

Fig. 1. Gene map of *Longpotamon kenliense* (Dai, 1997) mitochondrial genome.

Protein-coding genes and codon usage

The mitochondrial genome of *Longpotamon kenliense*, similar to the other brachyuran mitochondrial genomes, contained 13 PCGs. Among these genes, *ND1*, *ND4*, *ND4L* and *ND5* were encoded by the L-chain, and the remaining 9 PCGs were encoded by the H-chain, which is typical in the metazoan mitochondrial genome. According to GenBank data, the start codons of the PCGs in the Brachyura are relatively conserved, and most are ATN. Among the 13 PCGs in the mitochondrial genome of *L. kenliense*, the start codons of *COX1*, *COX2*, *COX3*, *ATP8*, *CYTB*, *ND2*, *ND4*, *ND4L* and *ND5* were all ATG, those of *ATP6* and *ND6* were ATA, that of *ND1* was ATT; and that of *ND3* was ATC. The stop codon of *ATP8* was TAG, those of *COX2*, *CYTB* and *ND5* were incomplete T, and those of all other genes were TAA. The special start codons could be converted to a common codon by RNA editing after transcription, and the translation process could be completed normally (Wilson et al., 2000). The incomplete stop codons could be complemented by polyadenylation after transcription. The average A + T content of the PCGs was 70.87%. Among the genes, *ATP8* had the highest A + T content of 79.87%, and *COX1* had the lowest A + T content of 65.37%.

RNA genes and noncoding regions

The tRNA genes of the brachyuran mitochondrial genome are approximately 60–80 bp in length and have a typical clover-leaf secondary structure. The lengths of the 22 tRNA genes were between 61 bp ($tRNA^{Arg}$) and 72 bp ($tRNA^{Val}$). Fourteen tRNA genes were encoded by the H-chain, and the rest were encoded by the L-chain. All tRNAs except $tRNA^{Ser(AGN)}$ exhibited a typical clover-leaf secondary structure, whereas $tRNA^{Ser(AGN)}$ lacked the dihydrouracil arm. This feature is also widely found in the mitochondrial genome of other metazoans (Wolstenholme, 1992; Ma et al., 2013). This special structure also has been found in mammals and insects (Ohtsuki et al., 2002).

Similar to those in other brachyuran mitochondrial genomes, the *16S rRNA* gene and 12*S rRNA* gene were encoded by the L-chain with lengths of 1310 bp and 811 bp, respectively. The *16S rRNA* gene was located between $tRNA^{Tyr}$ and $tRNA^{Val}$, whereas the *12S rRNA* was located between $tRNA^{Gln}$ and $tRNA^{Ile}$.

The mitochondrial genome of *Longpotamon kenliense* had 19 noncoding regions, the longest region of which was located between *12S rRNA* and $tRNA^{Ile}$ and had a length of 1664 bp. This region was a putative control region (CR), and its AT content was significantly higher than that of other regions (81.73%), exceeding the AT content of the entire genome (72.3%). The length of the CR in different species is generally derived from highly repetitive sequences, which can further form stem-loop structures, ranging from simple hairpin structures to complicated secondary structures (Jondeung et al., 2012). The CR in the mitochondrial genome of crustaceans contains conserved sequences such as AT-box and poly-T sequences, which are thought to be involved in mitochondrial DNA replication (Shi et al., 2015).

Gene rearrangement

Crustaceans have a highly rearranged mitochondrial gene structure in which rearrangements of tRNA genes are the most common, with fewer rearrangements of PCGs and rRNA genes. To determine whether the genome was rearranged, the brachyuran mitochondrial genome was used as a reference (fig. 2).

Fig. 2. Gene rearrangements of freshwater crabs.

Based on the current published sequences of mitochondrial genomes of freshwater crabs, members of the Brachyura, including *Longpotamon kenliense*, share a shifted *tRNA*^{His} (transfer from between *ND5* and *ND4* to between *tRNA*^{Glu} and *tRNA*^{Phe}). Translocation of *tRNA*^{Gln} occurred in *Longpotamon kenliense*, *Longpotamon xiushuiense*, *Longpotamon yangtsekiense*, *Sinopotamon yaanense*, *Huananpotamon lichuanense*, *Chinapotamon maolanense*, and *Geothelphusa dehaani* (from between *tRNA*^{Ile} and *tRNA*^{Met} to between *tRNA*^{Val} and *12S rRNA*); this rearrangement was found only in the Potamidae. *L. kenliense* shared the same genetic sequence with *L. yangtsekiense* and *S. yaanense*. The arrangement pattern was also very similar to that in *L. xiushuiense*. The only difference was in the position of the putative control region. Compared with the gene sequences of other freshwater crabs, the genus *Longpotamon* and genus *Sinopotamon* have a special genetic arrangement: the gene block *tRNA*^{Met}-ND2-*tRNA*^{Trp}-*tRNA*^{Cys}-*tRNA*^{Tyr} is transferred from between *tRNA*^{Gln} and *COX1* to between *tRNA*^{Leu} and *16S rRNA* intact. This forms a special gene sequence unique to the genus *Longpotamon* and genus *Sinopotamon* (Yuhui et al., 2017). This specific rearrangement supports a close genomic relationship between *Longpotamon* and *Sinopotamon*.

Phylogenetic relationships

The authors used 13 PCGs of the mitochondrial genome of *Longpotamon kenliense* and 67 other Brachyura available from the NCBI (*Clibanarius infraspinatus* and *Neopetrolisthes maculatus* were selected as outgroups) to establish ML and BI trees (fig. 3). The results were similar to the current molecular and morphological classifications, showing that the freshwater crabs were coming together in a single branch. *Somanniathelphusa boyangensis* of Gecarcinucoidea and eight species of Potamoidea (*Longpotamon kenliense*, *Longpotamon xiushuiense*, *Longpotamon yangtsekiense*, *Sinopotamon yaanense*, *Huananpotamon lichuanense*, *Chinapotamon maolanense*, *Potamiscus motuoensis* and *Geothelphusa dehaani*) formed two branches in the freshwater crab clade, respectively. *L. kenliense* clustered under Potamidae, together with *L. xiushuiense* and *L. yangtsekiense* they formed a clade of the genus *Longpotamon*, and together with *S. yaanense* formed a sister branch, which confirmed the hypothesis that the genera *Sinopotamon* and *Longpotamon* are closely related. According to the results, the molecular classification method of the mitochondrial PCGs has some degree of accuracy for the classification of Brachyura. However, the results of the ML and BI analyses greatly differ for the groups of individual species due to the lack of mitochondrial genome data for Brachyura, reflecting the long-branch attraction (LBA) effect (Straub et al., 2014), which suggested that data accumulation is of great significance for reconstructing the real phylogenetic process.

Divergence time of *Longpotamon kenliense*

The authors used the 13 PCGs from the mitochondrial genome of *Longpotamon kenliense* and 67 other brachyuran species available from the NCBI to estimate the divergence time of this species (fig. 4). The estimation of the divergence time showed that the freshwater crabs differentiated 109.56 Ma (million years from the present), and Gecarcinucoidea and Potamoidea differentiated 81.97 Ma. The divergence times of freshwater crabs and the genera in the Gecarcinucoidea and Potamoidea were similar to those reported by Tsang et al. (2014) (136.96 Ma/80 Ma) and Jia et al. (2018) (111.92 Ma/81.20 Ma). *Potamiscus motuoensis* differentiated 52.56 Ma. Four species of *Sinopotamon* and *Longpotamon* differentiated from *Geothelphusa dehaani*, *Chinapotamon maolanense* and *Huananpotamon lichuanense* 41.63 Ma. Within the clade of *Sinopotamon*/*Longpotamon*, *Sinopotamon yaanense* first differentiated from the genus *Longpotamon* 28.66 Ma ago. The divergence time of *L. kenliense* was 14.39 Ma, and the divergence time of *L. xiushuiense* and *L. yangtsekiense* was 10.40 Ma. The results also confirmed the hypothesis that *Sinopotamon* and *Longpotamon* are closely related and that the genus *Sinopotamon* has a longer history (Shih et al., 2016).

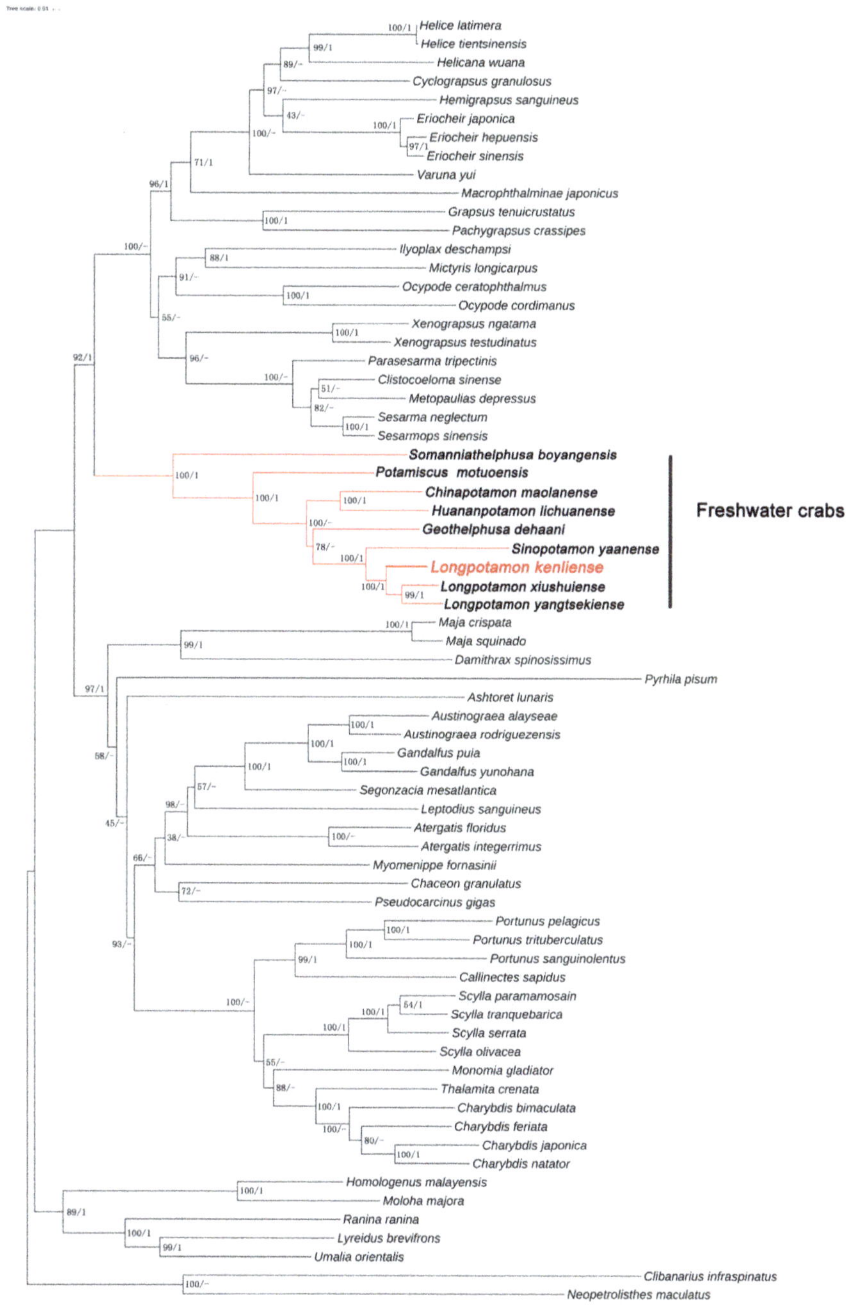

Fig. 3. Phylogenetic maximum likelihood (ML) tree of *Longpotamon kenliense* (Dai, 1997) and related brachyurans based on 13 PCGs sequences from mitochondrial genome; *Clibanarius infraspinatus* (Hilgendorf, 1869) and *Neopetrolisthes maculatus* (H. Milne Edwards, 1837) serve as the outgroups. The numbers at the internodes are maximum likelihood (ML) bootstrap proportions and Bayesian inference (BI) posterior proportions. The differences between the ML and BI trees are indicated by '–'. The species in boldface are freshwater crabs. Scale bars represent genetic distance.

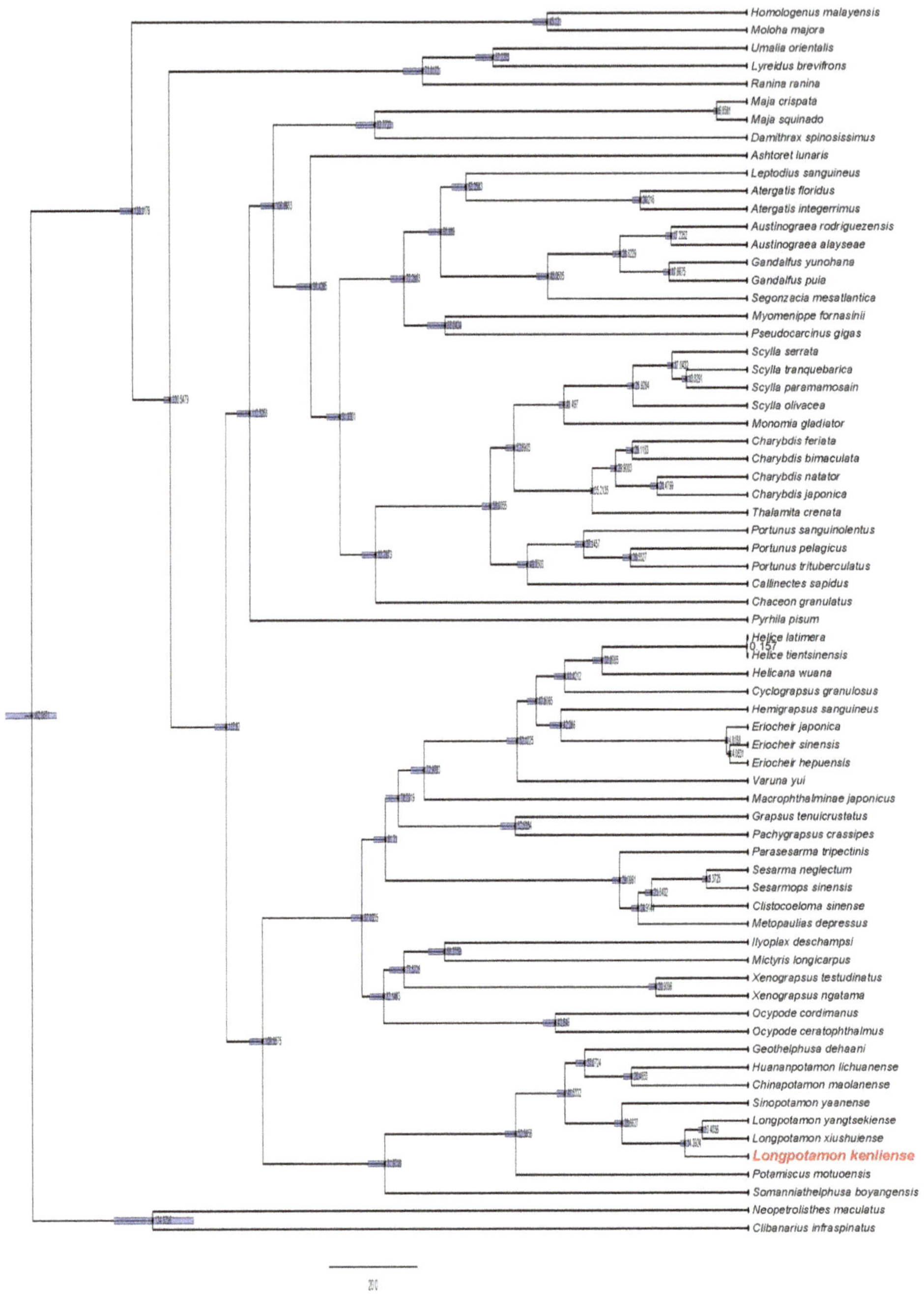

Fig. 4. Estimation of divergence time for *Longpotamon kenliense* (Dai, 1997) based on 13 PCGs nucleotide sequences of the mitochondrial genome. The scale bars represent millions of years from the present (Ma).

DISCUSSION

In this study, the authors obtained the mitochondrial genome sequence of *Longpotamon kenliense*, which had a total length of 18499 bp and contained 37 genes (13 protein-coding genes, 22 tRNA genes and 2 rRNA genes) and a control region. All tRNAs (except $tRNA^{Ser\,(AGN)}$, which lacked dihydrouracil arms) exhibited a typical clover structure. Compared with the gene sequences of other freshwater crabs, those of *Longpotamon*/*Sinopotamon* had a unique gene arrangement, as the gene block $tRNA^{Met}$-*ND2*-$tRNA^{Trp}$-$tRNA^{Cys}$-$tRNA^{Tyr}$ was transferred from between $tRNA^{Gln}$ and *COX1* to between $tRNA^{Leu}$ and *16S rRNA*. The results of the phylogenetic analyses were congruent with the current molecular and morphological classifications. The results showed that the freshwater crabs clustered together, and freshwater crabs in Potamoidea and Gecarcinucoidea formed two branches. The genera *Longpotamon* and *Sinopotamon* formed a single branch, which indicated that these two genera have a close evolutionary relationship. The divergence time of *L. kenliense* was estimated as 14.39 Ma, which was close to that of *Longpotamon yangtsekiense* and *Longpotamon xiushuiense* (10.40 Ma), whereas *Sinopotamon yaanense* differentiated earlier, 28.66 Ma. The differentiation of *Longpotamon* and *Sinopotamon* may be caused by the geographical isolation of the Sichuan Basin.

ACKNOWLEDGEMENTS

This work is supported by the National Natural Science Foundation of China (NSFC) http://www.nsfc.gov.cn/ (No. 21866020, 32060306), the National Parasitic Resources Center (NPRC-2019-194-30) and the Natural Science Foundation of Jiangxi (CN) http://www.jxstc.gov.cn/ (No. 20171BAB205108).

REFERENCES

BAI, J., S. X. XU, Z. H. NIE, Y. F. WANG, C. C. ZHU, Y. WANG, W. P. MIN, Y. X. CAI, J. X. ZOU & X. M. ZHOU, 2018. The complete mitochondrial genome of *Huananpotamon lichuanense* (Decapoda: Brachyura) with phylogenetic implications for freshwater crabs. Gene, **646**: 217-226. DOI:10.1016/j.gene.2018.01.015.

BURLAND, T. G., 2000. DNASTAR's Lasergene sequence analysis software. Methods Mol. Biol., **132**: 71-91.

DAI, A. Y., 1999. Fauna sinica (Arthropoda. Crustacea. Malacostraca. Decapoda. Parathelphusidae. Potamidae): 1-501. (Science Press, Beijing).

DRUMMOND, A. J. & A. RAMBAUT, 2007. BEAST: Bayesian evolutionary analysis by sampling trees. BMC Evol. Biol., **7**(214). DOI:10.1186/1471-2148-7-214.

GERNHARD, T., 2008. The conditioned reconstructed process. J. Theor. Biol., **253**: 769-778. DOI:10.1016/j.jtbi.2008.04.005.

IVEY, J. L. & S. R. SANTOS, 2007. The complete mitochondrial genome of the Hawaiian anchialine shrimp *Halocaridina rubra* Holthuis, 1963 (Crustacea: Decapoda: Atyidae). Gene, **394**: 35-44. DOI:10.1016/j.gene.2007.01.009.

JIA, X. N., S. X. XU, J. BAI, Y. F. WANG, Z. H. NIE, C. C. ZHU, Y. WANG, Y. X. CAI, J. X. ZOU & X. M. ZHOU, 2018. The complete mitochondrial genome of *Somanniathelphusa boyangensis* and phylogenetic analysis of genus *Somanniathelphusa* (Crustacea: Decapoda: Parathelphusidae). PLoS One, **13**(2): e0192601. DOI:10.1371/journal.pone.0192601.

JONDEUNG, A., W. KARINTHANYAKIT & J. KAEWKHUMSAN, 2012. The complete mitochondrial genome of the black mud crab, *Scylla serrata* (Crustacea: Brachyura: Portunidae) and its phylogenetic position among (pan)crustaceans. Mol. Biol. Rep., **39**: 10921-10937. DOI:10.1007/s11033-012-1992-2.

KARASAWA, H., C. E. SCHWEITZER & R. M. FELDMANN, 2008. Revision of Portunoidea Rafinesque, 1815 (Decapoda: Brachyura) with emphasis on the fossil genera and families. J. Crust. Biol., **28**: 82-127. DOI:10.1651/07-2882r.1.

KARASAWA, H., C. E. SCHWEITZER & R. M. FELDMANN, 2011. Phylogenetic analysis and revised classification of podotrematous Brachyura (Decapoda) including extinct and extant families. J. Crust. Biol., **31**: 523-565. DOI:10.1651/10-3423.1.

KATOH, K., K. KUMA, H. TOH & T. MIYATA, 2005. MAFFT version 5: improvement in accuracy of multiple sequence alignment. Nucleic Acids Res., **33**: 511-518. DOI:10.1093/nar/gki198.

KI, J. S., H. U. DAHMS, J. S. HWANG & J. S. LEE, 2009. The complete mitogenome of the hydrothermal vent crab *Xenograpsus testudinatus* (Decapoda, Brachyura) and comparison with brachyuran crabs. Comp. Biochem. Phys. D, **4**: 290-299. DOI:10.1016/j.cbd.2009.07.002.

LIU, Y. A. & Z. X. CUI, 2010. Complete mitochondrial genome of the Asian paddle crab *Charybdis japonica* (Crustacea: Decapoda: Portunidae): gene rearrangement of the marine brachyurans and phylogenetic considerations of the decapods. Mol. Biol. Rep., **37**: 2559-2569. DOI:10.1007/s11033-009-9773-2.

LOWE, T. M. & P. P. CHAN, 2016. tRNAscan-SE On-line: integrating search and context for analysis of transfer RNA genes. Nucleic Acids Res., **44**(W1): W54-W57. DOI:10.1093/nar/gkw413.

MA, H. Y., C. Y. MA, X. C. LI, Z. XU, N. N. FENG & L. B. MA, 2013. The complete mitochondrial genome sequence and gene organization of the mud crab (*Scylla paramamosain*) with phylogenetic consideration. Gene, **519**: 120-127. DOI:10.1016/j.gene.2013.01.028.

NG, P. K. L., D. GUINOT & P. J. F. DAVIE, 2008. Systema Brachyurorum: part I. An annotated checklist of extant brachyuran crabs of the world. Raffles Bull. Zool., (suppl.) **17**: 1-286.

OHTSUKI, T., G. KAWAI & K. WATANABE, 2002. The minimal tRNA: unique structure of *Ascaris suum* mitochondrial tRNA$^{(Ser)}$(UCU) having a short T arm and lacking the entire D arm. FEBS Letters, **514**: 37-43. DOI:10.1016/S0014-5793(02)02328-1.

RONQUIST, F., M. TESLENKO, P. VAN DER MARK, D. L. AYRES, A. DARLING, S. HOHNA, B. LARGET, L. LIU, M. A. SUCHARD & J. P. HUELSENBECK, 2012. MrBayes 3.2: efficient Bayesian phylogenetic inference and model choice across a large model space. Syst. Biol., **61**: 539-542. DOI:10.1093/sysbio/sys029.

SCHWEITZER, C. E. & R. M. FELDMANN, 2000. New fossil portunids from Washington, USA, and Argentina, and a re-evaluation of generic and family relationships within the Portunoidea Rafinesque, 1815 (Decapoda: Brachyura). J. Paleontol., **74**: 636-653. DOI:10.1017/S0022336000032765.

SEGAWA, R. D. & T. AOTSUKA, 2005. The mitochondrial genome of the Japanese freshwater crab, *Geothelphusa dehaani* (Crustacea: Brachyura): evidence for its evolution via gene duplication. Gene, **355**: 28-39. DOI:10.1016/j.gene.2005.05.020.

SHEN, X., J. REN, Z. CUI, Z. SHA, B. WANG, J. XIANG & B. LIU, 2007. The complete mitochondrial genomes of two common shrimps (*Litopenaeus vannamei* and *Fenneropenaeus chinensis*) and their phylogenomic considerations. Gene, **403**: 98-109. DOI:10.1016/j.gene.2007.06.021.

SHI, G. H., Z. X. CUI, M. HUI, Y. LIU, T. Y. CHAN & C. W. SONG, 2015. The complete mitochondrial genomes of *Umalia orientalis* and *Lyreidus brevifrons*: The phylogenetic position of the family Raninidae within brachyuran crabs. Mar. Genomics, **21**: 53-61. DOI:10.1016/j.margen.2015.02.002.

SHIH, H. T., C. HUANG & P. K. L. NG, 2016. A re-appraisal of the widely-distributed freshwater crab genus *Sinopotamon* Bott, 1967, from China, with establishment of a new genus (Crustacea: Decapoda: Potamidae). Zootaxa, **4138**: 309-331. DOI:10.11646/zootaxa.4138.2.5.

SUN, H. Y., K. Y. ZHOU & D. X. SONG, 2005. Mitochondrial genome of the Chinese mitten crab *Eriocheir japonica sinenesis* (Brachyura: Thoracotremata: Grapsoidea) reveals a novel gene order and two target regions of gene rearrangements. Gene, **349**: 207-217. DOI:10.1016/j.gene.2004.12.036.

TAMURA, K., G. STECHER, D. PETERSON, A. FILIPSKI & S. KUMAR, 2013. MEGA6: molecular evolutionary genetics analysis version 6.0. Mol. Biol. Evol., **30**: 2725-2729. DOI:10.1093/molbev/mst197.

TSANG, L. M., C. D. SCHUBART, S. T. AHYONG, J. C. LAI, E. Y. AU, T. Y. CHAN, P. K. L. NG & K. H. CHU, 2014. Evolutionary history of true crabs (Crustacea: Decapoda: Brachyura) and the origin of freshwater crabs. Mol. Biol. Evol., **31**: 1173-1187.

VAN BAKEL, B. W. M., D. GUINOT, P. ARTAL, R. H. B. FRAAIJE & J. W. M. JAGT, 2012. A revision of the Palaeocorystoidea and the phylogeny of raninoidian crabs (Crustacea, Decapoda, Brachyura, Podotremata). Zootaxa, **3215**: 1-216.

VAZQUEZ, G. E., I. PONZONI, M. C. SANCHEZ & N. B. BRIGNOLE, 2000. ModGen: a model generator for instrumentation analysis. Adv. Eng. Softw., **32**: 37-48.

WANG, Y., Q. W. ZENG, J. X. ZOU, C. C. ZHU, J. BAI, H. T. SHIH, S. L. CHENG & X. M. ZHOU, 2016. The complete mitochondrial genome of freshwater crab *Sinopotamon xiushuiense* (Decapoda: Brachyura: Potamoidea). Mitochondrial DNA B, **1**: 750-752. DOI:10.1080/23802359.2016.1209094.

WILSON, K., V. CAHILL, E. BALLMENT & J. BENZIE, 2000. The complete sequence of the mitochondrial genome of the crustacean *Penaeus monodon*: are malacostracan crustaceans more closely related to insects than to branchiopods? Mol. Biol. Evol., **17**: 863-874. DOI:10.1093/oxfordjournals.molbev.a026366.

WOLSTENHOLME, D. R., 1992. Animal mitochondrial-DNA — structure and evolution. Int. Rev. Cytol., **141**: 173-216. DOI:10.1016/S0074-7696(08)62066-5.

XIA, X. & Z. XIE, 2001. DAMBE: software package for data analysis in molecular biology and evolution. J. Hered., **92**: 371-373. DOI:10.1093/jhered/92.4.371.

YAMAUCHI, M. M., M. U. MIYA & M. NISHIDA, 2003. Complete mitochondrial DNA sequence of the swimming crab, *Portunus trituberculatus* (Crustacea: Decapoda: Brachyura). Gene, **311**: 129-135. DOI:10.1016/S0378-1119(03)00582-1.

YUHUI, X., Z. LIJUN, H. YUE, W. XIAOQI, Z. CHEN, H. ZHANG, W. RUORAN, P. DA & S. HONGYING, 2017. Complete mitochondrial genomes from two species of Chinese freshwater crabs of the genus *Sinopotamon* recovered using next-generation sequencing reveal a novel gene order (Brachyura, Potamidae). ZooKeys, **705**(6): 41-60. DOI:10.3897/zookeys.705.11852.

First received 27 February 2019.
Final version accepted 28 August 2020.

[When citing this volume, please refer to *Crustaceana* 93 (2020), 1233-1546]

Prof. Ruiyu Liu Memorial Issue

DESCRIPTIONS OF A NEW SPECIES OF *MINPOTAMON* DAI & TÜRKAY, 1997, AND A MONOTYPIC NEW GENUS OF AQUATIC FRESHWATER CRAB (BRACHYURA, POTAMIDAE) FROM EASTERN GUANGDONG, CHINA

BY

SIYING MAO[1]) and CHAO HUANG[1,2,3])

[1]) School of Life Sciences, Sun Yat-sen University, Guangzhou 510275, P.R. China

[2]) Australian Museum, 1 William St, Sydney NSW 2010, Australia

ABSTRACT

A new genus new species of freshwater crab and a new *Minpotamon* Dai & Türkay, 1997 species are described from eastern Guangdong, China, based on morphology and mitochondrial 16S rDNA sequences. The new genus, *Teoswamon* gen. nov., is established for *T. scolasticum* sp. nov., which resembles species from several other genera from Guangdong, but its unique combination of carapace, male gonopod and female vulvae characters set it apart from others. *Minpotamon kityang* sp. nov. is almost identical with *Minpotamon nasicum* (Dai, Chen, Song, Fan, Lin & Zeng, 1979) externally, but differs drastically in the male gonopod morphology. Molecular analysis supports the current taxonomic treatments. *Teoswamon scolasticum* sp. nov. forms its own lineage closely related to *Huananpotamon* Dai & Ng, 1994. *Minpotamon kityang* sp. nov. clusters with *Minpotamon nasicum*, but their genetic distance is of an interspecific level.

RÉSUMÉ

Un nouveau genre et nouvelle espèce de crabe d'eau douce ainsi qu'une nouvelle espèce de *Minpotamon* Dai & Türkay, 1997 sont décrits de l'est de Guangdong, Chine, à partir de la morphologie et des séquences de 16SrDNA mitochondrial. Le nouveau genre, *Teoswamon* gen. nov., est établi pour *T. scolasticum* sp. nov., qui ressemble aux espèces de plusieurs autres genres de Guangdong, mais la combinaison unique des caractères de la carapace, des gonopodes mâles et des vulves femelles le place à part des autres. *Minpotamon kityang* sp. nov. est presque identique à *Minpotamon nasicum* (Dai, Chen, Song, Fan, Lin & Zeng, 1979) par son aspect externe, mais en diffère radicalement par la morphologie du gonopode mâle. L'analyse moléculaire soutient le traitement taxonomique présent. *Teoswamon scolasticum* sp. nov. se révèle très proche de *Huananpotamon* Dai & Ng, 1994. *Minpotamon kityang* sp. nov. se regroupe avec *Minpotamon nasicum*, mais leur distance génétique est de niveau interspécifique.

[3]) Corresponding author; e-mail: chaohuang1777@yahoo.com

 | DOI:10.1163/9789004464353_006

INTRODUCTION

Guangdong is situated within the Huanan freshwater zoogeographic province (Huang et al., 2020) and is very rich in freshwater crabs with many new taxa being discovered in recent years, especially in the southern regions (Huang et al., 2012, Huang et al., 2014, Huang et al., 2016, Huang et al., 2017, Huang, 2018, Huang et al., 2018a, Huang et al., 2018b). The freshwater crabs of eastern Guangdong, however, remain poorly documented. In this paper we describe a new species of *Minpotamon* and a new genus and new species, *Teoswamon scolasticum* gen. nov. et sp. nov., from this region.

The first few *Minpotamon kityang* sp. nov. specimens the authors obtained in 2013 came from the Chinese aquarium trade, allegedly collected from Jieyang City, Guangdong. These specimens turned out to be a new species of the genus *Minpotamon* Dai & Türkay, 1997, but we refrained from describing it immediately because the exact locality had not been confirmed by us. On a field trip to eastern Guangdong in 2018, we managed to contact the original seller of the new *Minpotamon* species and he kindly agreed to take us to the locality in Jieyang City where we finally confirmed the locality and collected more specimens. The new species of *Minpotamon* is almost identical with *Minpotamon nasicum* (Dai, Chen, Song, Fan, Lin & Zeng, 1979) externally, but differs drastically in the male gonopod morphology. It is described herein as *Minpotamon kityang* sp. nov.

Another species of interest was collected from Puning City on the same 2018 field trip to eastern Guangdong. We first received this species from Jun-Hao Huang in 2017 when he kindly donated to us three specimens which he collected during a freshwater fish survey. Generic assignment of this species was initially difficult as its external characters were unremarkable. This species is strikingly similar to some species from the genus *Cantopotamon* Huang, Ahyong & Shih, 2017, but upon close examination and genetic analysis, it was found that this crab is a new species from a new genus. It has a unique combination of carapace, male gonopod and female vulvae characters and is not directly related to *Cantopotamon* genetically. As of such, we describe it as *Teoswamon scolasticum* sp. nov. herein.

MATERIAL AND METHODS

Specimens were collected by hand and preserved in 75% ethanol from Guangdong and deposited in the Sun Yat-sen Museum of Biology, Sun Yat-sen University, Guangzhou, China (SYSBM), Department of Parasitology of the Medical College of Nanchang University (NCU MCP) and the Zoological Reference Collection of the Lee Kong Chian Natural History Museum, Singapore (ZRC). Measurements, in millimetres, are of the carapace width and length, respectively. Other abbreviations are as follows: G1, male first gonopod; G2, male second gonopod; CW, carapace width. The terminology used primarily follows that of Dai (1999) and Davie et al. (2015).

16S sequences of *Minpotamon kityang* sp. nov. and *Teoswamon scolasticum* sp. nov. were obtained by Hsi-Te Shih following Shih et al. (2009) using the primers 1471 and 1472 (Crandall & Fitzpatrick, 1996) and given to the authors for use in this study. Comparative sequences were acquired from GenBank. Sequences were aligned with the aid of ClustalW (Thompson et al., 1994) in MEGA X (Kumar et al., 2018), after verification with the complementary strand. We followed Shih et al. (2009) in excluding the variable regions in loop regions of the 16S gene that could not be aligned adequately for phylogenetic analyses. Sequences of the new taxa have been deposited in GenBank (fig. 9). To confirm the systematic position of the new taxa, a preliminary analysis showed that the two taxa belong to the "South China #1" (Huang et al., 2018) clade of the "China-East Asia Islands" group (Shih et al., 2009). As a result, taxa from the closely related "East Asia Island Arc" (Huang et al., 2018) clade, i.e., *Amamiku amamensis* (Minei, 1973), *Apotamonautes hainanensis* (Parisi, 1916), *Candidiopotamon rathbunae* (De Man, 1914) and *Hainanpotamon fuchengense* Dai, 1995, were used as outgroups. The relationships between many clades within the "South China #1" clade cannot be resolved by only using the 16S gene (Huang et al., 2018), therefore, only the closely related taxa from eastern Guangdong, Fujian and Jiangxi (*Minpotamon*, *Minutomon* Huang, Mao & Huang, 2014 and *Huananpotamon* Dai & Ng, 1994) and some others (*Cantopotamon* and *Yarepotamon* Dai & Türkay, 1997) were selected for the final analysis. A total of 10 ingroup species from six genera and the abovementioned four outgroup species were genetically analysed. Bayesian inference (BI) analysis was performed with MrBayes (v. 3.2.7; Ronquist & Huelsenbeck, 2003). Model parameters were estimated during the analysis. The search was run with four chains for 10 million generations, with trees sampled every 1000 generations. Convergence of the analyses was assessed by the standard deviation of split frequencies (<0.01), the first 1000 trees were discarded as burn-in accordingly. A maximum parsimony (MP) consensus tree was constructed using MEGA X (Kumar et al., 2018), with 2000 bootstrap reiterations of a simple heuristic search, TBR branch-swapping (tree bisection-reconnection) (100 random-addition sequence replications; max no. of trees to retain = 10 000). The pairwise p-distances were calculated using MEGA X (Kumar et al., 2018).

TAXONOMY

Family POTAMIDAE Ortmann, 1896
Subfamily POTAMISCINAE Bott, 1970

Minpotamon Dai & Türkay, 1997
(figs. 1-3, 7A, B, 8A)

Minpotamon Dai & Türkay, 1997: 255, fig. 9, pl. II fig. 5.

Diagnosis.— Small sized (carapace width less than 30 mm). Carapace slightly broader than long, slightly swollen; epigastric cristae low, separated by narrow gap (figs. 1, 2A); postorbital cristae blunt, not confluent with epibranchial teeth and epigastric cristae (figs. 1, 2A). Maxilliped 3 with relatively broad ischium, exopod reaching beyond anterior edge of ischium, with long flagellum (fig. 3A). Chelipeds slightly subequal (fig. 3F-I). Ambulatory legs slender (fig. 1). Male pleon narrowly triangular (fig. 2C). G1 generally slender, slightly sinuous; subterminal segment basal region wide; terminal segment long, slender, sinistrally twisted on left G1, with vertical groove along entire length, tip pointing inwards in ventral view (figs. 3C-E, 7A, B). G2 with long flagellum-like terminal segment (fig. 3B). Female vulva medium sized, reaching the sutures of sternites 5/6 anteriorly, positioned closely to one another, opens inwards, with relatively wide outer rim (fig. 2F).

Remarks.— *Minpotamon* was a monotypic genus before the herein described new species. The original definition of *Minpotamon* by Dai and Türkay (1997) was based on the morphology of *Minpotamon nasicum* and reads "Male first pleopod with terminal segment more than half as long as subterminal segment, clearly bent laterally outwards, with subdistal projection prominent". As the new species, *M. kityang* sp. nov. does not possess a G1 terminal segment subdistal projection, the genus is redefined accordingly. The presence or absence of such a subdistal projection can merely be intraspecific variation and should be used with caution (Huang, 2018). We have also found that the terminal segment is in fact slightly shorter than half-length of the subterminal segment in both *M. nasicum* and *M. kityang* sp. nov. Furthermore, the G1 of both species is clearly bent inwards and not outwards. This error has likely arisen from mistaking the G1 figured in Dai & Türkay (1997: fig. 9(4)) as the left G1 whereas it is in fact the right G1.

Minpotamon kityang sp. nov.
(figs. 1-3, 7A, B, 8A)

Material examined.— Holotype: SYSBM 001283, male 19.7 × 16.1 mm, Xinheng County (23°39′19″N, 116°18′40″E), Jieyang City, Guangdong, coll. local, December 2013. Paratypes: SYSB 001284, male 18.6 × 15.8 mm, same data as holotype. SYSB 001285, female 16.6 × 14.3 mm, same data as holotype. SYSB 001849-001851, 3 males 17.5 × 14.9 mm, 13.7 × 11.8 mm,

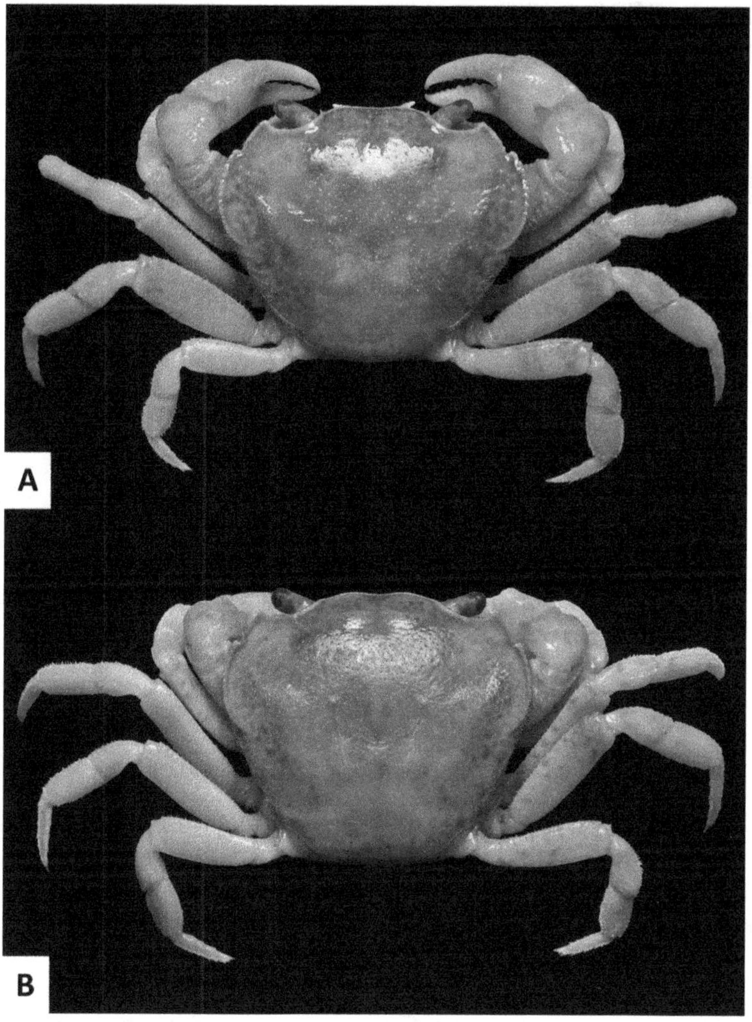

Fig. 1. *Minpotamon kityang* sp. nov., dorsal overall view: A, female paratype (SYSBM 001285); B, male holotype (SYSBM 001283).

13.3 × 11.5 mm, Xinheng County, Jieyang City, Guangdong, coll. C. Huang, November 2018. SYSB 001852, female 12.1 × 10.6 mm, same data as above. NCU MCP 416601, male 17.5 × 14.7 mm, Xinheng County, Jieyang City, Guangdong, coll. C. Huang, November 2018. NCU MCP 416602, female 12.8 × 11.2 mm, same data as above. ZRC 2017.1291, 3 males 18.8 × 15.6 mm, 17.2 × 15.1 mm, 15.9 × 14.0 mm, Xinheng County, Jieyang City, Guangdong, coll. local, December 2013. ZRC 2017.0811, 2 males 17.2 × 15.1 mm, 15.9 × 14.0 mm, female 16.1 × 13.5 mm; same data as above.

Description.— Male: carapace broader than long, width about 1.1-1.2 times as long as length ($n = 12$); regions indistinct, dorsal surface slightly convex; surface generally smooth, pitted, anterolateral region and front weakly rugose (fig. 1A).

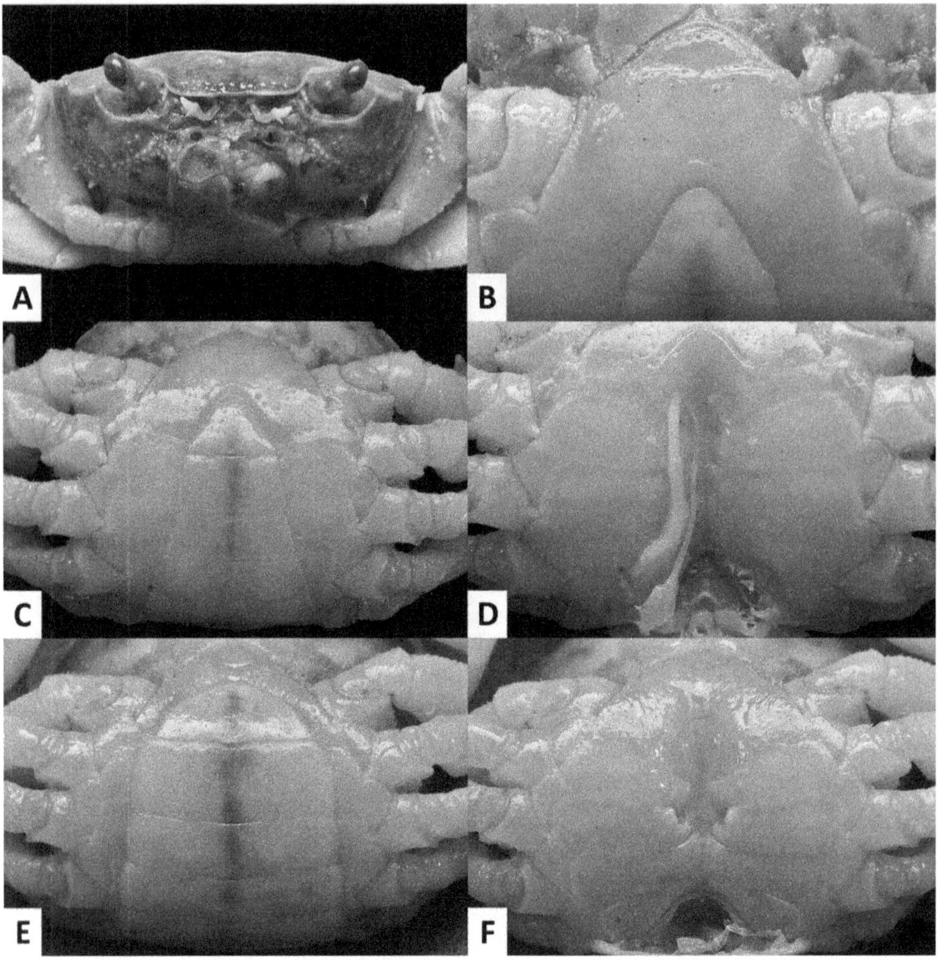

Fig. 2. *Minpotamon kityang* sp. nov., male holotype (SYSBM 001283): A, frontal view of cephalothorax; B, anterior thoracic sternum; C, ventral view of anterior thoracic sternum and pleon; D, ventral view showing sterno-pleonal cavity with right G1 in situ (left G1 removed). Female paratype (SYSBM 001285): E, ventral view of pleon; F, ventral view showing female vulvae.

Front sloping downwards, margin ridged (figs. 1A, 2A). Epigastric cristae low, separated by narrow gap (fig. 1A). Postorbital cristae blunt, laterally expanded, not confluent with epibranchial teeth and epigastric cristae (fig. 1A). Branchial regions slightly swollen (figs. 1A, 2A). Cervical groove shallow (fig. 1A). Mesogastric region slightly convex (fig. 1A). External orbital angle broadly triangular, outer margin gently convex, separated from anterolateral margin by conspicuous gap (fig. 1A). Epibranchial tooth granular (fig. 1A). Anterolateral margin cristate, lined with around 16-18 granules; bent inward posteriorly (fig. 1A). Posterolateral surface with low, oblique striae, converging towards posterior carapace margin (fig. 1A). Orbits large; supraorbital margins cristate, infraorbital margins lined

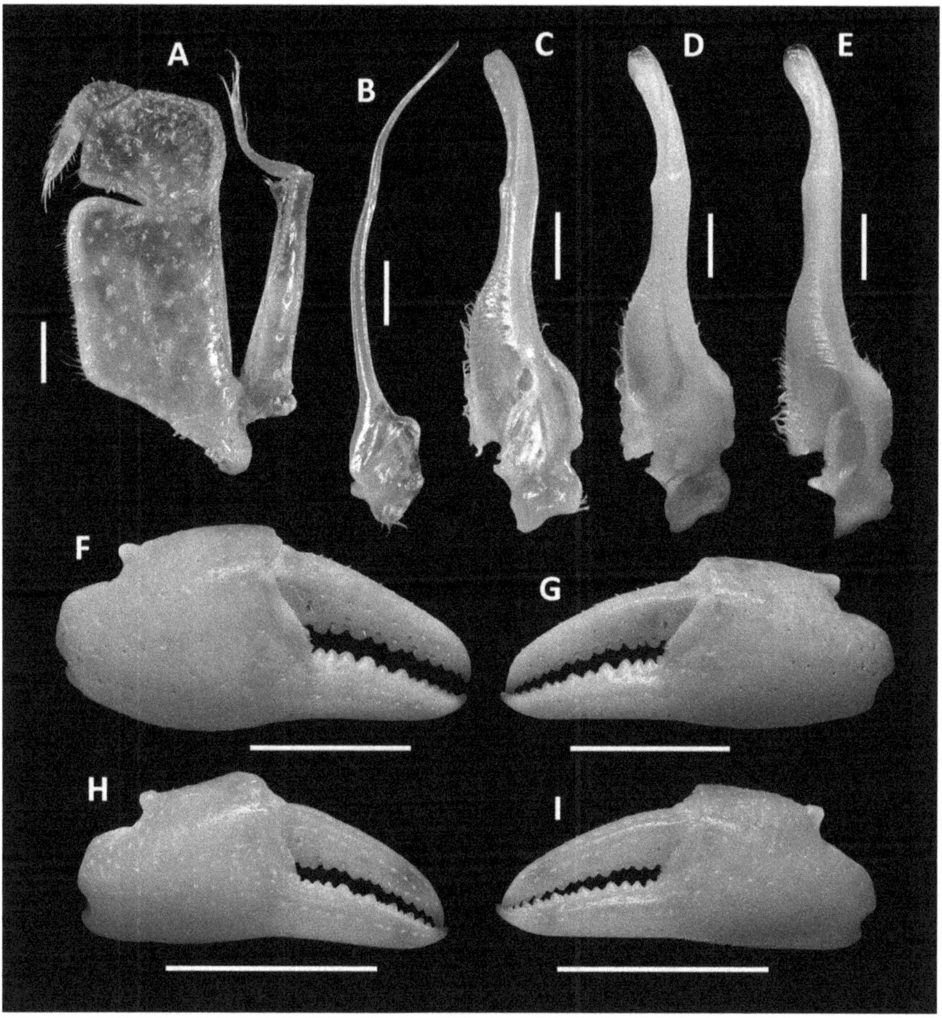

Fig. 3. *Minpotamon kityang* sp. nov., male holotype (SYSBM 001283): A, left third maxilliped; B, left G2; C, male left G1; F, male right cheliped; G, male left cheliped. Male paratype (SYSBM 001284): D, male left G. Male paratype (SYSBM 001849): E, male left G1. Female paratype (SYSBM 001285): H, female right cheliped; I, female left cheliped. Scale bars: A-E = 1.0 mm; F-I = 5.0 mm.

with granules (fig. 2A). Sub-orbital and pterygostomial regions covered with small, rounded granules; sub-hepatic regions rugose (fig. 2A). Median lobe of epistome posterior margin broadly triangular, lateral margins sinuous (fig. 2A).

Maxilliped 3 merus about 1.2 times as wide as long; ischium width about 0.8 as long as length; merus subtrapezoidal, with median depression; ischium subtrapezoidal, with distinct median sulcus, mesial margin rounded. Exopod reaching to proximal one-third of merus height; flagellum long (fig. 3A).

Chelipeds (pereiopod 1) unequal (fig. 3F, G). Merus trigonal in cross section; margins crenulated, surface rugose (fig. 1A). Carpus with sharp spine at inner-distal angle, spinule at base (fig. 1A). Major cheliped palm length about 1.2-1.3 times as long as height ($n = 3$); dactylus 0.9-1.0 times as long as palm length ($n = 3$) (fig. 3F). Palm surface pitted (fig. 3F, G). Dactylus as long as pollex (fig. 3F, G). Occlusal margin of fingers with 11-12 irregular blunt teeth; slight gape when closed (fig. 3F, G).

Ambulatory legs (pereiopods 2-5) slender with short setae (fig. 1A). Pereiopod 3 merus 0.6 times as long as carapace length ($n = 5$) (fig. 1A). Pereiopods 5 propodus 2.1-2.2 times as long as broad ($n = 5$), shorter than dactylus (fig. 1A).

Male thoracic sternum generally smooth; sternites 1-4 narrow, width about 1.6 times as long as length; sternites 1, 2 forming triangular structure; sternites 2, 3 fused, but demarcated by shallow transverse sulcus; sternites 3, 4 fused, demarcation inconspicuous (fig. 2B). Male sterno-pleonal cavity reaching anteriorly to level of midlength of chelipeds coxae base (fig. 2D); deep median longitudinal groove between sternites 7, 8 (fig. 2D). Male pleonal locking tubercle positioned at mid-length of sternite 5 (fig. 2D).

Male pleon triangular; somites 3-6 progressively narrower; somite 6 posterior margin almost same length as anterior margin; somite 6 width approximately 1.9 times as long as length; telson width 1.4 times as long as length; apex rounded (fig. 2C).

G1 slender, tip well exceeding pleonal locking tubercle in-situ, reaching beyond the suture between thoracic sternites 4/5 (fig. 2D). Subterminal segment length about 2.2 times as long as length of terminal segment ($n = 3$), base wide, tapering anteriorly. Terminal segment long, slender, sinistrally twisted on left G1, with vertical groove along entire length and large opening at tip, finger-shaped, bent inwards (figs. 3C-E, 7A, B). G2 subterminal segment about 1.5-1.6 times as long as length of flagelliform distal segment ($n = 3$) (fig. 3B).

Female: chelipeds slightly less inflated (fig. 3H, I). Major cheliped palm length about 1.2-1.4 times as long as height ($n = 3$); dactylus 1.0-1.1 times as long as palm length ($n = 3$) (fig. 3H). Pereiopod 3 merus 0.5-0.6 times as long as carapace length ($n = 2$) (fig. 1B). Pereiopods 5 propodus 2.1 times as long as broad ($n = 1$) (fig. 1B). Female vulvae small with wide rim, reaching the sutures of sternites 5/6 anteriorly, positioned closely to one another (fig. 2F). Female pleon ovate (fig. 2E).

Etymology.— The species name is a synonym for Jieyang, which is the city where it was collected. "Kityang" is the local Teochew dialect pronunciation of Jieyang. The name is used as a noun in apposition.

Habitat.— *Minpotamon kityang* sp. nov. lives in low-altitude hillstreams with sandy sediment in Jieyang, where they hide under rocks. Specimens were mostly collected by flipping over rocks in the shallows whereas a few were collected from submerged grass near the bank of the stream.

Remarks.— The carapace of *Minpotamon kityang* sp. nov. resembles that of *M. nasicum*; both have prominent epibranchial teeth and crested anterolateral margins lined with granules, slightly rugose anterolateral regions and a slightly bulging postgastric region (fig. 1, Dai, 1999). The main difference between *M. kityang* sp. nov. and *M. nasicum* is the G1 morphology. The G1 terminal segment of the new species is slender and finger-shaped with a large opening at the tip and is only slightly bent inwards (figs. 3C-E, 7A, B), whereas that of *M. nasicum* has a bifurcated tip and is strongly bent inwards (Dai, 1999: fig. 207 (6)). In addition, the G1 of the new species is obviously longer and reaches beyond the suture between thoracic sternites 4/5 (figs. 3C-E, 7A, B), whereas that of *M. nasicum* is shorter and barely reaches the suture (Dai, 1999: fig. 207 (6)).

Genus **Teoswamon** gen. nov.
(figs. 4-6, 7C, D, 8B)

Type species: *Teoswamon scolasticum* sp. nov. by present designation.

Diagnosis.— Very small sized (carapace width less than 20 mm). Carapace broader than long, relatively flat (figs. 4, 5A); postorbital cristae and epigastric cristae low, almost fused with each other (figs. 4, 5A); external orbital angle broadly triangular, separated from anterolateral margin by small gap (figs. 4, 5A). Median lobe of epistome broadly triangular (fig. 5A). Maxilliped III with relatively broad ischium, exopod reaching beyond anterior edge of ischium with long flagellum (fig. 6A). Cheliped fingers with slight gape when closed (fig. 6F-I). Male anterior thoracic sternum wide, width 1.9 times as long as length (fig. 5B). Male pleon broadly triangular (fig. 5C). G1 generally slender, subterminal segment tapering anteriorly; terminal segment long, slender, sinuous, sinistrally twisted on left G1, with visible vertical medial groove and distally positioned opening (figs. 6C-E, 7C, D). G2 distal segment relatively long (fig. 6B). Female vulvae large, reaching the sutures of sternites 5/6 anteriorly (fig. 5F).

Etymology.— The name of the new genus is a combination of "Teoswa", which refers to the locality of the type species, and the genus name *Potamon*, which is the type genus of the family. "Teoswa" (Teochew dialect pronunciation, equivalent to "Chaoshan" in Mandarin) refers to the cultural-linguistic region in eastern Guangdong, where the unique Teochew dialect is spoken. The Teo-Swa region is homeland or ancestral homeland of over 10 million Teochew people living around the globe whom take great pride in their cultural identity. Gender neuter.

Remarks.— *Teoswamon* gen. nov. is externally unremarkable and similar to several genera of freshwater crabs from southern China, including *Cantopotamon*,

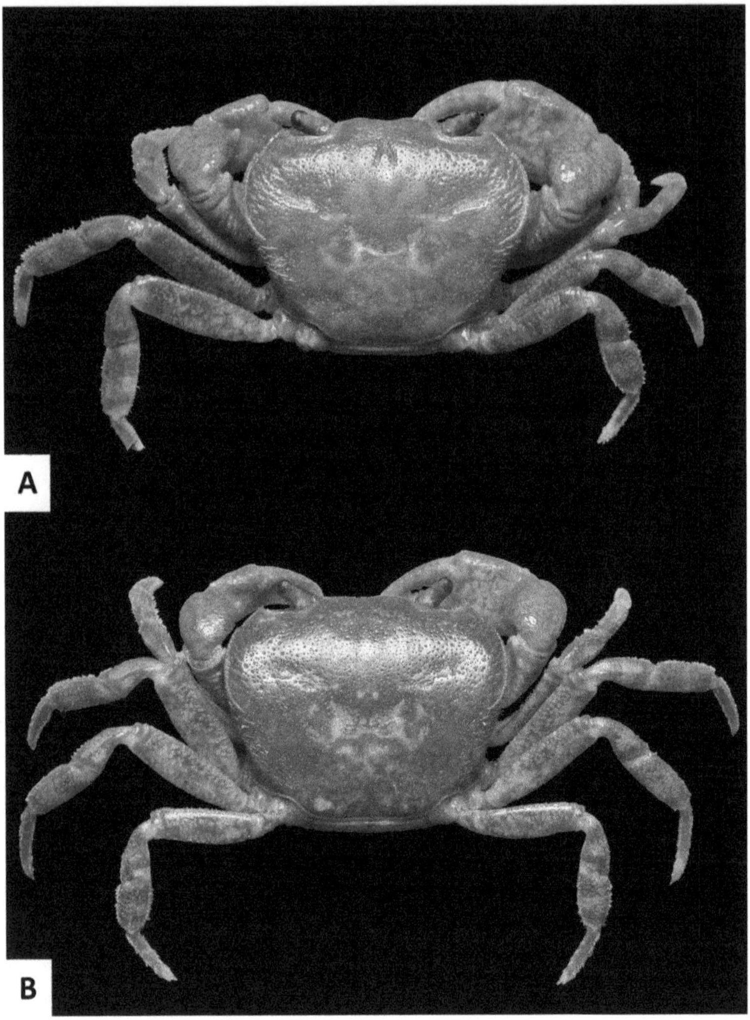

Fig. 4. *Teoswamon scolasticum* gen. nov., sp. nov., dorsal overall view. A, male holotype (SYSBM 001841); B, Female paratype (SYSBM 001847).

Eurusamon, *Minutomon* and *Yarepotamon*. Convergence in external morphology caused by similar evolutionary responses to similar habitats has increased the difficulty in trying to separate these taxa, even on the genus level (Huang, 2018).

Teoswamon gen. nov. is very similar to *Cantopotamon* with regards to the shape of the carapace and male G1 morphology. The anterolateral margins are lined with granules and curved inwards posteriorly in both genera (fig. 4, Huang et al., 2017: fig.1A). The type species of the new genus, *T. scolasticum* sp. nov., is particularly similar to *Cantopotamon shangchuanense* in having a G1 terminal segment that is bent inwards at the base and points anteriorly at the tip, however, the G1 of *T.*

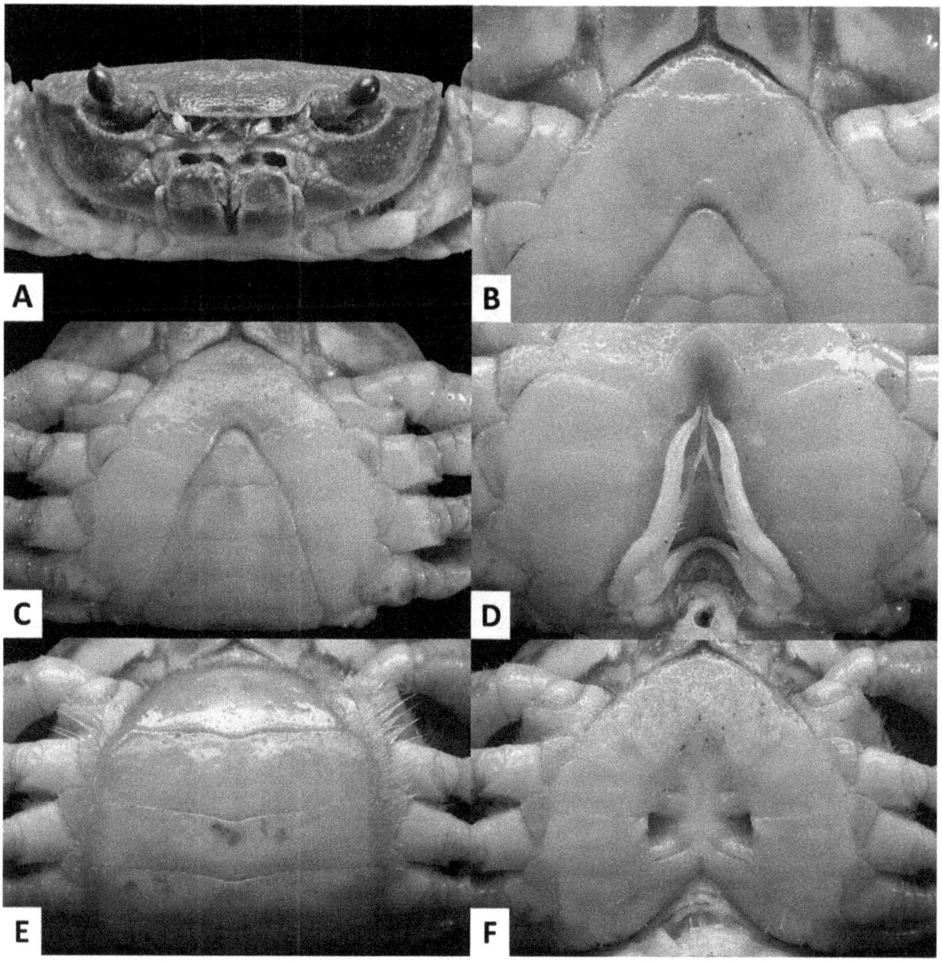

Fig. 5. *Teoswamon scolasticum* gen. nov., sp. nov., male holotype (SYSBM 001841): A, frontal cephalothorax; B, anterior thoracic sternum; C, ventral view of anterior thoracic sternum and pleon; D, ventral view showing sterno-pleonal cavity with right G1 in situ (left G1 removed). Female paratype (SYSBM 001847): E, ventral view of pleon; F, ventral view showing female vulvae.

scolasticum sp. nov. has a large opening at the tip (figs. 6C-E, 7C, D) whereas *C. shangchuanense*, as with other congeners, has a relatively small opening (Huang et al., 2017: fig. 4B). The G2 terminal segment is long in *Teoswamon* gen. nov. (fig. 6B) (length of subterminal segment: terminal segment = 1.9-2.0) when compared to *Cantopotamon* (Huang et al., 2017: fig. 2A) (length of subterminal segment: terminal segment = 2.6). The female vulvae of *Teoswamon* gen. nov. is large and reaches the suture of sternites 5/6 (fig. 5F); that of *Cantopotamon* species are consistently smaller and located within the boundaries of sternite 6 (Huang et al., 2017: fig. 11). The new genus is also smaller in size, carapace width not

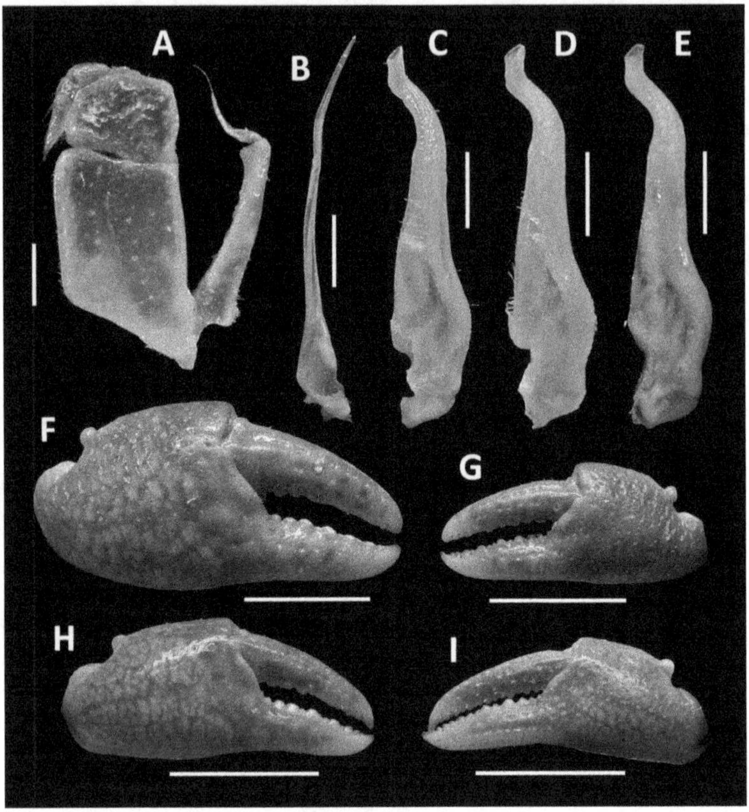

Fig. 6. *Teoswamon scolasticum* gen. nov., sp. nov., male holotype (SYSBM 001841): A, left third maxilliped; B, left G2; C, male left G1; F, male right cheliped; G, male left cheliped. Male paratype (SYSBM 001842): D, male left G1. Male paratype (NCU MCP 423001): E, male left G1. Female paratype (SYSBM 001847): H, female right cheliped; I, female left cheliped. Scale bars: A-E = 1.0 mm; F-I = 5.0 mm.

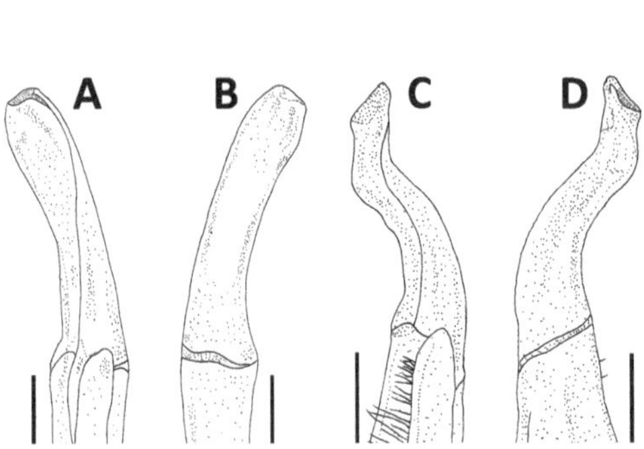

Fig. 7. *Minpotamon kityang* sp. nov., male holotype (SYSBM 001283), G1 terminal segment: A, ventral view; B, dorsal view. *Teoswamon scolasticum* gen. nov., sp. nov., male holotype (SYSBM 001841), G1 terminal segment: C, ventral view; D, dorsal view. Scale bars 1 mm.

exceeding 20 mm, whereas *Cantopotamon* can reach up to around 30 mm (Huang et al., 2017).

The G1 of *Teoswamon* gen. nov. is also similar to the monotypic *Eurusamon* in general form, but the G1 of the new genus exceeds well beyond the tubercle in situ and has a large opening at the tip (figs. 6C-E, 7C, D), whereas that of *Eurusamon* only slightly exceeds the pleonal locking tubercle and has a small opening (Huang, 2018: fig. 13D). The male pleonal somite 3 of the new genus (fig. 5C) is also much broader than *Eurusamon* (Huang, 2018: fig. 13C). The female vulvae of the new genus are large (fig. 5F), whereas those of *Eurusamon* are clearly smaller (Huang, 2018: fig. 16F). In addition, the new genus has a carapace width not exceeding 20 mm, which is much smaller than *Eurusamon*, which can reach up to 60 mm.

The new genus is also very similar to *Yarepotamon* externally, but can be separated by its postorbital cristae being almost fused with the epibranchial teeth and epigastric cristae (fig. 4A, B), whereas that of all four known *Yarepotamon* species are separate (Huang, 2018: figs. 2A, 3A, 4A, 6A). The third maxilliped exopod flagellum is long in the new genus (fig. 6A), whereas that of *Yarepotamon* is short to absent (Dai & Türkay, 1997: figs. 6(1), 7(4); Huang, 2018: figs. 5D, 7D). The new genus is also smaller in size, carapace width not exceeding 20 mm, whereas *Yarepotamon* can reach up to around 30 mm (Huang, 2018).

Teoswamon gen. nov. is superficially similar to the monotypic *Minutomon* in its small size, but the two can immediately be separated by their G1 morphology. The terminal segment of the new genus is long, slender and sinuous (figs. 6C-E, 7C, D), whereas that of *Minutomon* is much wider and knife-shaped (Huang et al., 2014: figs. 8B, C, 9A, B).

Teoswamon scolasticum sp. nov.
(figs. 4-6, 7C, D, 8B)

Material examined.— Holotype: SYSBM 001841, male 15.2 × 11.9 mm, Puning City (23°15′9″N, 116°8′38″E), Guangdong; coll. C. Huang, November 2018. Paratypes: SYSBM 001842-001846, 5 males 14.3 × 11.1 mm, 13.3 × 10.7 mm, 12.6 × 9.6 mm, 11.3 × 9.0 mm, 12.6 × 9.5 mm; same data as holotype. SYSBM 001847-001848, 2 females 15.9 × 12.5 mm, 13.9 × 11.0 mm; same data as holotype. NCU MCP 423001-423002, 2 males 14.1 × 11.3 mm, 10.7 × 8.9 mm, Puning City, Guangdong, coll. J. H. Huang, August 2017. NCU MCP 423003, female 15.0 × 11.8 mm; same data as above.

Description.— Male: carapace broader than long, width about 1.2-1.3 times as long as length ($n = 11$); regions indistinct, dorsal surface relatively flat, pitted, anterolateral and frontal regions rugose (fig. 4A). Front deflexed, margin ridged in dorsal view (figs. 4A, 5A). Epigastric cristae low, separated by narrow gap (figs. 4A, 5A). Postorbital cristae low, laterally expanded, almost fused with epibranchial teeth and epigastric cristae (fig. 4A). Branchial regions relatively flat (fig. 4A). Cervical groove discernible (fig. 4A). Mesogastric region slightly convex (fig. 4A). External orbital angle broadly triangular, outer margin gently convex, separated from anterolateral margin by small gap (fig. 4A). Epibranchial tooth small, granular (fig. 4A). Anterolateral margin cristate, lined with approximately 19-21 granules; bent inward posteriorly (fig. 4A). Posterolateral surface covered with low, oblique striae, converging towards posterior carapace margin (fig. 4A). Orbits large; supraorbital margins cristate, infraorbital margins lined with granules (fig. 5A). Suborbital and pterygostomial regions covered with small granules; sub-hepatic regions rugose, covered with fused small granules (fig. 5A). Median lobe of epistome posterior margin broadly triangular, lateral margins slightly sinuous (fig. 5A).

Maxilliped 3 merus about 1.2 times as wide as long; ischium width about 0.8 times as long as length; merus subtrapezoidal, with median depression; ischium subtrapezoidal, with distinct median sulcus, mesial margin rounded (fig. 6A). Exopod reaching to proximal one-third of merus; flagellum long (fig. 6A).

Chelipeds (pereiopod 1) unequal (fig. 6F, G). Merus trigonal in cross section, margins crenulated, surface rugose (fig. 4A). Carpus with sharp spine at inner-distal angle and spinule at base (fig. 4A). Major cheliped palm length about 1.2-1.5 times as long as height ($n = 3$); dactylus 0.8-0.9 times as long as palm length ($n = 3$) (fig. 6F). Palm surface pitted (fig. 6F). Dactylus as long as pollex (fig. 6F). Occlusal margin of fingers with irregular blunt teeth, slight gape when closed (fig. 6F).

Ambulatory legs (pereiopods 2-5) slender with short setae (fig. 4A). Pereiopod 3 merus 0.6-0.7 times as long as carapace length ($n = 4$) (fig. 4A). Pereiopods 5 propodus 1.9-2.2 times as long as broad ($n = 4$), slightly longer than dactylus (fig. 4A).

Male thoracic sternum generally smooth; sternites 1-4 narrow, width about 1.9 times as long as length; sternites 1, 2 forming triangular structure; sternites 2, 3 fused, but demarcated by shallow transverse sulcus; sternites 3, 4 fused without trace of sulcus (fig. 5B). Male sterno-pleonal cavity reaching anteriorly to level of midlength of chelipeds coxae base (fig. 5B). Male pleonal locking tubercle positioned slightly posterior to mid-length of sternite 5 (fig. 5D).

Male pleon triangular; somites 3-6 progressively narrower; somite 6 width approximately 2.2 times as long as length; telson width 1.4 times as long as length; apex rounded, posterior margin sinuous (fig. 5C).

G1 slender, tip of terminal segment exceeding pleonal locking tubercle in situ, reaching beyond the suture between thoracic sternites 4/5 (fig. 5D). Subterminal segment length about 2.2 ($n = 3$) times as long as length of terminal segment, tapering anteriorly. Terminal segment slender, tube-like, "S" shaped, curved inwards at base and abruptly bending anteriorly at around 1/3 length to the tip, with large opening directed inner-anteriorly (figs. 6C-E, 7C, D). G2 subterminal segment about 1.9-2.0 times as long as length of flagelliform distal segment ($n = 2$) (fig. 6B).

Female: chelipeds less inflated in females (fig. 6H, I). Major cheliped palm length about 1.4 times as long as height ($n = 3$); dactylus 0.9 times as long as palm length ($n = 3$) (fig. 6H). Pereiopod 3 merus 0.6-0.7 times as long as carapace length ($n = 3$) (fig. 4B). Pereiopod 5 propodus 2.0-2.2 times as long as broad ($n = 3$) (fig. 4B). Female vulvae large, reaches the sutures of sternites 5/6 anteriorly (fig. 5F). Female pleon broadly ovate (fig. 5E).

Etymology.— The species name is derived from Latin "scolasticus", meaning scholar, which is in memory of the late Prof. Rui-Yu Liu, an early Chinese carcinologist who contributed greatly to the research of the freshwater and marine crustaceans of China during his lifetime.

Habitat.— This species lives in low-altitude hillstreams in eastern Guangdong. They are most abundant in seeps where shallow water flows slowly through the substrate, but can also be found in shallow parts of the hillstream. Its small size appears to be an adaptation to this particular micro-habitat. An unidentified species of *Longpotamon* was found to be sympatric with the new species and appears to occupy the deeper parts of the streams and hides under large rocks.

Remarks.— See remarks for genus.

PHYLOGENETIC ANALYSIS AND DISCUSSION

A 528 bp segment, excluding the variable regions, of the 16S rDNA gene was amplified and aligned. Both phylogenetic analyses recovered the "South China #1" clade with high support (fig. 9). The generic placement of *Teoswamon scolasticum* sp. nov. is only supported by the BI analysis. It belongs in a clade with other eastern Guangdong potamids (*Minpotamon*, *Minutomon* and *Huananpotamon*) with *Huananpotamon* being its sister genus. Despite the striking similarity between *Teoswamon* gen. nov. and *Cantopotamon*, the two are evidently not immediately related. Further relationships between *Teoswamon* gen. nov. and other genera within the "South China #1" clade will have to be studied in the future using additional genetic markers.

The sister species of *Minpotamon kityang* sp. nov. is *Minpotamon nasicum* with very high support from both methods. Empirically, the species-level threshold of the nucleotide pairwise divergence of the 16S gene for potamid crabs is roughly around 1% (Shih et al., 2004; Shih et al., 2007). Despite obvious morphological

Fig. 8. A, *Minpotamon kityang* sp. nov. in natural habitat; B, *Teoswamon scolasticum* gen. nov., sp, nov. in natural habitat.

disparity, the 1.47% divergence between the two suggest a very close relationship, and is relatively small when compared to the average interspecific distance in *Cantopotamon* (5.80%, recalculated from Huang et al., 2017) and *Yarepotamon* (3.59%, recalculated from Huang, 2018), but is nevertheless larger than some other pairs of closely related species (e.g., 1.18% between *Nanhaipotamon formosanum* (Parisi, 1916) and *N. dongyinense* Shih, Chen & Wang, 2005; 0.93% between *N. nanriense* Dai, 1997 and *N. yongchuense* Dai, 1997, cf. Shih et al., 2011). *Minpotamon kityang* sp. nov. is thus also genetically supported as a new species.

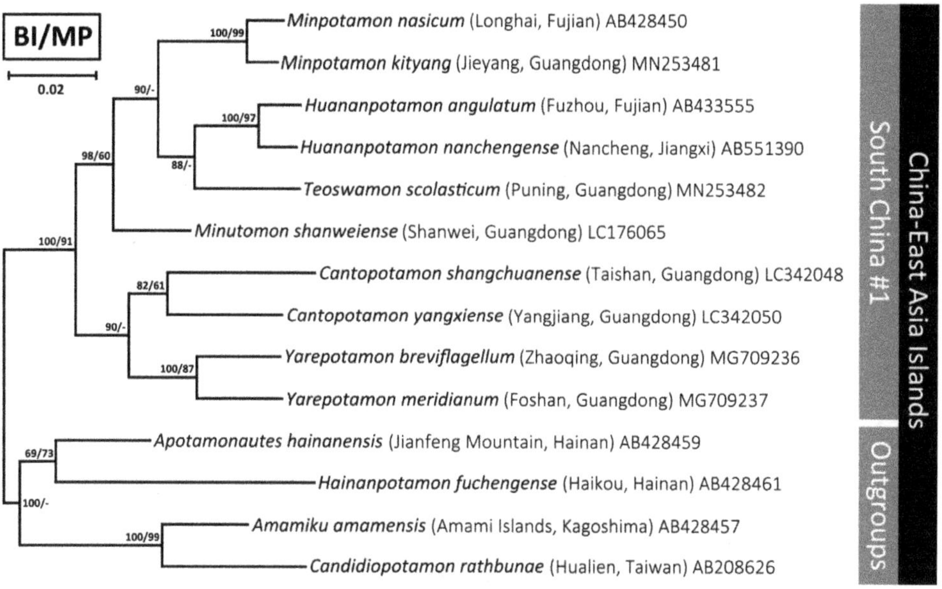

Fig. 9. A Bayesian inference (BI) tree based on the 16SrDNA sequence. Support values are represented at the nodes. Only ⩾50% are shown for maximum parsimony (-, not supported).

ACKNOWLEDGEMENTS

The authors wish to express gratitude towards Hsi-Te Shih (National Chung Hsing University) for his voluntary help in sequencing the specimens. A special thanks is also due to Dong Huang, Hui-Wen Xiao, Jun-Hao Huang and Zhuo-Cheng Zhou for their assistance in the discovery and collection of the new species. We would also like to thank the two anonymous reviewers who have helped to improve the manuscript.

REFERENCES

CRANDALL, K. A. & J. F. FITZPATRICK JR., 1996. Crayfish molecular systematics: using a combination of procedures to estimate phylogeny. Syst. Biol., **45**: 1-26.

DAI, A. Y., 1995. Study on freshwater crabs of a new genus *Hainanpotamon* from Hainan Island, China (Crustacea: Decapoda: Brachyura: Potamidae). Acta Zootaxon. Sin., **20**: 391-397. [In Chinese with English summary.]

DAI, A. Y., 1997. A revision of freshwater crabs of the genus *Nanhaipotamon* Bott, 1968 from China (Crustacea: Decapoda: Brachyura: Potamidae). Raffles B. Zool., **45**: 209-235.

DAI, A. Y., 1999. Fauna sinica (Arthropoda. Crustacea. Malacostraca. Decapoda. Parathelphusidae. Potamidae): 1-501, figs. 1-238, pls. 1-30. (Science Press, Beijing). [In Chinese with English summary.]

DAI, A. Y., G. X. CHEN, Y. Z. SONG, P. F. FAN, Y. G. LIN & Y. Q. ZENG, 1979. On new species of freshwater crabs harbouring metacercariae of lung flukes. Acta Zootaxon. Sin., **4**: 122-131, pl. 1. [In Chinese with English summary.]

DAI, A. Y. & P. K. L. NG, 1994. Establishment of a new genus of freshwater crab, *Huananpotamon* (Crustacea: Decapoda: Brachyura: Potamidae) from southern China. Raffles B. Zool., **42**: 657-661.

DAI, A. Y. & M. TÜRKAY, 1997. Revision of the Chinese freshwater crabs previously placed in the genus *Isolapotamon* Bott, 1968 (Crustacea: Decapoda: Brachyura: Potamidae). Raffles B. Zool., **45**: 237-264.

DAVIE, P. J., D. GUINOT & P. K. L. NG, 2015. Systematics and classification of Brachyura. In: P. CASTRO, P. J. F. DAVIE, D. GUINOT, F. R. SCHRAM & J. C. VON VAUPEL KLEIN (eds.), Treatise on zoology — anatomy, taxonomy, biology. The Crustacea, Vol. **9C-I**: 1049-1130. (Brill, Leiden, The Netherlands).

DE MAN, J. G., 1914. Note sur quelques crustacés décapodes brachyures terrestres et d'eau douce appartenant au Musée Civique de Gènes. Ann. Mus. Civ. stor. Nat. Giacomo Doria, **6**: 122-138.

HUANG, C., 2018. Revision of *Yarepotamon* Dai & Türkay, 1997 (Brachyura: Potamidae), freshwater crabs endemic to southern China, with descriptions of two new genera and four new species. J. Crust. Biol., **38**: 173-189.

HUANG, C., S. T. AHYONG & H. T. SHIH, 2017. *Cantopotamon*, a new genus of freshwater crabs from Guangdong, China, with descriptions of four new species (Crustacea: Decapoda: Brachyura: Potamidae). Zool. Stud., **56**: 1-21.

HUANG, C., M. C. EBACH & S. T. AHYONG, 2020. Bioregionalisation of the freshwater zoogeographical areas of mainland China. Zootaxa, **4742**: 271-298.

HUANG, C., J. R. HUANG & P. K. L. NG, 2012. A new species of *Nanhaipotamon* Bott, 1968 (Crustacea: Decapoda: Brachyura: Potamidae) from Zhuhai, Guangdong Province, China. Zootaxa, **3588**: 55-63.

HUANG, C., S. Y. MAO & J. R. HUANG, 2014. Two new potamid crabs, *Yuexipotamon arcophallus* new genus, new species and *Minutomon shanweiense* new genus, new species, (Crustacea: Decapoda: Brachyura: Potamidae) from southern China. Zootaxa, **3764**: 455-466.

HUANG, C., H.-T. SHIH & S. T. AHYONG, 2018a. Two new genera and two new species of narrow-range freshwater crabs from Guangdong, China (Decapoda: Brachyura: Potamidae). J. Crust. Biol., **38**: 614-624.

HUANG, C., K. C. WONG & S. T. AHYONG, 2018b. The freshwater crabs of Macau, with the description of a new species of *Nanhaipotamon* Bott, 1968 and the redescription of *Nanhaipotamon wupingense* Cheng, Yang, Zhong & Li, 2003 (Crustacea, Decapoda, Potamidae). ZooKeys, **810**: 91-111.

HUANG, C., H.-T. SHIH & S. Y. MAO, 2016. *Yuebeipotamon calciatile*, a new genus and new species of freshwater crab from southern China (Crustacea, Decapoda, Brachyura, Potamidae). ZooKeys, **615**: 61-72.

KUMAR, S., G. STECHER, M. LI, C. KNYAZ & K. TAMURA, 2018. MEGA X: molecular evolutionary genetics analysis across computing platforms. Mol. Biol. Evol., **35**: 1547-1549.

MINEI, H., 1973. Potamoid crabs of the Ryukyu Islands, with descriptions of five new species (Crustacea, Decapoda, Potamoidea). J. Fac. Agric. Kyushu Univ., **17**: 203-226.

PARISI, B., 1916. I Decapodi Giapponesi del Museo di Milano. IV. Cyclometopa. Atti Soc. Ital. Sci. Nat., Mus. Civ. Stor. Nat. Milano, **55**: 153-190, Pls. 7-11.

RONQUIST, F. & J. P. HUELSENBECK, 2003. MrBayes 3: Bayesian phylogenetic inference under mixed models. Bioinformatics, **19**: 1572-1574.

SHIH, H.-T., G. X. CHEN & L. M. WANG, 2005. A new species of freshwater crab (Decapoda: Brachyura: Potamidae) from Dongyin Island, Matsu, Taiwan, defined by morphological and molecular characters, with notes on its biogeography. J. Nat. Hist., **39**: 2901-2911.

SHIH, H.-T., P. K. L. NG & H. W. CHANG, 2004. Systematics of the genus *Geothelphusa* (Crustacea, Decapoda, Brachyura, Potamidae) from southern Taiwan: a molecular appraisal. Zool. Stud., **43**: 561-570.

SHIH, H.-T., D. C. J. YEO & P. K. L. NG, 2009. The collision of the Indian plate with Asia: molecular evidence for its impact on the phylogeny of freshwater crabs (Brachyura: Potamidae). J. Biogeogr., **36**: 703-719.

SHIH, H.-T., P. K. L. NG, C. D. SCHUBART & H.-W. CHANG, 2007. Phylogeny and phylogeography of the genus *Geothelphusa* (Crustacea: Decapoda, Brachyura, Potamidae) in southwestern Taiwan based on two mitochondrial genes. Zool. Sci., **24**: 57-66.

SHIH, H.-T., X. M. ZHOU, G. X. CHEN, I. C. CHIEN & P. K. L. NG, 2011. Recent vicariant and dispersal events affecting the phylogeny and biogeography of East Asian freshwater crab genus *Nanhaipotamon* (Decapoda: Potamidae). Mol. Phylogenet. Evol., **58**: 427-438.

THOMPSON, J. D., D. G. HIGGINS & T. J. GIBSON, 1994. CLUSTAL W: improving the sensitivity of progressive multiple sequence alignment through sequence weighting, position-specific gap penalties and weight matrix choice. Nucleic Acids Res., **22**: 4673-4680.

First received 11 September 2019.
Final version accepted 2 June 2020.

[When citing this volume, please refer to *Crustaceana* 93 (2020), 1233-1546]

Prof. Ruiyu Liu Memorial Issue

RUIYUPOTAMON, A NEW GENUS AND FOUR NEW SPECIES OF FRESHWATER CRABS FROM NORTHWESTERN YUNNAN, CHINA (DECAPODA, BRACHYURA, POTAMIDAE)

BY

PENGFEI WANG, ZEWEI ZHANG and HONGYING SUN[1])
Jiangsu Key Laboratory for Biodiversity and Biotechnology, College of Life Sciences, Nanjing Normal University, Nanjing 210023, P.R. China

ABSTRACT

A new genus *Ruiyupotamon* gen. nov. and four new species, *R. banqiaopingense* sp. nov., *R. zhongjiangense* sp. nov., *R. jinjiangense* sp. nov., and *R. zhongyiense* sp. nov. belonging to the subfamily Potamiscinae (Potamidae), collected from northwestern Yunnan Province, P.R. China, are described. Molecular phylogenetic data derived from both mitochondrial (16S), and nuclear (28S and ITS2) markers supports the establishment of the new genus and four new species.

Key words. — Potamiscinae, *Ruiyupotamon*, new genus, new species, molecular markers, phylogeny

RÉSUMÉ

Un nouveau genre *Ruiyupotamo* nov. gen. et quatre nouvelles espèces, *R. banqiaopingense* nov. sp., *R. zhongjiangense* nov. sp., *R. jinjiangense* nov. sp. et *R. zhongyiense* nov. sp. Appartenant à la sous famille des Potamiscinae (Potamidae), récoltés dans le nord ouest de la province du Yunnan, R.p. de Chine, ont été décrits. Les données de phylogénie moléculaire proviennent à la fois de marqueurs mitochondriaux (16S) et nucléaires (28S et ITS2) supportent l'établissement du nouveau genre et des quatre nouvelles espèces.

Mots clés. — Potamiscinae, *Ruiyupotamon*, nouveau genre, nouvelles espèces, marqueurs moléculaires, phylogénie

INTRODUCTION

Yunnan Province has the highest species richness of freshwater crabs in China, with 60 species in 17 genera described (Dai, 1999; Chu et al., 2018a, b, Naruse et al., 2018). Of these, 58 species of 16 genera have been assigned to the subfamily

[1]) Corresponding author; e-mail: sunhongying@njnu.edu.cn

 | DOI:10.1163/9789004464353_007

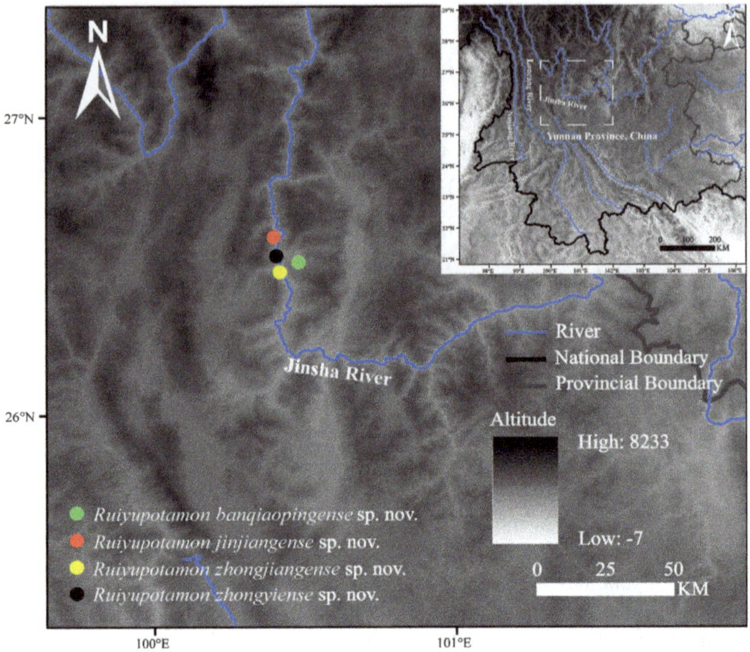

Fig. 1. A map showing the distribution of *Ruiyupotamon banqiaopingense* sp. nov. (green circle), *R. zhongjiangense* sp. nov. (yellow circle), *R. jinjiangense* sp. nov. (red circle), *R. zhongyiense* sp. nov. (black circle) from northwestern Yunnan Province, P.R. China.

Potamiscinae Bott, 1970 (Potamidae Ortmann, 1896). In recent years, regular field investigations of freshwater crab fauna were conducted in northwestern Yunnan Province, P.R. China. In this paper, we describe a new genus, *Ruiyupotamon* gen. nov. and four new species from the specimens collected from the dry-hot valley of Jinsha River. This new genus differs from other genera recorded in western and northern Yunnan, P.R. China, in the characters of the carapace and male first gonopod. Molecular phylogenetic analyses based on sequences of the mitochondrial 16S rRNA gene (16S), the nuclear 28S rRNA gene (28S) and the internal transcribed spacer 2 (ITS2), provide independent evidence for the recognition of the new genus and four new species. A key to the four new species of *Ruiyupotamon* gen. nov. is provided herein.

MATERIAL AND METHODS

Sampling and morphological comparison

Specimens were collected by hand from 4 locations in northwestern Yunnan (fig. 1), preserved in 95% ethanol and identified under a stereo dissection microscope (Nikon SMZ645). In this paper, the G1 terminal morphological characters of the four new species were described by using scanning electron microscopy (Phenom Pure+). All specimens examined are deposited in the Jiangsu Key Laboratory for Biodiversity and Biotechnology, College of Life Sciences, Nanjing Normal University (NNU), Nanjing, P.R. China. The following abbreviations are used: G1, male first gonopod; G2, male second gonopod; asl, altitude above sea level.

Molecular data

Genomic DNA was extracted from leg muscle tissue using the standard proteinase K and Phenol/chloroform techniques (Sambrook & Russell, 2001). DNA fragments were amplified: (i) 590 bp fragment of 16S; (ii) 1000 bp fragment of nuclear 28S; (iii) 1000 bp fragment of ITS2 (table I). To investigate the phylogenetic relationships among the new genus and other Asian potamid taxa, a total of 65 available partial sequences of 16S from 62 species and 50 genera (Shih et al., 2009; Huang et al., 2016, 2017; Ji et al., 2016; Chu et al., 2017, 2018a), including the seven new sequences from the four new species obtained in this study and an unclassified species of Potamiscinae (voucher-specimen number NNU170502), were used.

Phylogenetic analyses

All sequences were aligned using MAFFT 7.215 (Katoh & Standley, 2013) with the iterative refinement method G-INS-i (accurate alignment), in which the gap opening and extension penalties were 1.53 and 0.123 respectively. The primer sequences were removed from alignments. Ambiguous or randomly similar sites in 16S alignments were removed by Aliscore 2.0 (Misof & Misof, 2009; Kück et al., 2010) using default settings with gaps treated as a fifth character. A NJ tree was generated as the initial guiding tree. After the analysis, no ambiguous or randomly similar sites were found and the original 491 bp segments with 197 bp parsimony-informative sites and 254 bp variable sites were used for further analysis. The best-fitting model for sequence evolution of the 16S dataset was determined by JModeltest vers. 2.1.1 (Darriba et al., 2012), selected by the Bayesian information criterion (BIC). The best model obtained was HKY + I + G, and was subsequently applied for Bayesian inference (BI). Sequences of the different haplotypes of 16S, 28S and ITS2 have been deposited in the GenBank (accession numbers are listed in table II).

Maximum likelihood (ML) analysis for the dataset was performed using RAxML-HPC2 on XSEDE 8.1.12 (Stamatakis, 2006, 2014) through the Cipres Science Gateway (Miller et al., 2010) with GTR + CAT model. CAT approximation of rate heterogeneity was used for rapid bootstrapping, but the Gamma model

TABLE I

Primer sequences used in the present study

Gene	Primer name	Sequence (5′ → 3′)	Length (bp)	Product length (bp)	Annealing temperature (°C)	Reference
16S	H16S	GCCTGCTTATCAAAAACAT	19	approx. 590	50	Ji et al. (2016)
	L16S	AAGAGATAGAAATCAACCTGG	21			
28S	F28S	CAGCCCTAAGCAGGTGGTAAACT	23	approx. 1000	52	Ji et al. (2016)
	R28S	CCATCTTTCGGGTCCCAACAT	21			
ITS2	FITS	AAGCTGCGTGTCGGTATGTGAAT	23	approx. 1000	54	This study
	RITS	CAAGCATAACTTAGGCGGTGAGA	23			

TABLE II

Sequence GenBank-accession number of the 16S, 28S and ITS2 haplotypes for the four new species of the genus *Ruiyupotamon* gen. nov. from Yunnan Province, P.R. China

Species	Localities in Yunnan, P.R. China	Museum catalogue no.	Accession No.		
			16S	28S	ITS2
R. banqiaopingense	Banqiaoping Village, Yongsheng County, Lijiang City	NNU427103	MT661600	MT668527	MT622755
		NNU427110	MT668534	MT668528	MT649478
R. jinjiangense	Jinjiang Town, Gucheng District, Lijiang City	NNU428408	MT668537	MT668531	MT663743
R. zhongjiangense	Zhongjiang Village, Heqing County, Dali City	NNU427204	MT668535	MT668529	MT663745
		NNU427205	MT668536	MT668530	MT663744
R. zhongyiense	Zhongyi Village, Heqing County, Dali City	NNU4210305	MT668538	MT668532	MT677854
		NNU4210306	MT668539	MT668533	MT677869

was used for the final ML search. Node reliability was obtained after 1000 bootstrap replicates.

MrBayes 3.2.6 (Huelsenbeck & Ronquist, 2001; Ronquist et al., 2012) was used to perform the Bayesian inference (BI) analysis. Two parallel runs were created. In each run, three heated and one cold MCMC chains were run for ten million generations, with every one thousand generations being sampled. The first 25% of samples from the cold chain were discarded as burn-in. The average standard deviation (SD) of split frequencies between both runs was checked (0.003 < 0.01). The sampled parameters were investigated using Tracer 1.6 (Rambaut et al., 2014). The effective sampling sizes (ESS) for all parameters were more than 200. The maximum clade credibility tree with the nodal posterior probability was generated by TreeAnnotator 1.8.2 (Drummond et al., 2012). The bootstrap support (BS) and Bayesian posterior probability (BPP) were used to assess statistical support. Sequences of the 28S and ITS2 gene were concatenated with the 16S gene and analyzed by the BI and ML methods. PartitionFinder 1.1.1 (Lanfear et al., 2012) was used to find the best-fit partitioning schemes and models using a greedy search based on the Akaike information criterion (AIC). Subsequently, the best partitioning schemes for ML and BI analyses correspond to the three gene regions, and the best models found for BI are HKY + I, GTR + I, HKY + G for 16S, 28S and ITS2 respectively. The parameters for the ML and BI analyses were found following the method mentioned above.

RESULTS

Systematic account
Family POTAMIDAE Ortmann, 1896
Subfamily POTAMISCINAE Bott, 1970
Genus **Ruiyupotamon** gen. nov.

Type species.— *Ruiyupotamon banqiaopingense* sp. nov., present designation.

Comparative material.— *Aparapotamon inflomanum* Dai & Chen, 1985: 1 male (adult), 16.8 × 13.6 mm, NNU 142201, Xianggelila, Yunnan, Sep 2015; *Artopotamon compressum* Dai & Chen, 1985: holotype, male (adult), 23.3 × 18.6 mm, IZCAS CB00733, Lushui county, Yunnan, 1981; *Semicirculara lincangensis* Chu, Wang & Sun, 2018: holotype, male (adult), 18.8 × 14.4 mm, NNU 160562, Cangyuan County, Yunnan, May 2016.

Diagnosis.— Carapace broader than long, subquadrate, slightly convex, branchial regions relatively flat (fig. 2A); anterolateral margin convex, lined with distinct granules, cervical groove shallow, inconspicuous (fig. 2A); H-shaped groove between gastric and cardiac regions deep and distinct (fig. 2A); epigastric cristae slightly convex, separated medially by a shallow "Y" shaped groove (fig. 2A, B); external orbital angle sharply triangular (fig. 2A). Third maxilliped

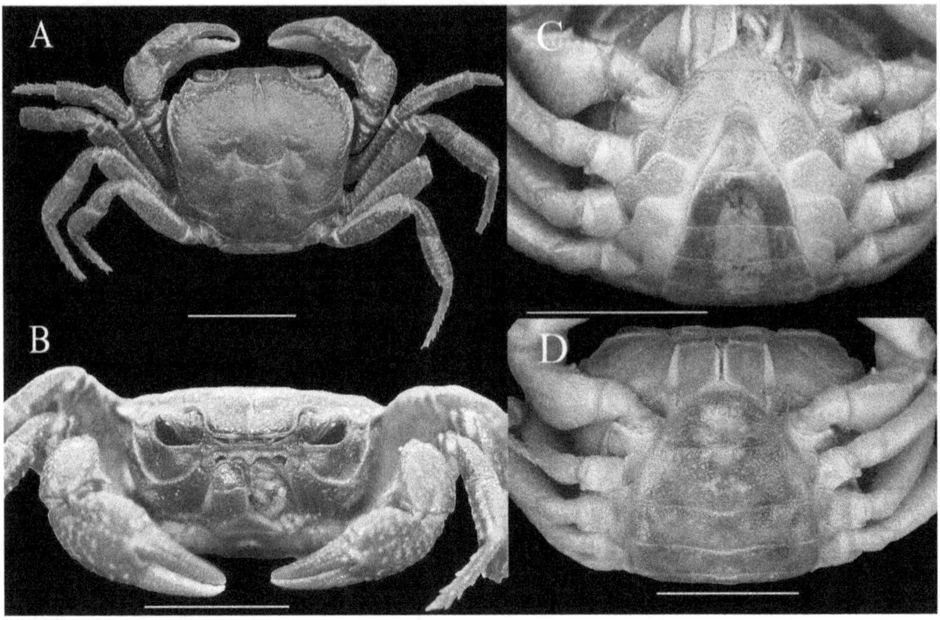

Fig. 2. *Ruiyupotamon banqiaopingense* sp. nov. A-C, male holotype, 17.4 × 14.5 mm, NNU 427101; D, female paratype, 21.6 × 17.9 mm, NNU 427105. A, Dorsal view; B, frontal view of cephalothorax; C & D, ventral view of thoracic sternum and pleon. Scale bars = 1.0 cm.

exopod reaching proximal one-third of merus length, without flagellum (fig. 3A). Male pleon broadly triangular (fig. 2C). G1 reaching beyond tubercle of male abdominal locking mechanism (fig. 3B); tip of terminal segment with protrusion, ventral lobe broader than dorsal lobe (fig. 11B); G2 longer than G1, terminal segment slightly curved to the median longitudinal suture in situ (fig. 3B, G). Female vulvae ovate, on suture between thoracic sternites 5/6 (fig. 12A).

Etymology.— The name of the genus is an arbitrary combination of Prof. Liu Ruiyu's given name (Rui-Yu) and *Potamon* in order to honour Prof. Liu, a senior academician of the Institute of Oceanology, Chinese Academy of Sciences, who devoted his life to studying the classification and systematics of Chinese marine crustaceans.

Remarks.— Although the new genus is similar to *Aparapotamon* Dai & Chen, 1985, *Semicirculara* Chu, Wang & Sun, 2018 and *Artopotamon* Dai & Chen, 1985 by having a third maxilliped exopod without a flagellum, it can be easily distinguished from the three genera by the carapace being slightly convex (fig. 2A), (versus carapace distinctly convex in *Aparapotamon* and *Semicirculara*, and relatively flat in *Artopotamon* (cf. Dai, 1999: pl. 342, 344; Chu et al., 2018a: fig. 2A)); tip of the G1 terminal segment reaching beyond tubercle of abdominal locking mechanism in situ (fig. 3B), (versus tip of the G1 terminal segment

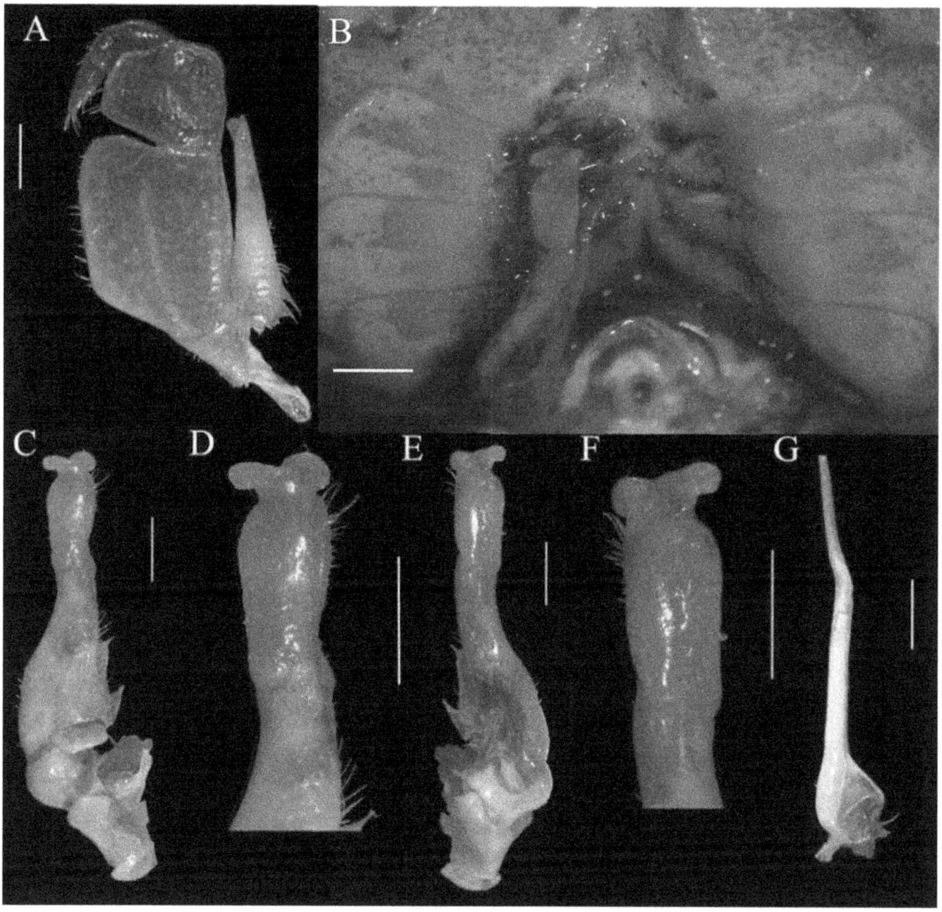

Fig. 3. *Ruiyupotamon banqiaopingense* sp. nov. male holotype, 17.4 × 14.5 mm, NNU 427101. A, Left third maxilliped; B, ventral view showing sterno-abdominal cavity with right G1 in situ (left G1 removed); C, left G1 (dorsal view); D, left G1 terminal segment (dorsal view); E, left G1 (ventral view); F, left G1 terminal segment (ventral view); G, left G2. Scale bars = 1.0 mm.

reaching thoracic sternites 4/5 in situ in *Aparapotamon* and *Semicirculara*, not reaching tubercle of abdominal locking mechanism in situ in *Artopotamon* (cf. Dai, 1999: fig. 186(3) and 196(3); Chu et al., 2018a: fig. 4A)); G1 slender, straight (fig. 4D) (versus G1 slender, distal part of terminal segment curved dorsally and inwards in *Aparapotamon*; terminal segment generally slenderer than the distal part of subterminal segment in *Artopotamon* (fig. 4A, B; cf. Dai, 1999: fig. 186(4) and 196(4))); ventral lobe of terminal segment broader than dorsal lobe with distinct ventral flap in middle part (figs. 4D, 11A) (versus the two lobes almost with the identical width in *Aparapotamon* and *Artopotamon* (fig. 4A, B; cf. Dai, 1999: fig. 186(4) and 196(4))); tip of G1 with a disc-like protrusion pointing outwards (fig. 4D) (versus tip of dorsal lobe convex as a discoid shape in

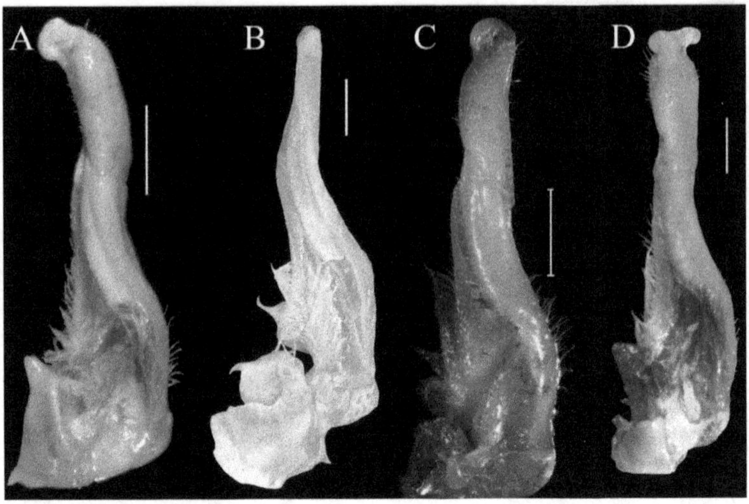

Fig. 4. Ventral view of G1 of *Ruiyupotamon banqiaopingense* sp. nov. and allied genera. A, *Aparapotamon inflomanum* Dai & Chen, 1985 (male, 16.8 × 13.6 mm, NNU 142201); B, *Artopotamon compressum* Dai & Chen, 1985 (male holotype, 23.3 × 18.6 mm, IZCAS CB00733); C, *Semicirculara lincangensis* (male holotype, 18.8 × 14.4 mm, NNU 160562); D, *Ruiyupotamon banqiaopingense* sp. nov. (male holotype, 17.4 × 14.5 mm, NNU 427101). Scale bars = 1.0 mm.

Aparapotamon; tip of dorsal lobe elongated with a ventrally directed semicircular lobe in *Semicirculara*; flattened in *Artopotamon* (fig. 4A, B, C; cf. Dai, 1999: fig. 196(4))). Descriptions of these differences are listed in table III.

Ruiyupotamon banqiaopingense sp. nov.
(figs. 2, 3, 4D, 11A, B, 12A, 13A)

Material examined.— Holotype: male (adult), 17.4 × 14.5 mm, NNU 427101, Banqiaoping Village (26°30′51″N 100°28′43″E, 2188 m asl), Shunzhou Town, Yongsheng County, Lijiang City in Yunnan Province, coll. Pengfei Wang, Zewei Zhang, Qiang Zhao & Hongying Sun, 6 May 2017. Paratype: female (allotype), 21.6 × 17.9 mm, NNU 427105, same data as holotype. One male (adult), 21.1 × 17.2 mm, NNU 427111, same data as holotype. One female (adult), 20.6 × 16.9 mm, NNU 427107, same data as holotype.

Etymology.— This species is named after the type locality, Banqiaoping Village, Shunzhou Town, Yongsheng County in Yunnan Province, P.R. China.

Diagnosis.— H-shaped groove between gastric and cardiac regions deep and distinct (fig. 2A). Male pleon smooth, generally pitted, broadly triangular, somites 3-6 progressively broader longitudinally (fig. 2C); telson broadly triangular, lateral margins almost straight, tip rounded (fig. 2C); ventral lobe of G1 terminal segment broader than dorsal lobe (figs. 3D, 11B); tip of G1 reaches to thoracic sternites 4/5 in situ (fig. 3B), with a blunt disc-like protrusion, almost as wide as the terminal segment transversely (figs. 3C-F; 11A, B); female vulvae on suture between thoracic sternites 5/6, ovate, margin nearly flat, opening ventrally (fig. 12A).

TABLE III

Morphological comparisons among *Aparapotamon* Dai & Chen, 1985, *Artopotamon* Dai & Chen, 1985, *Semicirculara* Chu, Wang & Sun, 2018 and *Ruiyupotamon* gen. nov.

Character	*Aparapotamon inflomanum*[1]	*Artopotamon compressum*[2]	*Semicirculara lincangensis*[3]	*Ruiyupotamon* gen. nov.
Carapace	Distinctly convex	Relatively flat	Distinctly convex	Slightly convex
Outline of G1	Slender, distal part of terminal segment curved dorsally and inwards (fig. 4A)	Terminal segment generally slenderer than the distal part of subterminal segment (fig. 4B)	Slender, straight; ventral lobe of terminal segment broader than dorsal lobe with distinct ventral flap in middle part (fig. 4C)	Slender, straight; ventral lobe of terminal segment broader than dorsal lobe with distinct ventral flap in middle part (fig. 4D)
Tip of G1	Tip of dorsal lobe convex as a discoid shape (fig. 4A)	Flattened (fig. 4B)	Tip of dorsal lobe elongated with a ventrally directed semicircular lobe (fig. 4C)	With a disc-like protrusion pointing outwards (fig. 4D)
G1 in situ	Reaching thoracic sternites 4/5	Not reaching tubercle of abdominal locking mechanism	Reaching thoracic sternites 4/5	Reaching beyond tubercle of abdominal locking mechanism (fig. 3B)

[1]. cf. Dai (1999): pl. 363; fig. 196.
[2]. cf. Dai (1999): pl. 343; fig. 186.
[3]. cf. Chu et al. (2018a): fig. 2A, B, 3C, D, 4A.

Description.— Carapace about 1.2 times broader than long, subquadrate, punctate, dorsal surface slightly convex, anterolateral region with granules (fig. 2A). Cervical groove shallow, indistinct, H-shaped groove between gastric and cardiac regions deep and distinct (fig. 2A). Epigastric cristae slightly convex, rugose, separated medially by a deep "Y" shaped groove, just anterior of postorbital cristae (fig. 2A, B); postorbital region concave and epibranchial region with obvious granules (fig. 2A). Frontal region slightly convex, surface covered with granules, anterior border emarginated medially (fig. 2A, B). Dorsal orbital margin ridged, glabrous (fig. 2A). Median part of epistome with triangular median lobe, tip rounded, inferior margin sinuous, upper margin of epistome almost straight (fig. 2B). Pterygostomian region comparatively smooth (fig. 2B); external orbital angle triangular (fig. 2A); epibranchial tooth granular, distinct, clearly demarcated from external orbital tooth with a small notch (fig. 2A); anterolateral margin distinctly convex, without obvious granules (fig. 2A); posterolateral margin comparatively smooth, lined with oblique striae (fig. 2A).

Third maxilliped merus about 1.2 times as broad as long, trapezoidal (fig. 3A); ischium about 1.3 times as long as broad, rectangular, with distinct median sulcus (fig. 3A); exopod reaching proximal 1/3 of merus length, without flagellum (fig. 3A).

Chelipeds slightly unequal (fig. 2A, B); merus trigonal in cross-section, margin crenulated (fig. 2A); carpus with sharp spine on inner distal angle, with granules at base (fig. 2A, B); manus of larger cheliped about 1.4 times as long as high, outer surface with granules, movable finger as long as fixed finger (fig. 2A, B).

Ambulatory legs slender, glabrous (fig. 2A); dactylus slender, margin sharp, spinous (fig. 2A). Last leg with propodus about 2 times as long as broad, slightly shorter than dactylus (fig. 2A).

Male pleon smooth, pitted, broadly triangular, somites 3-6 progressively broader longitudinally (fig. 2C); somite 6 width 2.2 times length (fig. 2C); telson broadly triangular, tip rounded, about 1.4 times as broad as long, thoracic sternum pitted (fig. 2C). Sternites 1, 2 completely fused, triangular (fig. 2C); suture between sternites 2, 3 continuous, distinct (fig. 2C); sternites 3, 4 fused with shallow median suture (fig. 2C); interruption between sutures of sternites 4/5, 5/6, 6/7 broad, median groove of thoracic sternum deep (fig. 3B); median longitudinal suture between sternites 7, 8 short (fig. 3B).

Tip of G1 terminal segment reaching beyond male abdominal locking mechanism, reaching to suture between thoracic sternites 4/5 in situ (fig. 3B); distal part of subterminal segment almost as wide as basal part of terminal segment (fig. 3C-F); subterminal segment about 1.8 times as long as terminal segment (fig. 3E). Terminal segment almost straight, tip of terminal segment with a blunt, disc-like protrusion, almost as wide as diameter of terminal segment (figs. 3C-F,

11A, B). G2 longer than G1, subterminal segment about 2.6 times as long as distal segment (fig. 3B, G).

The largest adult paratype female specimen resembles the holotype in most non-sexual characters. Female vulvae ovate; sixth somite about 2.9 times as broad as long, telson semicircular, about 2.2 times as broad as long (fig. 2D). Female vulvae on suture between thoracic sternites 5/6, ovate, margin flat, opening ventrally (fig. 12A).

Remarks.— *R. banqiaopingense* sp. nov. is closest to *R. jinjiangense* sp. nov. in most of its external morphology and tip of G1 terminal segment reaching to the suture between thoracic sternites 4/5 (figs. 3B, 8B), but can be distinguished by its tip of G1 with a blunt, disc-like protrusion (figs. 3C-F, 11A, B) (versus a beak-shaped protrusion in *R. jinjiangense* sp. nov.; figs. 8C-F, 11E, F). The new species is also similar to *R. zhongyiense* sp. nov. in most external morphology, but can be distinguished by its G1 reaching to the suture between thoracic sternites 4/5 (fig. 3B) (versus not reaching to the suture between thoracic sternites 4/5 in *R. zhongyiense* sp. nov.; fig. 10B); tip of G1 with a blunt disc-like protrusion (figs. 3C-F, 11A, B) (versus a swollen protrusion in *R. zhongyiense* sp. nov.; figs. 10C-F, 11G, H). Integral morphological comparisons are listed in table IV.

Distribution and habitat.— *R. banqiaopingense* sp. nov. was found in Banqiaoping Village (26°30′51″N 100°28′43″E, 2188 m asl), Shunzhou Town, Yongsheng County in Yunnan Province, P.R. China. This species is narrowly distributed in small hillstreams, hiding under rocks or small burrow near the east coast of Jinsha River.

Live coloration.— Carapace is usually dark brown, while chelipeds and ambulatory legs are usually brown in life (fig. 13A).

Ruiyupotamon zhongjiangense sp. nov.
(figs. 5, 6, 11C, D, 12B, 13B)

Material examined.— Holotype: male (adult), 17.7 × 14.8 mm, NNU 427207, Zhongjiang Village (26°28′51″N 100°24′58″E, 1300 m asl), Longkaikou Town, Heqing County, Dali City in Yunnan Province, coll. Pengfei Wang, Zewei Zhang, Qiang Zhao & Hongying Sun, 6 May 2017. Paratype: female (adult), 20.7 × 16.7 mm, NNU 427206, same data as holotype. One male (adult), 17.7 × 14.4 mm, NNU 427205, same data as holotype. One male (adult), 17.8 × 15.1 mm, NNU 427208, same data as holotype. One female (adult), 18.5 × 14.7 mm, NNU 427216, same data as holotype.

Etymology.— This species is named after the type locality, Zhongjiang Village, Longkaikou Town, Heqing County in Yunnan Province, P.R. China.

Diagnosis.— H-shaped groove between gastric and cardiac regions deep and distinct (fig. 5A). Male pleon broadly triangular, lateral margins slightly curved, tip of telson rounded (fig. 5C). G1 not reaching suture between sternites 4/5 in situ

TABLE IV

Morphological comparisons among the four new species of *Ruiyupotamon* gen. nov., *R. banqiaopingense* sp. nov., *R. zhongjiangense* sp. nov., *R. jinjiangense* sp. nov. and *R. zhongyiense* sp. nov.

Character	*R. banqiaopingense* sp. nov.	*R. zhongjiangense* sp. nov.	*R. jinjiangense* sp. nov.	*R. zhongyiense* sp. nov.
G1 in situ	Reaching to the suture between thoracic sternites 4/5 (fig. 3B)	Not reaching to the suture between thoracic sternites 4/5 (fig. 6B)	Reaching to the suture between thoracic sternites 4/5 (fig. 8B)	Not reaching to the suture between thoracic sternites 4/5 (fig. 10B)
Tip of G1	Blunt and disc-like, almost as wide as the diameter of terminal segment (fig. 11A, B)	Tongue-shaped, planiform, curved downwards (fig. 11C, D)	Beak-shaped, with a sharp terminal (fig. 11E, F)	Swollen (fig. 11G, H)
Lateral margin of male telson	Almost straight (fig. 2C)	Slightly curved (fig. 5C)	Slightly curved (fig. 7C)	Distinctly curved (fig. 9C)

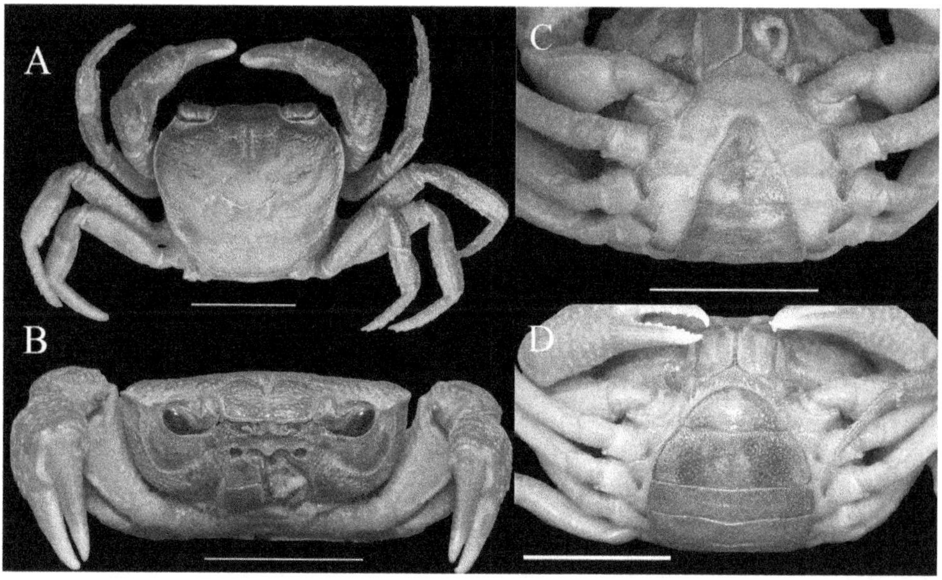

Fig. 5. *Ruiyupotamon zhongjiangense* sp. nov. A-C, male holotype, 17.7 × 14.8 mm, NNU 427207; D, female paratype, 20.7 × 16.7 mm, NNU 427206. A, Dorsal view; B, frontal view of cephalothorax; C, D, ventral view of thoracic sternum and pleon. Scale bars = 1.0 cm.

(fig. 6B), tip of terminal segment with a linguiform protrusion, curved downwards, ventral lobe of terminal segment broader than dorsal lobe (figs. 6C-F, 11C, D). Female vulvae on suture between thoracic sternites 5/6, widely ovate (fig. 12B); inferior margin slightly arched, opening ventrally (fig. 12B).

Description.— Carapace about 1.2 times broader than long, dorsal surface pitted, gently convex (fig. 5A); cervical groove shallow, H-shaped groove between gastric and cardiac regions deep and distinct (fig. 5A). Epigastric cristae with rugae, separated medially by a deep "Y" shaped groove (fig. 5A, B); postorbital cristae distinctly convex (fig. 5A); postorbital region concave and epibranchial region with obvious granules (fig. 5A). Frontal region distinctly convex, surface covered with granules, anterior border emarginated medially (fig. 5A, B). Orbital margin ridged, smooth (fig. 5B). Medial part of epistome with triangular medial lobe, tip rounded, inferior margin sinuous (fig. 5B). External orbital angle sharp triangle, epibranchial tooth blunt (fig. 5A); anterolateral margin with indistinct granules (fig. 5A).

Third maxilliped with merus about 1.2 times as broad as long, trapezoidal (fig. 6A); ischium about 1.2 times as long as broad (fig. 6A); ischium rectangular, with distinct median sulcus (fig. 6A); exopod reaching to proximal 1/3 of merus, without flagellum (fig. 6A).

Chelipeds slightly unequal (fig. 5A, B); merus trigonal in cross-section, carpus with sharp spine on inner distal angle, with granules at base (fig. 5A, B); outer

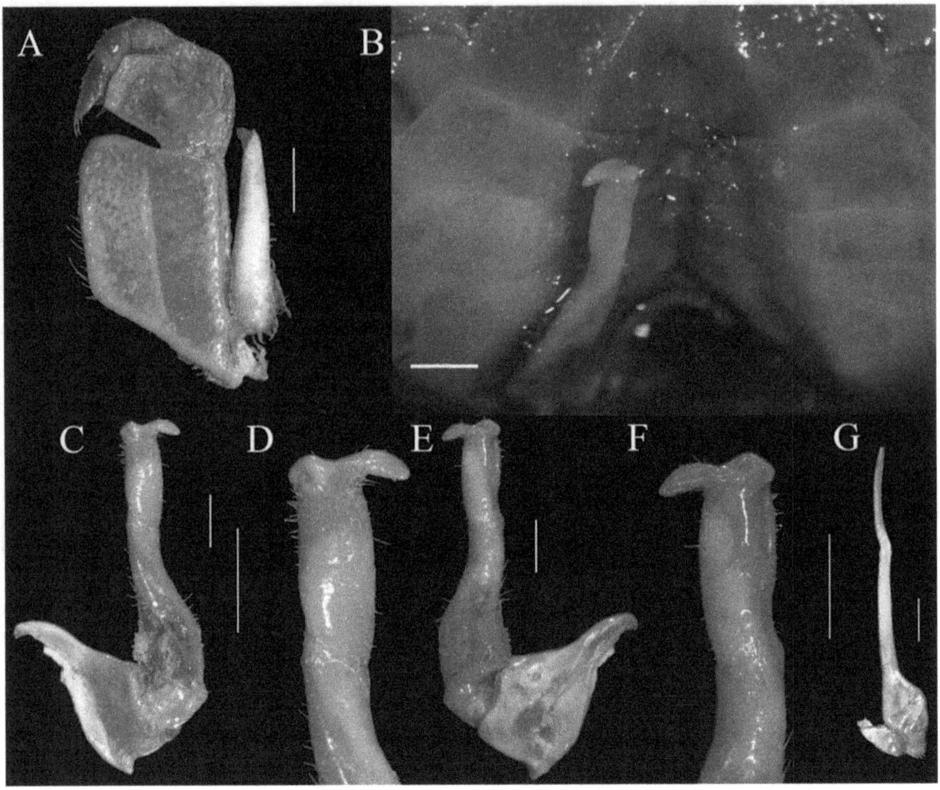

Fig. 6. *Ruiyupotamon zhongjiangense* sp. nov. male holotype, 17.7 × 14.8 mm, NNU 427207. A, Left third maxilliped; B, ventral view showing sterno-abdominal cavity with right G1 in situ; C, left G1 (ventral view); D, left G1 terminal segment (ventral view); E, left G1 (dorsal view); F, left G1 terminal segment (dorsal view); G, left G2. Scale bars = 1.0 mm.

surface of manus lined with distinctly convex granules, about 1.4 times as long as high (fig. 5B); fixed finger nearly as long as movable finger (fig. 5B).

Ambulatory legs slender (fig. 5A); dactylus slender, with dense spinules (fig. 5A); last leg with propodus about 2 times as long as broad, slightly shorter than dactylus (fig. 5A).

Male pleon broadly triangular, surface pitted (fig. 5C); somite 6 width 2.6 times length, telson broadly triangular, tip rounded, with about 1.4 times as broad as long, thoracic sternum pitted (fig. 5C); sternites 1, 2 completely fused to form triangular structure, sternites 2, 3 separated by continuous suture (fig. 5C); sternites 3, 4 fused with shallow median suture (fig. 5C). Median groove of thoracic sternum deep, interruption between sutures of sternites 4/5, 5/6, 6/7 gently broad (fig. 6B); median longitudinal suture between sternites 7/8 moderately short (fig. 6B).

Tip of G1 terminal segment not reaching to suture between thoracic sternites 4/5 in situ (fig. 6B); subterminal segment about 1.6 times as long as terminal segment (fig. 6C, E); distal part of subterminal segment almost as wide as the basal part of terminal segment (fig. 6D); terminal segment straight, tip of G1 with a linguiform protrusion curved downwards (figs. 6C-F, 11C, D). G2 subterminal segment about 2.6 times as long as terminal segment (fig. 6G).

The largest adult paratype female specimen resembles the holotype in most non-sexual characters. Pleon ovate, surface pitted, covering major parts of thoracic sternum, sixth somite about 2.7 times as broad as long, telson semicircular, about 2.3 times as broad as long (fig. 5D). Female vulvae on suture between thoracic sternites 5/6, widely ovate (fig. 12B); inferior margin slightly arched, opening ventrally (fig. 12B).

Remarks.— *R. zhongjiangense* sp. nov. is closest to *R. zhongyiense* sp. nov. in most of its external morphology and in the tip of the G1 terminal segment not reaching to the suture between thoracic sternites 4/5 (figs. 6B, 10B), but can be distinguished by its tip of the G1 having a tongue-shaped protrusion curved downwards (figs. 6C-F, 11C, D) (versus a swollen protrusion in *R. zhongyiense* sp. nov.; figs. 10C-F, 11G, H). Integral morphological comparisons are listed in table IV.

Distribution and habitat.— *R. zhongjiangense* sp. nov. was found in Zhongjiang Village (26°28′51″N 100°24′58″E, 1300 m asl), Longkaikou Town, Heqing County in Yunnan Province, P.R. China. This species is distributed in shallow moist burrows or under rocks of farmland around the mountain stream near the west coast of Jinsha River.

Live coloration.— Carapace is usually dark brown, while chelipeds and ambulatory legs are usually reddish brown in life (fig. 13B).

Ruiyupotamon jinjiangense sp. nov.
(figs. 7, 8, 11E, F, 12C, 13C)

Material examined.— Holotype: male (adult), 16.8 × 13.7 mm, NNU 428401, Luoxuan Village (26°35′55″N 100°23′42″E, 2151 m asl), Jinjiang Town, Gucheng District, Lijiang City in Yunnan Province, coll. Pengfei Wang, Zewei Zhang, Qiang Zhao & Hongying Sun, 6 May 2017. Paraype: female (allotype), 18.5 × 15.4 mm, NNU 428403, same data as holotype. One male (adult), 15.1 × 12.3 mm, NNU 428410, same data as holotype. One male (adult), 15.5 × 13.6 mm, NNU 428406, same data as holotype. One female (adult), 18.3 × 14.6 mm, NNU 428427, same data as holotype.

Etymology.— This species is named after the type locality, Jinjiang Town, Gucheng District, Lijiang City in Yunnan Province, P.R. China.

Diagnosis.— H-shaped groove between gastric and cardiac regions deep and distinct (fig. 7A). Male pleon broadly triangular, lateral margins slightly curved, tip of telson rounded (fig. 8C). G1 long, straight, reaches to suture between sternites 4/5 in situ (fig. 7B), ventral lobe of terminal segment broader than dorsal lobe (figs. 8F, 11F); tip with a protrusion pointing outwards, beak-shaped, terminally cuspidal

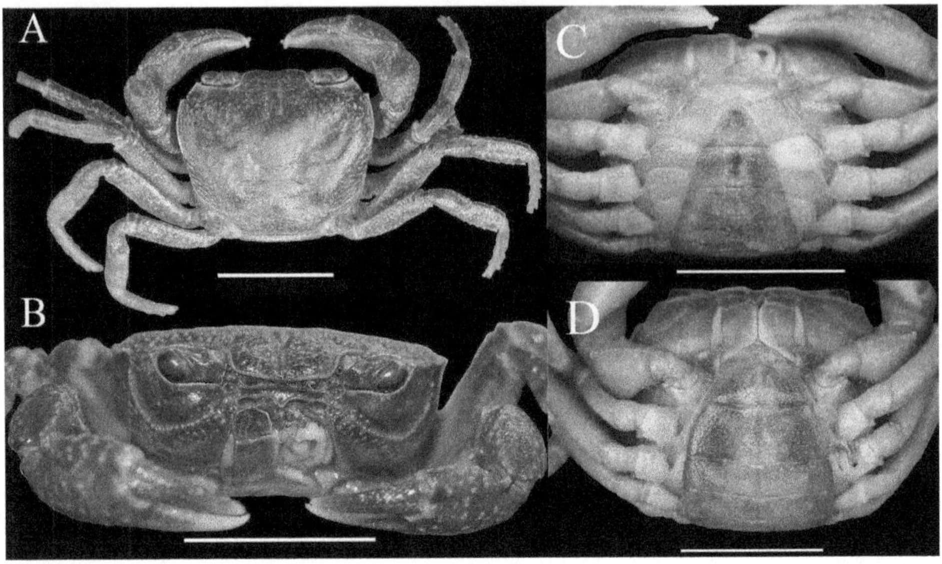

Fig. 7. *Ruiyupotamon jinjiangense* sp. nov. A-C, male holotype, 16.8 × 13.7 mm, NNU 428401; D, female paratype, 18.5 × 15.4 mm, NNU 428403. A, Dorsal view; B, frontal view of cephalothorax; C, D, ventral view of thoracic sternum and pleon. Scale bars = 1.0 cm.

(figs. 8C-F, 11E, F). Female vulvae on the suture between thoracic sternites 5/6, widely ovate (fig. 12C); lower margin slightly arched, opening ventrally (fig. 12C).

Description.— Carapace subquadrate, about 1.2 times broader than long, slightly convex, dorsal surface pitted (fig. 7A). Cervical groove distinct (fig. 7A). H-shaped groove between gastric and cardiac regions deep and distinct (fig. 7A). Epigastric cristae slightly convex, divided medially by a deep "Y" shaped groove (fig. 7A, B); postorbital region concave and epibranchial region with obvious granules (fig. 7A); external orbital angle triangular (fig. 7A, B); epibranchial tooth granular, clearly demarcated from external orbital angle with a notch (fig. 7A). Anterolateral margin gently convex, lined with indistinct granules (fig. 7A); posterolateral margin rugose (fig. 7A); frontal region slightly convex, surface covered with granules, anterior border gently emarginated medially (fig. 7A, B); orbital margin smooth, ridged (fig. 7A, B). Medial part of epistome with triangular medial lobe, tip rounded, inferior margin of epistome sinuous (fig. 7B). Pterygostomian regions comparatively smooth (fig. 7B).

Third maxilliped merus about 1.2 times as broad as long, trapezoidal (fig. 8A); ischium about 1.4 times as long as broad, rectangular, with distinct median sulcus (fig. 8A); exopod reaching proximal 1/3 of merus length, without flagellum (fig. 8A).

Chelipeds slightly unequal (fig. 7A, B); merus trigonal in cross-section (fig. 7A). Carpus with sharp spine at inner-distal angle, with granules at base (fig. 7A, B).

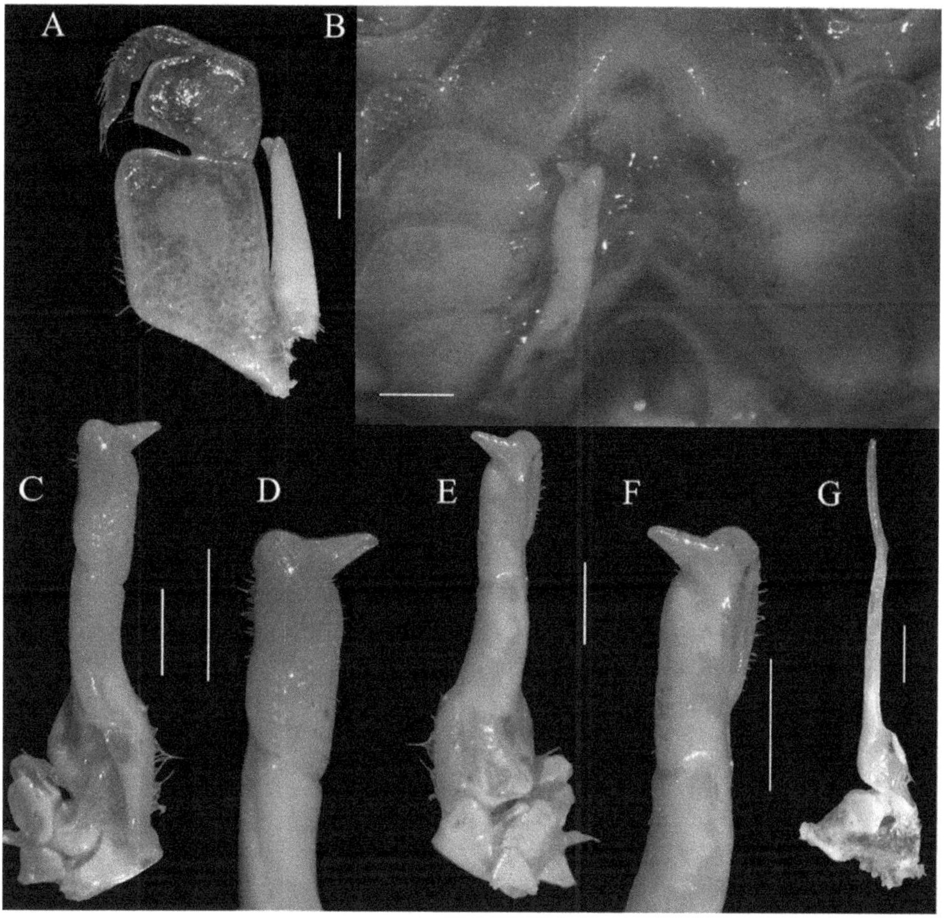

Fig. 8. *Ruiyupotamon jinjiangense* sp. nov. male holotype, 16.8 × 13.7 mm, NNU 428401. A, Left third maxilliped; B, ventral view showing sterno-abdominal cavity with right G1 in situ; C, left G1 (ventral view); D, left G1 terminal segment (ventral view); E, left G1 (dorsal view); F, left G1 terminal segment (dorsal view); G, left G2. Scale bars = 1.0 mm.

Outer surface of manus with distinct granules, about 1.5 times as long as high, length of movable finger nearly equal to fixed finger (fig. 7A, B).

Ambulatory legs moderately slender, surface gently pitted (fig. 7A); dactylus slender, with dense spinules; last leg with propodus about 2.1 times as long as broad, slightly shorter than dactylus (fig. 7A).

Male pleon triangular, third somite widest (fig. 7C); somite 6 width 2.2 times length (fig. 7C); telson about 1.5 times as broad as long, with a rounded tip, lateral margins of telson slightly curved (fig. 7C). Thoracic sternum generally smooth, pitted (fig. 7C); sternites 1, 2 completely fused to form triangular structure (fig. 7C); suture between thoracic sternites 2/3 complete (fig. 7C); sternites 3, 4 fused with shallow median suture (fig. 7C). Median groove of male thoracic sternum deep (fig. 8B); interruption between sutures of sternites 4/5, 5/6, 6/7 broad (fig. 8B); median longitudinal suture between sternites 7, 8 short (fig. 8B).

Tip of G1 terminal segment reaching to the suture between thoracic sternites 4/5 in situ (fig. 8B); subterminal segment about 1.8 times as long as terminal segment (fig. 8C); distal part of subterminal segment almost as wide as the basal part of terminal segment (fig. 8C-F); terminal of G1 with a beak-shaped protrusion, sticking laterally upwards at an angle of about 40° to parallel (figs. 8C-F, 11E, F); G2 subterminal segment about 2.3 times as long as terminal segment (fig. 8G).

The largest adult paratype female specimen resembles the holotype in most non-sexual characters. Pleon broadly ovate, surface pitted, sixth somite about 2.5 times as broad as long (fig. 7D); telson semicircular, about 2.2 times as broad as long (fig. 7D). Female vulvae on the suture between thoracic sternites 5/6, ovate, inferior margin slightly arched, opening ventrally (fig. 12C).

Remarks.— *R. jinjiangense* sp. nov. is similar to *R. banqiaopingense* sp. nov. in most of its external morphology and in the tip of the G1 terminal segment reaching to the suture between thoracic sternites 4/5 (figs. 3B, 8B), but can be separated by a unique combination of characters as shown in the Remarks section for *R. banqiaopingense* sp. nov. and table IV.

Distribution and habitat.— *R. jinjiangense* sp. nov. was found in Jinjiang Town (26°35′55″N 100°23′42″E, 2151 m asl), Gucheng District, Lijiang City in Yunnan Province, P.R. China. The new species lives under rocks near a hillstream which is near the west coast of Jinsha River.

Live coloration.— Carapace is usually dark brown, while chelipeds and ambulatory legs are usually reddish brown in life (fig. 13C).

Ruiyupotamon zhongyiense sp. nov.
(figs. 9, 10, 11G, H, 12D, 13D)

Material examined.— Holotype: male (adult), 19.1 × 15.3 mm, NNU 4210301, Zhongyi Village (26°32′6″N 100°24′13″E, 1628 m asl), Duomei Town, Heqing County, Dali City in Yunnan Province, coll. Pengfei Wang, Zewei Zhang, Qiang Zhao & Hongying Sun, 6 May 2017. Paratypes: one female (allotype), 20.3 × 16.5 mm, NNU 4210303, same data as holotype. One male (adult), 17.0 × 13.5 mm, NNU 4210307, same data as holotype. One male (adult), 18.0 × 14.9 mm, NNU 4210308, same data as holotype. One female (adult), 18.2 × 15.0 mm, NNU 4210304, same data as holotype.

Etymology.— This species is named after the type locality, Zhongyi Village, Duomei Town, Heqing County in Yunnan Province, P.R. China.

Diagnosis.— H-shaped groove between gastric and cardiac regions distinct (fig. 9A). Male pleon broadly triangular (fig. 9C), lateral margins of telson distinctly curved, tip rounded (fig. 9C). G1 not reaching suture between sternites 4/5 in situ (fig. 10B); tip of terminal segment with protrusion, ventral lobe of

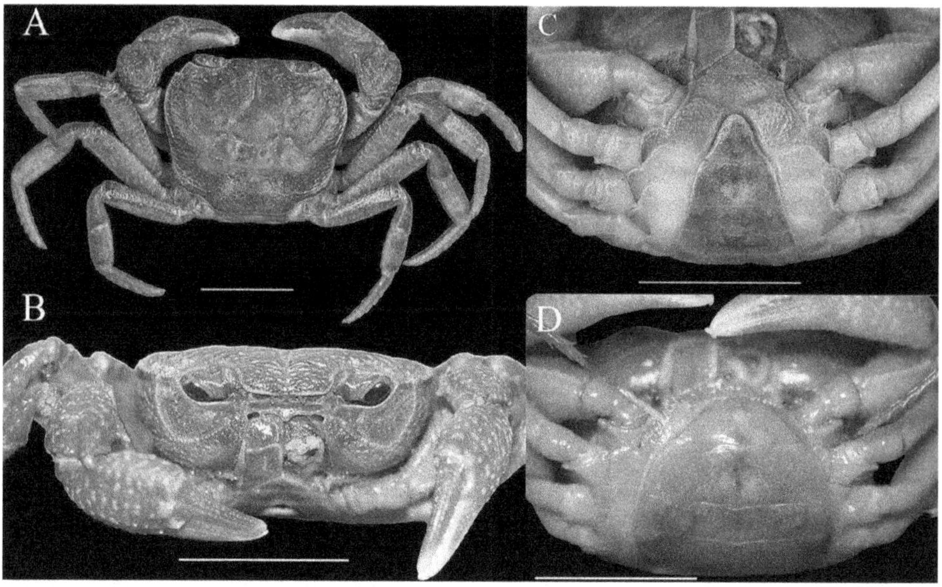

Fig. 9. *Ruiyupotamon zhongyiense* sp. nov. A-C, male holotype, 19.1 × 15.3 mm, NNU 4210301; D, female paratype, 20.3 × 16.5 mm, NNU 4210303. A, Dorsal view; B, frontal view of cephalothorax; C, D, ventral view of thoracic sternum and pleon. Scale bars = 1.0 cm.

terminal segment broader than dorsal lobe (figs. 10E, F, 11G, H). Female vulvae on the suture between thoracic sternites 5/6, widely ovate (fig. 12D); inferior margin slightly arched, opening ventrally (fig. 12D).

Description.— Carapace about 1.2 times broader than long, subquadrate (fig. 9A); cervical groove shallow, inconspicuous (fig. 9A). H-shaped groove between gastric and cardiac regions distinct (fig. 9A). Epigastric cristae slightly convex, separated medially by deep "Y" shaped groove (fig. 9A); postorbital region concave and epibranchial region with obvious granules (fig. 9A); external orbital angle obtuse triangle (fig. 9A); epibranchial tooth granular, clearly demarcated from external orbital angle with a distinct notch (fig. 9A); anterolateral margin convex with indistinct granules (fig. 9A). Posterolateral margin comparatively smooth, lined with oblique striae (fig. 9A). Frontal region slightly convex, surface covered with granules, anterior border gently emarginated medially (fig. 9A, B). Supraorbital margin sinuous, ridged, infraorbital margin with tiny granules (fig. 9A, B). Median part of epistome with triangular lobe, tip rounded, inferior margin sinuous, upper margin almost straight (fig. 9B). Pterygostomial region relatively smooth, with several rounded granules (fig. 9B).

Third maxilliped merus about 1.1 times as broad as long, trapezoidal (fig. 10A); ischium about 1.5 times as long as broad, rectangular, with distinct median sulcus (fig. 10A); exopod reaching proximal 1/3 of merus length, without flagellum (fig. 10A).

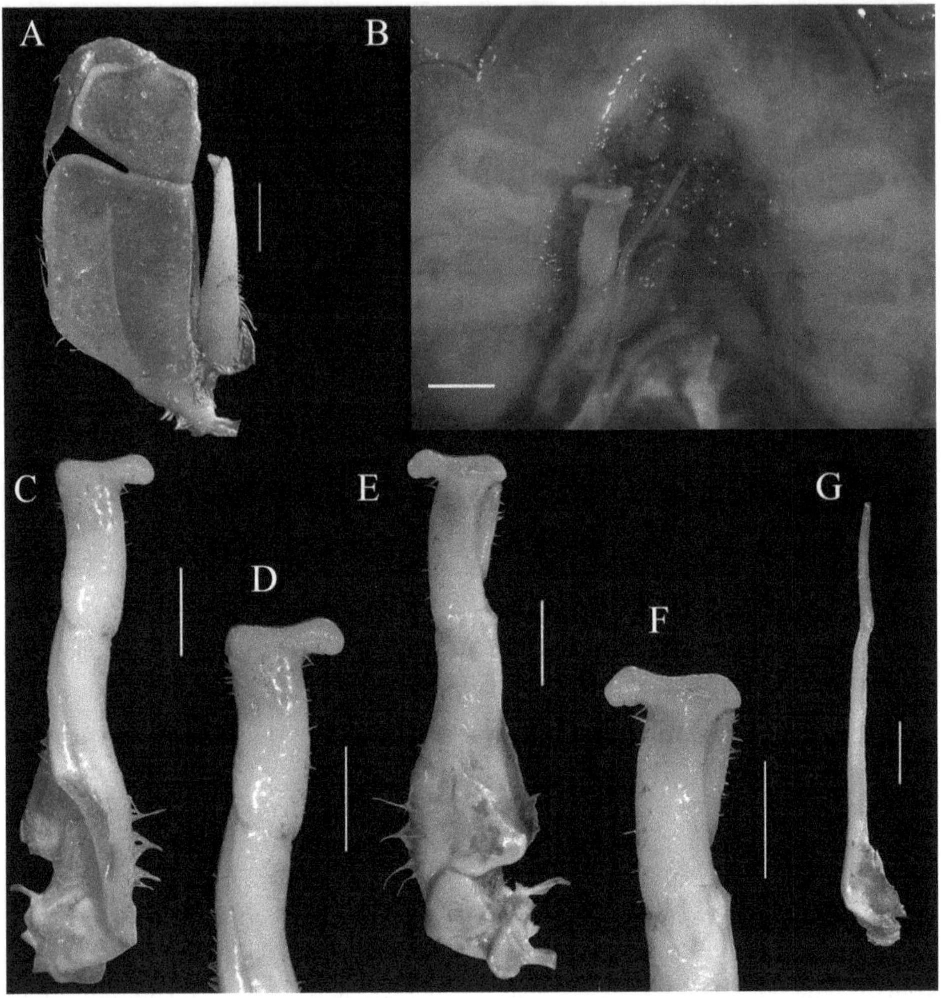

Fig. 10. *Ruiyupotamon zhongyiense* sp. nov. male holotype, 19.1 × 15.3 mm, NNU 4210301. A, Left third maxilliped; B, ventral view showing sterno-abdominal cavity with right G1 in situ; C, left G1 (ventral view); D, left G1 terminal segment (ventral view); E, left G1 (dorsal view); F, left G1 terminal segment (dorsal view); G, left G2. Scale bars = 1.0 mm.

Chelipeds slightly unequal (fig. 9A, B); merus trigonal in cross-section (fig. 9A). Carpus with sharp spine at inner-distal angle, with granules at base (fig. 9A, B). Outer surfaces of manus with distinct granules (fig. 9B), about 1.5 times as long as high, length of movable finger nearly equal to fixed finger (fig. 9A, B).

Ambulatory legs long, slender, dactylus with dense spinules (fig. 9A); last leg with propodus about 2.4 times as long as broad, slightly shorter than dactylus (fig. 9A).

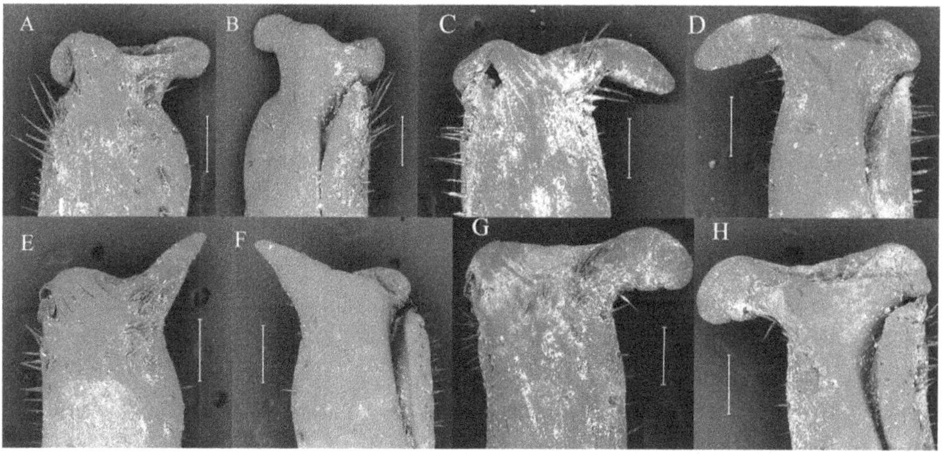

Fig. 11. Ventral and dorsal view of distal part of G1 terminal segment. A, B, *Ruiyupotamon banqiaopingense* sp. nov. (male holotype, NNU 427101); C, D, *R. zhongjiangense* sp. nov. (male holotype, NNU 427207); E, F, *R. jinjiangense* sp. nov. (male paratype, NNU 428407); G, H, *R. zhongyiense* sp. nov. (male holotype, NNU 4210301). A, C, E, G, Ventral view; B, D, F, H, dorsal view. Scale bars = 300 μm.

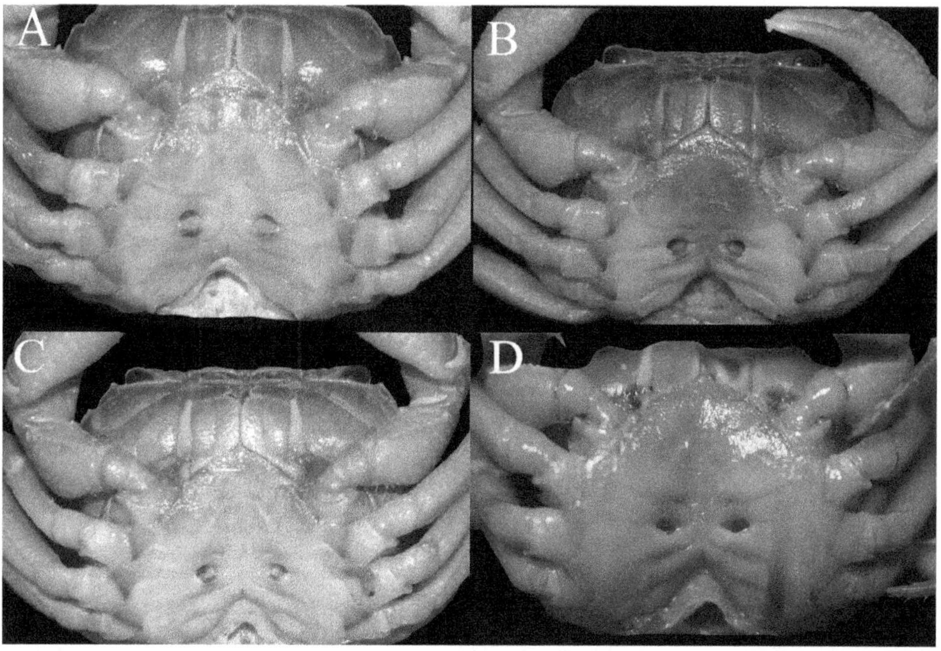

Fig. 12. Female vulvae. A, *Ruiyupotamon banqiaopingense* sp. nov. female paratype (21.6 × 17.9 mm, NNU 427105); B, *Ruiyupotamon zhongjiangense* sp. nov. female paratype (20.7 × 16.7 mm, NNU 427206); C, *Ruiyupotamon jinjiangense* sp. nov. female paratype (18.5 × 15.4 mm, NNU 428403); D, *Ruiyupotamon zhongyiense* sp. nov. female paratype (20.3 × 16.5 mm, NNU 4210303).

Male pleon widely triangular, third somite widest (fig. 9C); somite 6 width 2.1 times length, telson about 1.4 times as broad as long, with a rounded tip, lateral margins of telson distinctly curved (fig. 9C). Thoracic sternum generally pitted (fig. 9C); sternites 1, 2 completely fused to form triangular structure (fig. 9C); suture between thoracic sternites 2/3 distinct (fig. 9C); sternites 3, 4 fused with shallow median suture (fig. 9C). Median groove of male thoracic sternum deep (fig. 10B); interruption between sutures of sternites 4/5, 5/6, 6/7 broad (fig. 10B), median longitudinal suture between sternites 7/8 short (fig. 10B).

Tip of G1 terminal segment not reaching to suture between thoracic sternites 4/5 in situ (fig. 10B); subterminal segment about 1.7 times as long as terminal segment (fig. 10C, E); terminal of G1 with a swollen protrusion, extending outwards (figs. 10C, D, 11G, H); G2 distinctly longer than G1, subterminal segment about 2.3 times as long as distal segment (fig. 10G).

The largest adult paratype female specimen resembles the holotype in most non-sexual characters. Pleon broadly ovate, surface pitted (fig. 9D); sixth somite about 2.9 times as broad as long (fig. 9D); telson semicircular, about 2.7 times as broad as long (fig. 9D). Female vulvae on the suture between thoracic sternites 5/6, ovate, inferior margin indistinctly arched, opening ventrally (fig. 12D).

Remarks.— *R. zhongyiense* sp. nov. is very similar to *R. zhongjiangense* sp. nov. in most of its external morphology and in the tip of the G1 terminal segment not reaching to the suture between thoracic sternites 4/5 (figs. 6B, 10B), but can also be distinguished by a unique character as shown in the Remarks section for *R. zhongjiangense* sp. nov. and table IV. The new species is also similar to *R. banqiaopingense* sp. nov. in most of its external morphology, but can be distinguished by a combination of characters as noted in the Remarks section for *R. banqiaopingense* sp. nov. and table IV.

Distribution and habitat.— *R. zhongyiense* sp. nov. was found in Zhongyi Village (26°32′6″N 100°24′13″E, 1628 m asl), Duomei Town, Heqing County in Yunnan Province, P.R. China. The new species lives under rocks and in moist burrows near the small mountain stream of the west coast of Jinsha River.

Live coloration.— Carapace is usually dark brown or black, while chelipeds and ambulatory legs are usually reddish brown in life (fig. 13D).

Key to the Four Described Species of the Genus *Ruiyupotamon*

1. G1 relatively long, tip of terminal segment reaching to the suture between thoracic sternites 4/5 2
– G1 relatively short, tip of terminal segment not reaching to the suture between thoracic sternites 4/5 3

2. Tip of G1 terminal segment with a round and blunt protrusion . . . . *R. banqiaopingense*
– Tip of G1 terminal segment with a beak-shaped protrusion . . . . *R. jinjiangense*

3. Tip of G1 terminal segment with a flat and tongue-shaped protrusion . . . . *R. zhongjiangense*
– Tip of G1 terminal segment with a swollen protrusion . . . . *R. zhongyiense*

Fig. 13. Coloration of the four new species of the new genus *Ruiyupotamon*. A, *R. banqiaopingense* sp. nov.; B, *R. zhongjiangense* sp. nov.; C, *R. jinjiangense* sp. nov.; D, *R. zhongyiense* sp. nov.

Phylogenetic relationships

In the present phylogenetic analyses, 62 species from 50 potamid genera were included. The BI and ML analyses based on 16S resulted in similar topologies (fig. 14). The phylogenetic tree indicates that the four new species cluster together forming a monophyletic group with strong support (BPP/BS = 1/91). This new genus is nested within the seven known genera (*Aparapotamon*, *Pararanguna*, *Tenuipotamon*, *Trichopotamon*, *Arquatopotamon*, *Semicirculara*, *Artopotamon*) to form a 'SW China' clade with strong support (BPP/BS = 1/100). However, the phylogenetic relationships among the four new species of *Ruiyupotamon* gen. nov. and among the eight genera in the 'SW China clade' are still unresolved in the tree topologies inferred from the available 16S sequences.

According to the combined analysis of the 16S, 28S and ITS2 data, individuals from the same species clustered together with strong to moderate support (fig. 15). Within this genus, a sister group relationship was recovered (BPP/BS = 1/100), *R. banqiaopingense* sp. nov., and the other three, *R. zhongjiangense* sp. nov., *R. jinjiangense* sp. nov. and *R. zhongyiense* sp. nov.

DISCUSSION

Morphological and molecular evidences support the recognition of the new genus *Ruiyupotamon* gen. nov., comprising the four new species, *R. banqiaopingense* sp. nov., *R. zhongjiangense* sp. nov., *R. jinjiangense* sp. nov. and *R.*

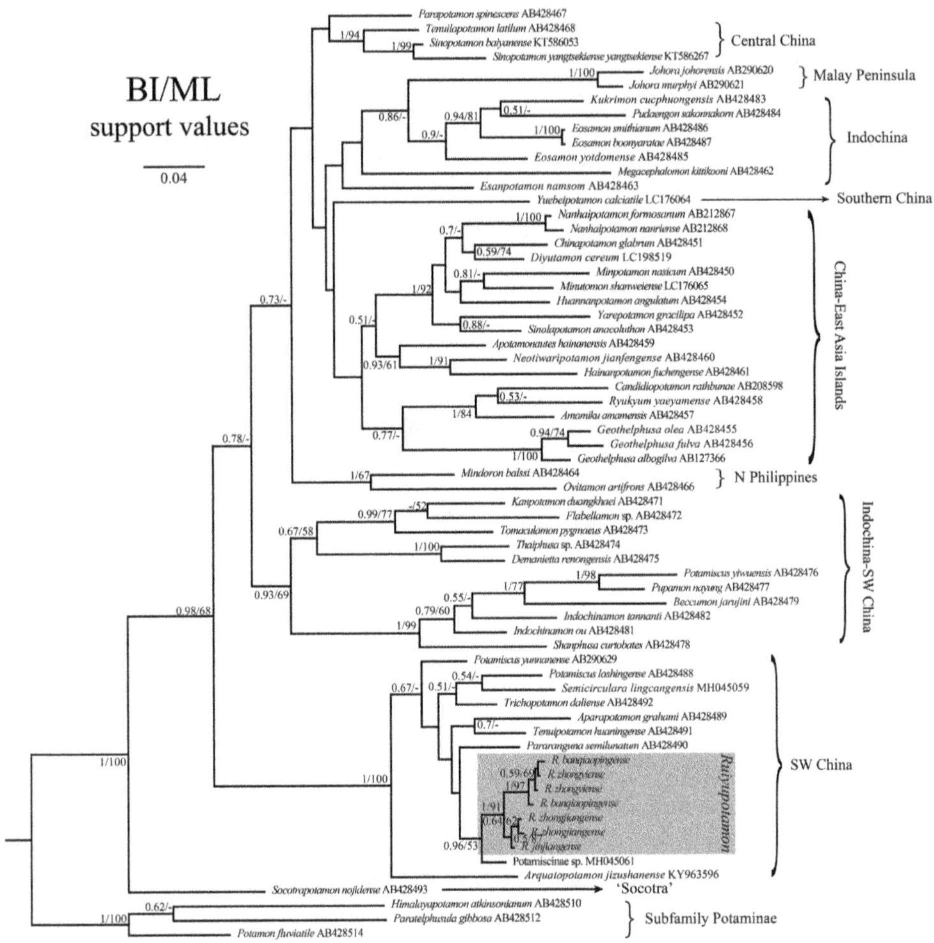

Fig. 14. Bayesian inference (BI) tree of the new genus and four new species, and the remaining Asian potamid crabs with accession numbers based on 16S rDNA sequences. Branch lengths and topologies from BI tree. Probability values at the nodes represent support values for BI and ML. Only values >50% are shown. The new species are highlighted in gray.

zhongyiense sp. nov.. Similar to the other genera within the 'SW China' clade, species of this new genus do not possess a flagellum on the third maxilliped exopod. The phylogenetic tree based on 16S validates the establishment of *Ruiyupotamon* gen. nov., but is unable to resolve species-level relationships (fig. 14). With multilocus markers, however, the phylogentic relationships among the four new species were clearly resolved (fig. 15) and provide evidence for the establishment of these new species. The split of *R. banqiaopingense* sp. nov. with the other three species might be due to geographic isolation. *R. banqiaopingense* sp. nov. is distributed in the east bank of the Jinsha River, while the other three, *R. zhongjiangense* sp. nov., *R. jinjiangense* sp. nov. and *R. zhongyiense* sp. nov. are distributed

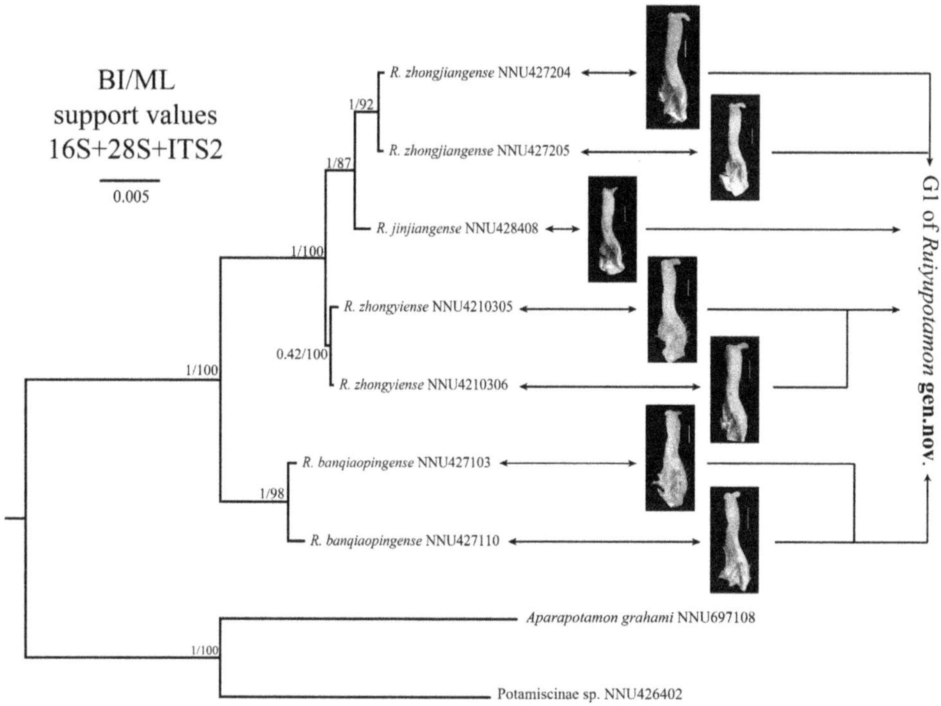

Fig. 15. Bayesian inference (BI) tree of the four new species of *Ruiyupotamon* gen. nov. based on the combined dataset of 16S rDNA, 28S rDNA and ITS2 with comparisons of G1. Bayesian posterior probabilities and maximum likelihood (ML) bootstrap values are indicated at nodes.

in the west bank of Jinsha River. Our results implicate that the dry-hot valley at low altitude (<1000 m) and the varying climatic environment among different altitudes in this region form a natural sky-island driving the fast speciation of these narrowly distributed freshwater crabs with weak diffusion capacity.

Yunnan Province is a global biodiversity hotspot region (Myers et al., 2000) and a possible centre of origin for the Potamidae of China and even East Asia (Shih & Ng, 2011). With the additional a new genus and four new species presently described, there are now 18 genera and 64 species known to occur in Yunnan Province. Even so, the explorations for freshwater crabs in the dry-hot valley of Jinsha River located in northwestern Yunnan are still in the initial stages. Due to the highly complex terrain and rich biodiversity of freshwater crabs in the Hengduan Mountains Region, further field investigations are expected.

ACKNOWLEDGEMENTS

We thank Peter K. L. Ng (National University of Singapore) for important suggestion about nomination of new species, and Neil Cumberlidge (Northern

Michigan University) for polishing the early version of this manuscript. We sincerely appreciate Dr. Wenliang Liu (School of Ecological & Environmental Science, East China Normal University) for providing the scanning electron microscopy platform and instruction that made possible the electron microscopy observation of G1. We thank Qiang Zhao (College of Life Sciences, Nanjing Normal University) for assistance with collecting specimens; and we also thank Research Fellow, Dr. Jun Chen and Mr. Kaibayier Meng (National Zoological Museum of China, Institute of Zoology, Chinese Academy of Sciences, IZCAS) for permission to examine type material deposited in National Zoological Museum of China, IZCAS. We are grateful to two reviewers, Dr. Wenliang Liu and Dr. Wei Jiang (Institute of Oceanology, Chinese Academy of Sciences), whose comments and suggestions improved our manuscript greatly. This work was supported by the National Natural Science Foundation of China (No. 31772427) to SHY, and Biodiversity Survey Observation and Assessment Program of Ministry of Ecology and Environment of China. P. F. Wang and Z. W. Zhang contributed equally to this work.

REFERENCES

CHU, K. L., P. F. WANG & H. Y. SUN, 2018a. A new genus and species of primary freshwater crab and a new species of *Artopotamon* Dai & Chen, 1985 (Crustacea, Brachyura, Potamidae) from western Yunnan, China. Zootaxa, **4422**: 115-131.

CHU, K. L., X. P. MA, Z. W. ZHANG, P. F. WANG, L. N. LÜ, Q. ZHAO & H. Y. SUN, 2018b. A checklist for the classification and distribution of China's freshwater crabs. Biodiversity Science, **26**: 274-282.

CHU, K. L., L. J. ZHOU & H. Y. SUN, 2017. A new genus and new species of freshwater crab (Decapoda: Brachyura: Potamidae Ortmann, 1896) from Yunnan Province, China. Zootaxa, **4286**: 241-253.

DAI, A. Y., 1999. Fauna sinica: Arthropoda Crustacea Malacostraca Decapoda Parathelphusidae Potamidae: 1-501, figs. 1-238, pls. 1-30. (Science Press, Beijing, China). [in Chinese with English summary.]

DARRIBA, D., G. L. TABOADA, R. DOALLO & D. POSADA, 2012. jModelTest 2: more models, new heuristics and parallel computing. Nat. Methods, **9**: 772.

DRUMMOND, A. J., M. A. SUCHARD, D. XIE & A. RAMBAUT, 2012. Bayesian phylogenetics with BEAUti and the BEAST 1.7. Mol. Biol. Evol., **29**: 1969-1973.

HUANG, C., H. T. SHIH & S. Y. MAO, 2016. *Yuebeipotamon calciatile*, a new genus and new species of freshwater crab from southern China (Crustacea, Decapoda, Brachyura, Potamidae). ZooKeys, **615**: 61-72.

HUANG, C., H. T. SHIH & P. K. L. NG, 2017. A new genus and new species of Potamidea (Crustacea: Decapoda: Brachyura: Potamoidae), the first stygomorphic cave crab known from China and East Asia. Zootaxa, **4232**: 71-84.

HUELSENBECK, J. P. & F. RONQUIST, 2001. MRBAYES: Bayesian inference of phylogenetic trees. Bioinformatics, **17**: 754-755.

JI, Y. K., Y. F. SUN, W. GAO, K. L. CHU, R. C. WANG, Q. ZHAO & H. Y. SUN, 2016. Out of the Sichuan Basin: rapid species diversification of the freshwater crabs in *Sinopotamon*, (Decapoda: Brachyura: Potamidae) endemic to China. Mol. Phylogenet. Evol., **100**: 80-94.

KATOH, K. & D. M. STANDLEY, 2013. MAFFT multiple sequence alignment software version 7: improvements in performance and usability. Mol. Biol. Evol., **30**: 772-780.

KÜCK, P., K. MEUSEMANN, J. DAMBACH, B. THORMANN, B. M. V. REUMONT, J. W. WÄGELE & B. MISOF, 2010. Parametric and non-parametric masking of randomness in sequence alignments can be improved and leads to better resolved trees. Front. Zool., **7**: 1-12.

LANFEAR, R., B. CALCOTT, S. Y. W. HO & S. GUINDON, 2012. PartitionFinder: combined selection of partitioning schemes and substitution models for phylogenetic analyses. Mol. Biol. Evol., **29**: 1695-1701.

MILLER, M. A., W. PFEIFFER & T. SCHWARTZ, 2010. Creating the CIPRES Science Gateway for inference of large phylogenetic trees. Proceedings of the Gateway Computing Environments Workshop (GCE), 14 Nov. 2010, New Orleans, LA: 1-8.

MISOF, B. & K. MISOF, 2009. A Monte Carlo approach successfully identifies randomness in multiple sequence alignments: a more objective means of data exclusion. Syst. Biol., **58**: 21-34.

MYERS, N., R. A. MITTERMEIER, C. G. MITTERMEIER, G. A. B. FONSECA & J. KENT, 2000. Biodiversity hotspots for conservation priorities. Nature, **403**: 853-858.

NARUSE, T., J. E. CHIA & X. M. ZHOU, 2018. Biodiversity surveys reveal eight new species of freshwater crabs (Decapoda: Brachyura: Potamidae) from Yunnan Province, China. PeerJ, **6**: e5497.

RAMBAUT, A., M. A. SUCHARD, D. XIE & A. J. DRUMMOND, 2014. Tracer v1.6. Available from: http://beast.bio.ed.ac.uk/Tracer.

RONQUIST, F., M. TESLENKO, P. VAN DER MARK, D. L. AYRES, A. DARLING, S. HÖHNA, B. LARGET, L. LIU, M. A. SUCHARD & J. P. HUELSENBECK, 2012. MrBayes 3.2: efficient Bayesian phylogenetic inference and model choice across a large model space. Syst. Biol., **61**: 539-542.

SAMBROOK, J. & D. W. RUSSELL, 2001. Rapid isolation of yeast DNA. In: J. SAMBROOK & D. W. RUSSELL (eds.), Molecular cloning, a laboratory manual: 631-632. (Cold Spring Harbor Laboratory Press, New York).

SHIH, H. T. & P. K. L. NG, 2011. Diversity and biogeography of freshwater crabs (Crustacea: Brachyura: Potamidae, Gecarcinucidae) from east Asia. System. Biodivers., **9**: 1-16.

SHIH, H. T., D. C. J. YEO & P. K. L. NG, 2009. The collision of the Indian plate with Asia: molecular evidence for its impact on the phylogeny of freshwater crabs (Brachyura: Potamidae). J. Biogeogr., **36**: 703-719.

STAMATAKIS, A., 2006. RAxML-VI-HPC: maximum likelihood-based phylogenetic analyses with thousands of taxa and mixed models. Bioinformatics, **22**: 2688-2690.

STAMATAKIS, A., 2014. RAxML version 8: a tool for phylogenetic analysis and post-analysis of large phylogenies. Bioinformatics, **30**: 1312-1313.

First received 3 October 2019.
Final version accepted 2 June 2020.

[When citing this volume, please refer to *Crustaceana* 93 (2020), 1233-1546]

Prof. Ruiyu Liu Memorial Issue

THE IDENTITY OF *HYASTENUS PLEIONE* (HERBST, 1803) AND DESCRIPTION OF A NEW SPECIES FROM CHINA (DECAPODA, BRACHYURA, MAJOIDEA, EPIALTIDAE)

BY

BEE YAN LEE[1,2,3]) and PETER K. L. NG[1])

[1]) Tropical Marine Science Institute, National University of Singapore, 18 Kent Ridge Road, Singapore 119227, Republic of Singapore

[2]) Lee Kong Chian Natural History Museum, National University of Singapore, 2 Conservatory Drive, Singapore 117377

ABSTRACT

The poorly known epialtid crab, *Hyastenus pleione* (Herbst, 1803), from the Indian Ocean is re-described on the basis of the type material. Large sized specimens of a similar looking species from China are here assigned to a new species. The new species differs from *H. pleione* in having fewer large granules on the carapace, a blunt distal angle of the basal antennal article with the outer margin slightly constricted medially, possessing only one branchial granule along the lateral margin of the carapace, and has a relatively more elongate male first gonopod with the subdistal part weakly dilated.

Key words. — Spider crab, Epialtidae, Majoidea, taxonomy, new species, China, Indian Ocean

RÉSUMÉ

L'espèce peu connue de crabe épialtidé *Hyastenus pleione* (Herbst, 1803), de l'Océan Indien est re-décrite sur la base du matériel type. Des spécimens ressemblants de grande taille de Chine sont assignés ici à une nouvelle espèce. La nouvelle espèce diffère de *H. pleione* par de plus larges granules sur la carapace, un angle distal aplati de l'article basal de l'antenne avec le bord externe légèrement rétréci au milieu, la présence d'un seul granule branchial le long du bord latéral de la carapace et le premier gonopode mâle relativement plus allongé avec une part sous-distale légèrement dilatée.

Mots clés. — Crabe araignée, Epialtidae, Majoidea, taxonomie, espèce nouvelle, Chine, Océan Indien

[3]) Corresponding author; e-mail: beeyan06@gmail.com; beeyan_lee@u.nus.edu

 | DOI:10.1163/9789004464353_008

INTRODUCTION

There are currently 40 recognised species in the majoid genus *Hyastenus* White, 1847 (family Epialtidae MacLeay, 1838) (Ng et al., 2008; Windsor & Ahyong, 2013; Lee & Ng, 2019). Comparison of the type material of *Hyastenus pleione* (Herbst, 1803) with what has long been identified as this species from China showed that two different species were involved. The present paper re-describes *H. pleione* based on the type material and describes the Chinese taxon as new.

MATERIAL AND METHODS

Specimens examined are deposited in the Zoological Reference Collection of the Lee Kong Chian Natural History Museum (formerly the Raffles Museum of Biodiversity Research), National University of Singapore (ZRC); Museum für Naturkunde, Berlin, Germany (ZMB); Institute of Zoology, Chinese Academy of Sciences, Beijing, China (IZCAS); and Institute of Oceanology, Chinese Academy of Sciences, Qingdao, China (IOCAS). Measurements, in millimetres, are of the maximum carapace length (excluding the pseudorostral spine) and the maximum carapace width (excluding the lateral branchial spines). The following abbreviations are used: coll. = collected by; G1 = male first gonopod; IWP = Indo-West Pacific; P2 = first walking leg; and stn. = station.

TAXONOMY

Superfamily MAJOIDEA Samouelle, 1819
Family EPIALTIDAE MacLeay, 1838
Genus *Hyastenus* White, 1847
Hyastenus pleione (Herbst, 1803)
(figs. 1A, B, 2A-D, 3A-C)

Cancer plejone Herbst, 1803: 52, pl. 58 fig. 5 (type locality: East Indies).
Naxia pleione — Gerstaecker, 1856: 114. 115, pl. 5 figs. 1, 2 [new combination].
Hyastenus pleione — A. Milne-Edwards, 1872: 250 [new combination]. — Miers, 1886: 56 (list). — De Man, 1887a: 5 (list), 15. — De Man, 1887b: 225, pl. 7 fig. 3. — Henderson, 1893: 343. — Ortmann, 1894: 41 (key). — Alcock, 1895: 207 (key), 208. — Gravely, 1927: 151, 151 (key). — Balss, 1935: 123 (list). — Buitendijk, 1939: 247 (in part). — Pillai, 1951: 7. — Tirmizi & Serène, 1971: 24-27, text-fig. 2, pl. 2 fig. A. — Griffin & Tranter, 1986: 124 (list), 125 (key), 127, 148, 153. – Dai & Yang, 1991: 143 (in part). — Kazmi, 1997: 79 (list), 87 (list). – Sakai, 1999: 24, pl. 11E. — Jeyabaskaran et al., 2000: 11 (list), 36 (key), 74, 75, pl. 63 fig. b. — Dev Roy & Nandi, 2005: 222 (table). — Dev Roy & Nandi, 2007: 174 (table), 185 (table). — Dev Roy, 2008: 41, 42, 166 (table). — Ng et al., 2008: 103 (list). — Gopalakrishnan et al., 2012: 42 (list). — Dev Roy & Nandi, 2012: 196 (table). — Ng et al., 2017: 134 (list). — Al-Khafaji et al., 2017: 464 (list). — Vidhya et al., 2017: 2333 (list), 2334 (table). — Trivedi et al., 2018: table 1.

Halimus pleione — Laurie, 1906: 376 [new combination].

Material examined.— Lectotype (herein designated): male (22.5 × 19.0 mm) (ZMB 0055), East Indies, no other data. — Paralectotypes: 2 females (22.9 × 15.3 mm, 26.7 × 17.5 mm), 1 carapace only (34.7 × 23.6 mm) (ZMB 0056), East Indies, no other data. — Other material: 1 ovigerous female (20.5 × 15.9 mm) (ZRC 2019.0491), reef at Vaan Island, Tuticorin, Tamil Nadu, India coll. local students, 2019.

Diagnosis.— Pseudorostral spines straight, diverging. Supraorbital eave forming rectangular lobe; preorbital angle sharp, outer margin slightly constricted, antorbital angle blunt; postorbital lobe cup-like. Carapace with granules, regions defined; gastric region slightly swollen with 6 granules; 1 urogastric granule; cardiac region slightly swollen forming large granule; 1 metabranchial granule on margin of swollen cardiac; 3 branchial granules in straight row, 1 epibranchial above 2 mesobranchial granules; lateral branchial spine pointing outwards; intestinal spine medially on posterior region of carapace (figs. 2A, 3A); 3 granules along lateral margin of carapace (figs. 2D, 3C). Basal antennal article almost as wide as long, distal angle distinct, pointed outwards, outer margin constricted medially, basal angle round. Pterygostomial region with 3 granules on outer margin (figs. 2B, 3B). Chelipeds with propodus longer than fingers (fig. 2A). Ambulatory legs slender (figs. 2A, 3A). Male thoracic sternum slightly depressed anteriorly (in ventral view); sternites 3, 4 narrow, with lateral margin constricted. Male pleon triangular, telson triangular (fig. 2C). G1 relatively short, slightly curved, distal tip sharp, subdistal portion prominently dilated on lateral margin (see Tirmizi & Serène, 1971, text-fig. 2; reproduced here as fig. 1B).

Remarks.— Little is known about *Hyastenus pleione*. This species was originally described as *Cancer plejone* by Herbst (1803) from East Indian Seas, with one specimen figured (cf. Herbst, 1803, pl. 58 fig. 5; reproduced here as fig. 1A). No measurements of *H. pleione* were provided, but Herbst (1803) stated that the figure provided shows the actual size of the specimen. It was transferred to the genus *Naxia* Latreille, 1825, by Gerstaecker (1856) with no explanation, and then referred to *Hyastenus* by A. Milne-Edwards (1872), where it has remained ever since. All the literature after Herbst (1803) uses a different spelling from the original description without comment, i.e. "*pleione*" rather than the original "*plejone*" (see Griffin & Tranter, 1986: 153; Sakai, 1999: 24; Ng et al., 2008: 103). The species was clearly named after Pleione, an oceanid nymph in Greek mythology so the use of the former spelling is expected. The Latin "i" and "j" are often used interchangeably in old texts and the change in spelling is thus not surprising. While the species itself is not widely reported, being one of the earliest named species, its name has been used in comparisons and discussions frequently (e.g., see Griffin & Tranter, 1986). As such, in order to maintain the usage and stability of the

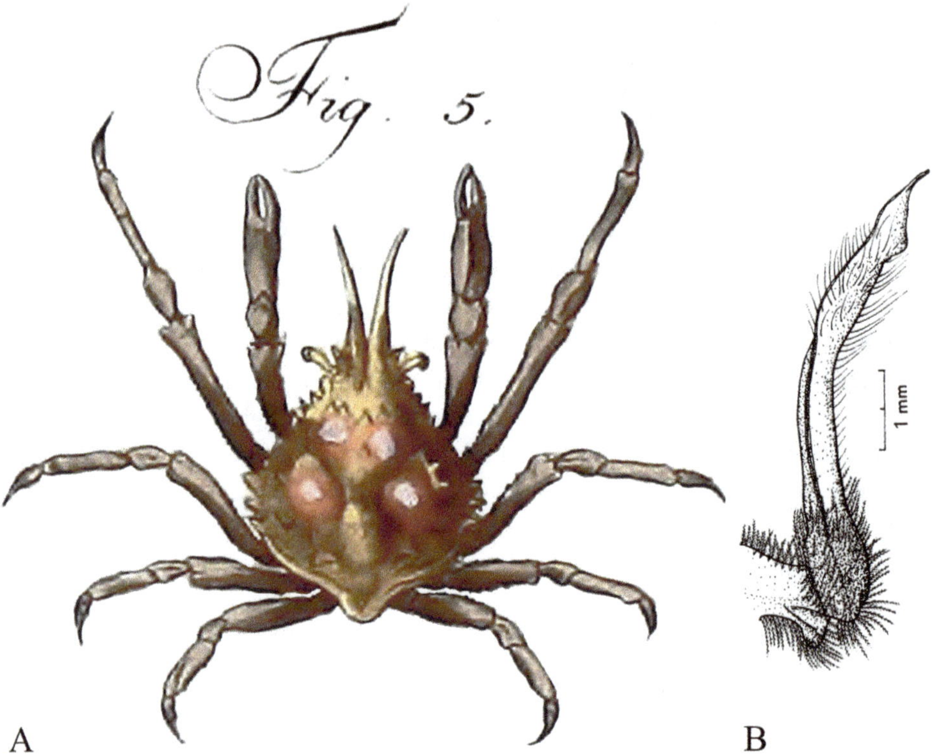

A B

Fig. 1. A, Illustration of *Hyastenus pleione* (Herbst, 1803), East Indies (after Herbst, 1803: pl. 58 fig. 5); B, illustration of *H. pleione* (Herbst, 1803), Arabian Sea, left G1 (after Tirmizi & Serène, 1971: text-fig. 2).

name, we follow Article 33.3.1 of the zoological code (ICZN, 1999: 43) to retain the spelling of the species as *Hyastenus pleione* (Herbst, 1803).

Sakai (1999: 24, pl. 11E) stated that the holotype (and only) specimen of *H. pleione* is deposited in the ZMB (catalogue number ZMB 0055) and was a male measuring 22 × 18.6 mm in length and width (without spines). Although one specimen was figured by Herbst (1803, pl. 58 fig. 5; fig. 1A), the actual number of specimens examined by Herbst is not known. The carapace figured by Herbst (1803, pl. 58 fig. 5; fig. 1A) measures approximately 25 mm in length (excluding the pseudorostral spines) and matches the "holotype" specimen identified by Sakai (1999) as well as the largest male examined here (22.5 × 19.0 mm (excluding the spines), ZMB 0055).

In the ZMB, however, there are two lots of specimens (ZMB 0055, ZMB 0056) from Herbst's collection, all of which match the description and figure, and were obtained at the same locality. As such, all should be treated as syntypes. The specimen reported and figured by Sakai (1999) is therefore not the holotype. The male specimen (22.5 × 19.0 mm, ZMB 0055) that was identified as "holotype"

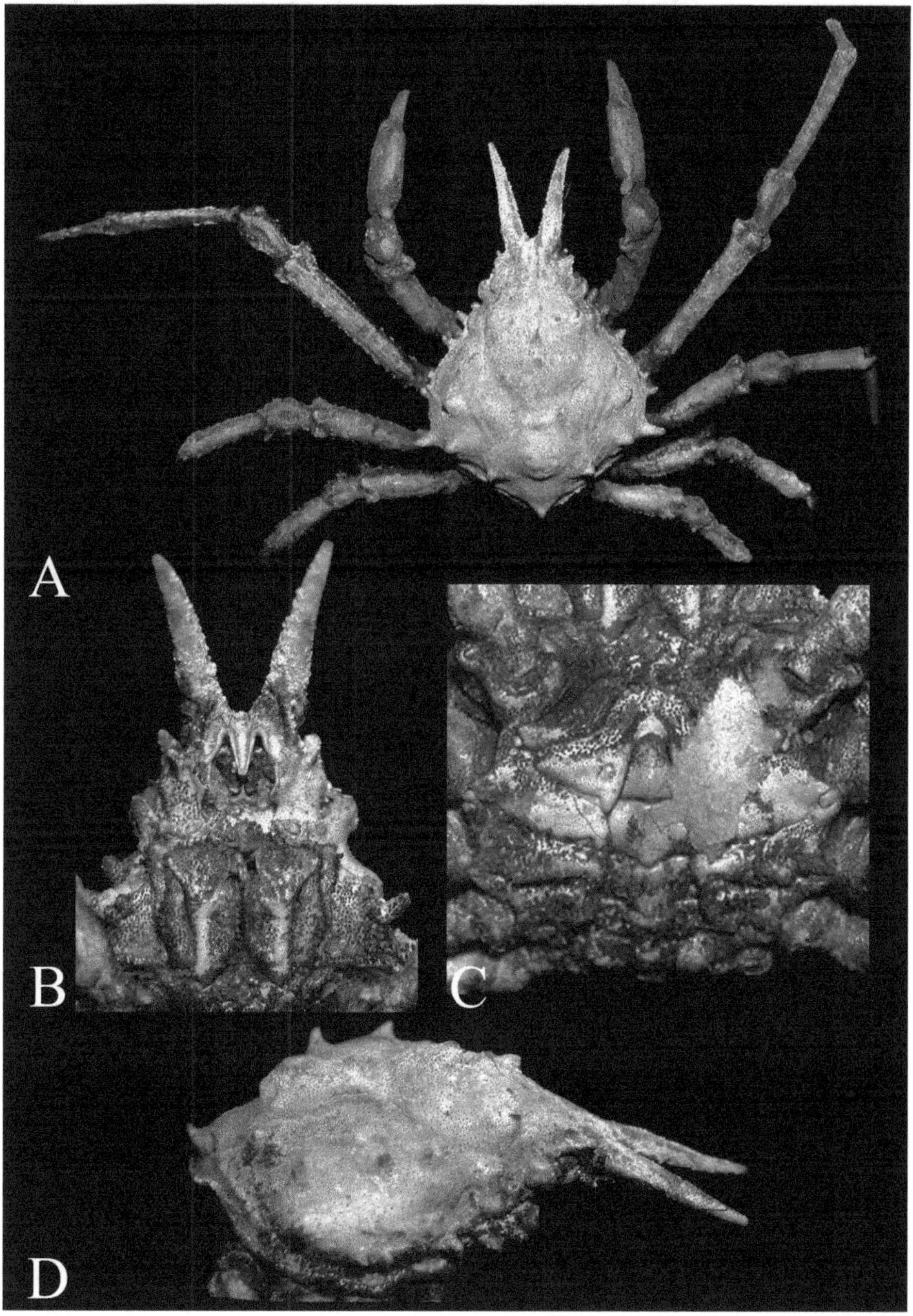

Fig. 2. *Hyastenus pleione* (Herbst, 1803), lectotype male (22.5 × 19.0 mm) (ZMB 0055), East Indies. A, Overall dorsal view of carapace; B, ventral frontal surface of carapace; C, male thoracic sternum view; D, lateral view of carapace.

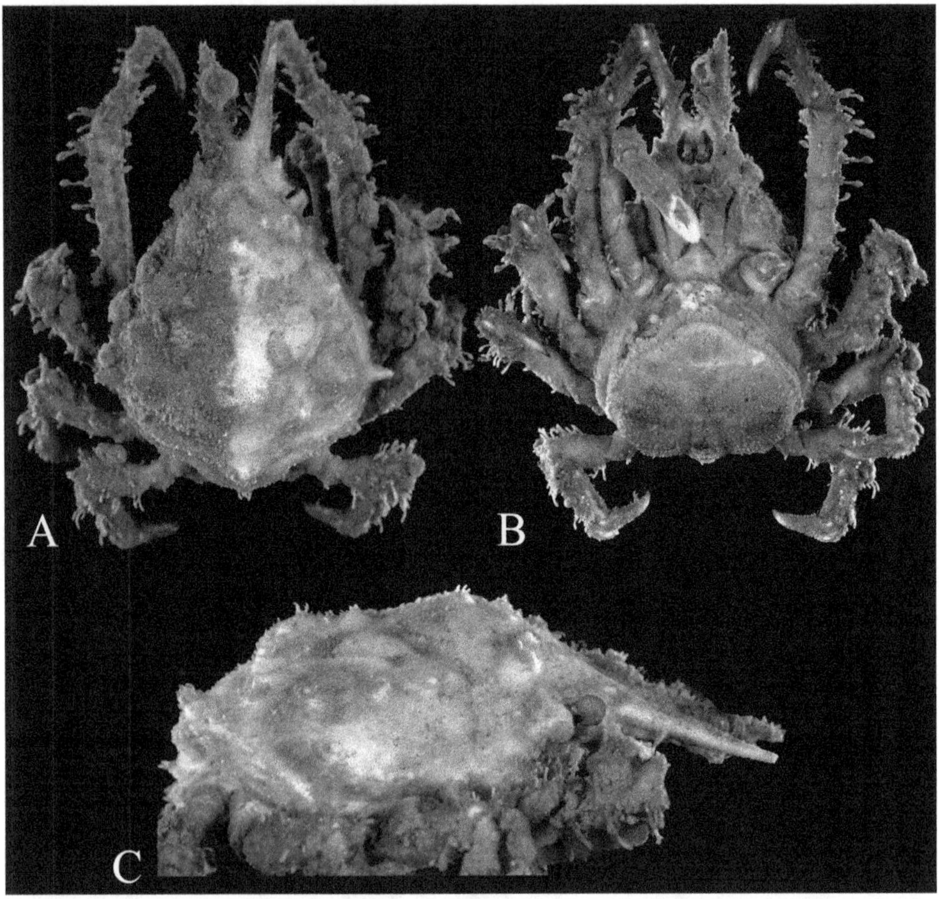

Fig. 3. *Hyastenus pleione* (Herbst, 1803), female (20.5 × 15.9 mm) (ZRC 2019.0491), Tuticorin, India. A, Overall dorsal view of carapace; B, overall ventral view; C, lateral view of carapace.

by Sakai (1999) is here designated as the lectotype. As the specimen is preserved dried, the G1 could not be examined. We have only been able to examine one recent female specimen of this species from India.

Gerstaecker (1856) examined four specimens (almost certainly the syntypes) from the Herbst collection and figured the top half of the dorsal and ventral view of one specimen (Gerstaecker, 1856, pl. 5 figs. 1, 2). Gerstaecker (1856) also compared *H. pleione* to *H. diacanthus* (De Haan, 1837) commenting that both species are similar but that *H. diacanthus* differed in having a slightly narrower and elongated carapace, and less granules on the carapace compared to *H. pleione*.

One specimen of *H. pleione* was examined by Gravely (1927) in his study on the fauna of Krusadai Island in the Gulf of Mannar although no description or figure was provided and the exact collecting locality was not stated. Based on the figure by Jeyabaskaran et al. (2000, pl. 63 fig. b), it seems likely that his specimen(s) examined are *H. pleione* as it matches the types in most aspects (figs. 2A, 3A). It is unlikely to be *H. aries* (Latreille, 1825), which has a similar geographic range, because *H. aries* lacks the numerous granules as seen on *H. pleione* (figs. 2A, 3A). Other records of *H. pleione* from India (see Dev Roy & Nandi, 2005; Dev Roy & Nandi, 2007; Dev Roy, 2008; Gopalakrishnan et al., 2012; Dev Roy & Nandi, 2012; Vidhya et al., 2017) only list the species from Gulf of Mannar, Tamil Nadu, Andaman Islands, and Indian Ocean (see summary in Trivedi et al., 2018), but without descriptions, diagnoses, figures or any clear indication if specimens were examined for their studies. Therefore, we cannot be sure if these identifications are correct.

Tirmizi & Serène (1971) examined four males and two ovigerous females that were collected from the Arabian Sea and sold at the Karachi fish market. The figure of a male specimen by Tirmizi & Serène (1971, pl. 2 fig. A) matches *H. pleione* s. str. (fig. 2) but is slightly larger than the lectotype. The G1 from one male specimen with carapace length of 57 mm (inclusive of pseudorostral spines) was figured by Tirmizi & Serène (1971, text-fig. 2; reproduced here as fig. 1B), and they commented that it was similar to the G1 of "*H. pleione*" illustrated by Buitendijk (1939, text-fig. 14).

Buitendijk (1939) examined specimens from Endeh, Flores, and Koepang, Timor, and figured a G1 from a specimen from Flores, and the G1 is superficially similar to that in Tirmizi & Serène (1971, text-fig. 2; reproduced here as fig. 1B). The G1 of the Flores specimen, however, has only a weak angle on the subdistal region (Buitendijk, 1939, text-fig. 14). Buitendijk (1939) commented that the strength and number of spines on the lateral margin of the carapace varies to some degree between specimens from Flores and Timor. Buitendijk (1949) later recorded "*H. pleione*" from Singapore, and subsequent workers have followed suit (see Ow-Yang, 1963; Lovett, 1981). Examination of specimens in the ZRC that had been identified as "*H. pleione*" from Singapore showed that they are all *H. hilgendorfi* (De Man, 1887) instead (unpublished data). The G1s of the Singapore specimens agree well with that figured of *H. hilgendorfi* by Griffin & Tranter (1986, fig. 42h, i), with the same shape and the subdistal portion weakly dilated on the lateral margin and with the distal tip less slender. They also agree well with the G1 of "*H. pleione*" figured by Buitendijk (1939, text-fig. 14). As such, we believe the specimens identified as "*H. pleione*" from Flores, Timor and Singapore by Buitendijik (1939, 1949) are *H. hilgendorfi* instead.

Hyastenus pleione is morphologically close to *H. hilgendorfi*, but in addition to the differences in G1 structure, their carapace can also be distinguished as follow: there are six tubercles on the swollen gastric region in *H. pleione* (figs. 2A, 3A) (versus three tubercles on swollen gastric region on *H. hilgendorfi*; Tirmizi & Serène, 1971, pl. 2 fig. B) (see also Tirmizi & Serène, 1971; Griffin & Tranter, 1986).

As discussed earlier, comparison of the types to specimens previously identified as *H. pleione* from East China Seas shows that the latter actually belong to a new species (see under *H. ducator* n. sp. for details). The known range of *H. pleione* s. str. thus remains in the Indian Ocean.

Based on the groupings by Griffin & Tranter (1986), *H. pleione* s. str. belongs to "Group 1 species" (see Griffin & Tranter, 1986: 124). Griffin & Tranter (1986) used the carapace granules on the carapace to differentiate *H. pleione* from the other *Hyastenus* species (see Griffin & Tranter, 1986: 124, 125). *Hyastenus pleione* is characterised by the relatively wide posterolateral half of the carapace (figs. 2A, 3A); distinct granules with more swollen carapace regions (fig. 2A); the outer margin of the basal antennal article is strongly constricted medially, with a rounded distal angle (fig. 2B); with distinct granules on the pterygostomial region (fig. 2B); and the male thoracic sternites 3 and 4 relatively shorter in length (fig. 2C).

Distribution.— *Hyastenus pleione* is known with certainty only from southern India, Sri Lanka (Griffin & Tranter, 1986), Gulf of Mannar (Jeyabaskaran et al., 2000), and Arabian Sea (Tirmizi & Serène, 1971).

Hyastenus ducator n. sp.
(figs. 4A, B, 5A-C, 6A, B, 7A-D)

Hyastenus pleione — Shen, 1932: 55–58, 284 (list), 286 (list), 289 (table), text-figs. 31-32, pl. 2 fig. 3. — Kamita, 1941: 3 (list), 4 (list), 74–76, 250 (list), 253 (list), text-fig. 34. — Kikuchi, 1959: 51 (list). — Miyake et al., 1962: 127 (list). — Shen & Dai, 1964: 41, text-fig. — Kim, 1973: 538, pl. 103 fig. 199. — Kim & Kim, 1982: 147, 154 (list). — Dai et al., 1986: 129 (in part), fig. 72 (2), pl. 16 fig. 7. — Dai & Yang, 1991: 142, 143 (in part), fig. 72 (2), pl. 16 figs. 7, 8. — Huang et al., 1993: 127 (list). — Ko, 1997: 1 (list). — Kim & Hong, 1999: 786 (list) [not *Cancer plejone* Herbst, 1803].

Material examined.— Holotype: male (53.1 × 40.0 mm) (IOCAS), seashore outside Institute of Oceanology, Qingdao beach, China, in tangle nets, coll. P.K.L. Ng & H.-L. Chen, 22 August 2002. Paratypes: 3 males (51.9 × 39.0 mm, 51.2 × 37.6 mm, 50.2 × 37.5 mm) (ZRC 2002.0497), same locality and collection information as Holotype. — 1 male (42.4 × 32.3 mm) (IZCAS CB 27038), Yingdao (= Hongdao), Qingdao, Shandong, China, East Ocean, coll. 2 June 1935 — 2 males (36.3 × 27.6 mm, 34.3 × 24.8 mm) (ZRC 2019.1511), Qingdao, China, in tangle nets, coll. P.K.L. Ng, 20 June 2010. — 2 males (47.3 × 35.2 mm, 46.2 × 33.8 mm) (MBM 163487), Qingdao small port, Qingdao, Shandong, China, coll. 29 June 1951. — 1 male (46.1 × 34.9 mm) (MBM 163482), Qingdao small port, Qingdao, Shandong, China, coll. 27 June 1951. — 1 female (49.2 × 36.8 mm) (MBM 163485), Qingdao, Shandong, China, coll. 30 September 1955. — 1 ovigerous female (45.6 × 33.4 mm) (MBM 163488), Qingdao front seas, Qingdao, Shandong, China, coll. Y.-S. Chen, 1 August 1952. — Other material: 1 female (52.3 × 40.3 mm) (IZCAS CB 27040), Jiesheng, Shanwei, Guangdong, China, coll. 25 February 1956. — 1 male (47.0 × 36.5 mm) (IZCAS CB 27039), Jiesheng, Shanwei, Guangdong Province, China, coll. 25 February 1956. —

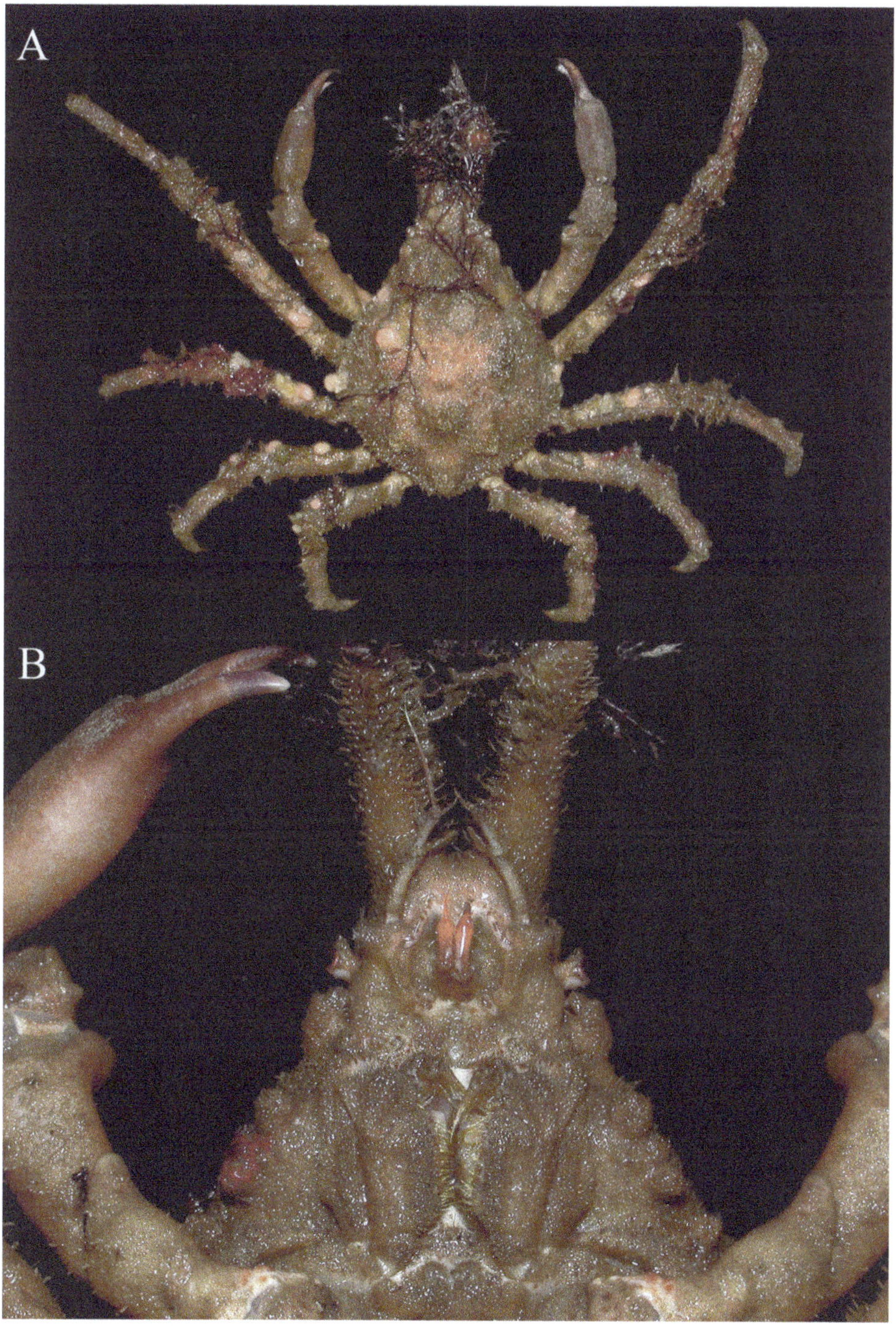

Fig. 4. *Hyastenus ducator* n. sp., holotype male (53.1 × 40.0 mm) (IOCAS), Qingdao, China, color in life. A, Overall view of carapace; B, ventral frontal surface of carapace.

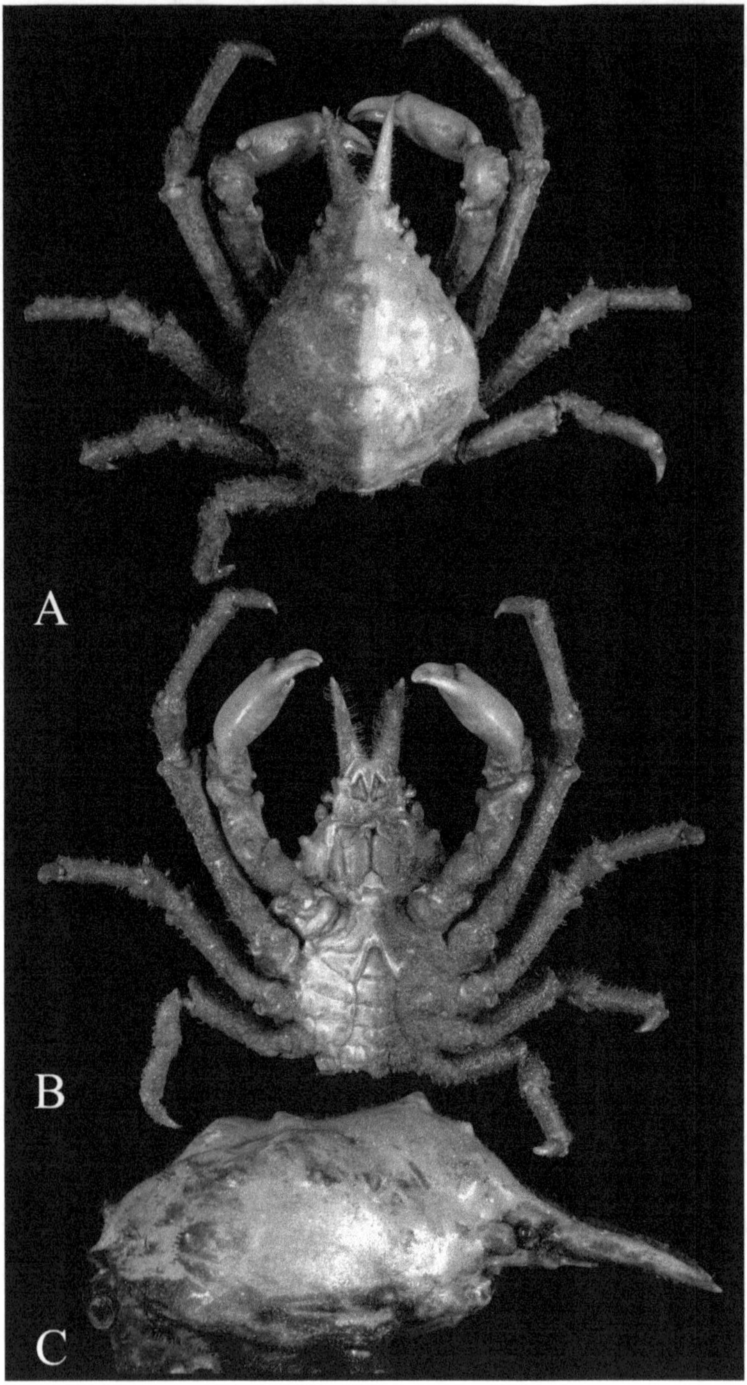

Fig. 5. *Hyastenus ducator* n. sp., holotype male (53.1 × 40.0 mm) (IOCAS), Qingdao, China. A, Overall dorsal view of carapace; B, overall ventral view; C, lateral view of carapace.

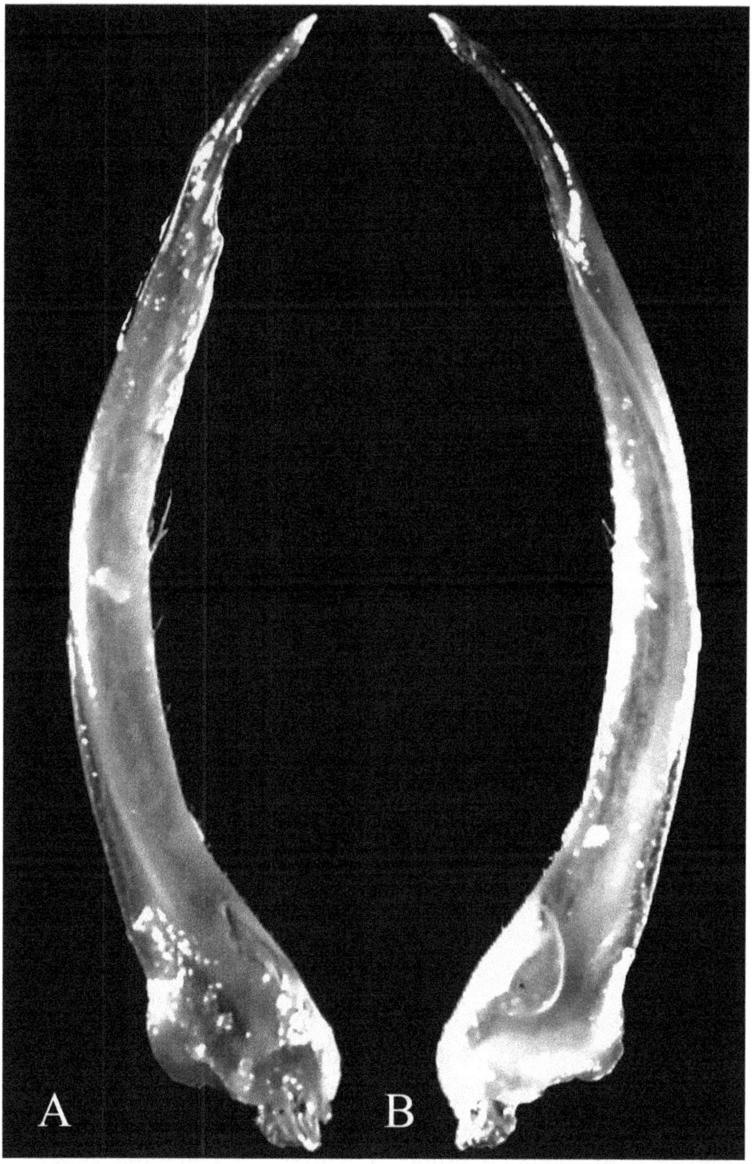

Fig. 6. *Hyastenus ducator* n. sp., paratype male (42.4 × 32.3 mm) (IZCAS CB27038), China, left G1. A, Ventral view; B, dorsal view.

1 ovigerous female (51.9 × 39.6 mm) (IZCAS CB 27041), Jimei, Fujian Province, China, coll. June 1976.

Diagnosis.— Carapace pyriform, covered with thin layer of setae, smooth when denuded. Pseudorostral spines stout, straight, diverging. Supraorbital eave forming rectangular lobe; preorbital angle blunt, outer margin slightly constricted medially,

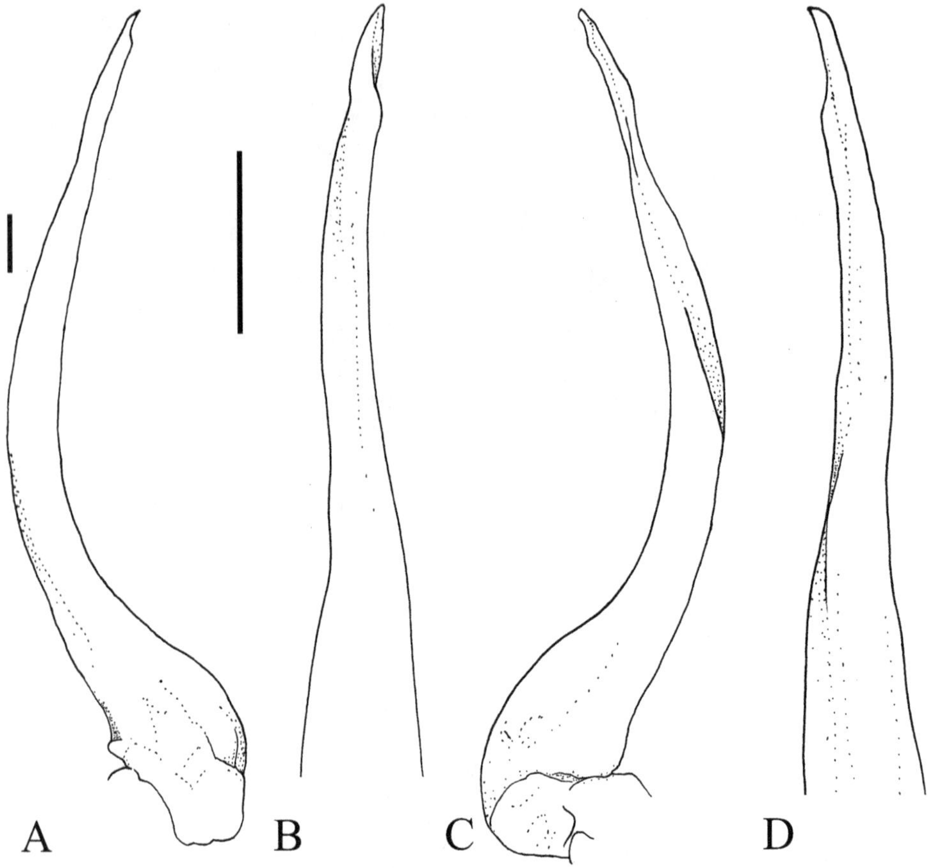

Fig. 7. *Hyastenus ducator* n. sp., holotype male (53.1 × 40.0 mm) (IOCAS), Qingdao, China, left G1. A, Ventral view; B, ventral view of distal portion; C, dorsal view; D, dorsal view of distal portion. Scale bars: 1 mm.

antorbital angle rounded; orbital hiatus between supraorbital and postorbital lobes form narrow U-shape; postorbital lobe cup-like. Carapace with regions defined; gastric region with 6 granules, 1 small mesogastric granule above 1 large mesogastric granule, 4 protogastric granules; small hepatic granule; 1 urogastric granule; row of 3 branchial granules, 1 epibranchial granule, 2 mesobranchial granules; metabranchial region slightly swollen; short lateral branchial spine pointing outwards; 1 intestinal granule medially along posterior region of carapace (figs. 4A, 5A); 1 branchial granule along lateral margin of carapace (fig. 5C). Antennal flagellum shorter than pseudorostral spines. Basal antennal article longer than broad, distal angle blunt, outer margin slightly constricted, basal angle weak. Buccal frame with distal angle slightly protruded. Pterygostomial region with 2 granules on outer margin, first granule largest (figs. 4B, 5B). Chelipeds covered with setae except on fingers, smooth when denuded; propodus slightly swollen, longer than fingers; carpus with 5 granules; merus with 2 granules along dorsal margin, row of 2 or 3 granules along ventral margin (figs. 4A, 5A). Ambulatory legs covered with setae except on dactylus (figs. 4A, 5A). Male thoracic sternum depressed anteriorly (in ventral view); sternites 3, 4 narrow, concave, lateral margin slightly constricted. Male pleon triangular, telson triangular (fig. 5B). G1 elongate, slightly curved, distal tip pointed, slender, subdistal portion weakly dilated (figs. 6A, B, 7A-D).

Etymology.— The new species name, *ducator*, is Latin for leader or chief. The name alludes to and honours Prof. Liu Riuyu, a good friend for many years and leader of the crustacean community in China. His friendship and kindness will be sadly missed. The name is used as a noun in apposition.

Type locality.— Qingdao, China.

Remarks.— *Hyastenus ducator* is one of the largest species in the genus, reaching sizes of 53 mm in carapace length (excluding pseudorostral spines) and 40 mm in carapace width. The present study shows that specimens from China previously identified as *H. pleione* (see Shen, 1932; Shen & Dai, 1964; Dai et al., 1986; Dai & Yang, 1991; Huang et al., 1993) actually belong to a separate species, here named *H. ducator* n. sp. *Hyastenus ducator* was first recorded as *H. pleione* by Shen (1932) from a single female specimen "dredged from the muddy bottom" off Shantung Peninsula, China (Shen, 1932: 58, text-fig. 31, 32, pl. 2 fig. 3). Shen (1932), however, did not indicate if he had any comparative material. In Japan, Kikuchi (1959) was the first to record *H. ducator* (as *H. pleione*) from Sado Island and adjacent waters but there was no indication of specimens examined or the exact location. Subsequently, Miyake et al. (1962) listed *H. ducator* (as *H. pleione*) from Toyama and Niigata in Japan, but again, without figures or information on the number of specimens examined. It is interesting to note that even though *H. ducator* is relatively large, there are no other recent records of this species from Japan and the taxon was not listed or even discussed in subsequent monographs of the brachyuran fauna of Japan (see Sakai, 1976; Miyake, 1983). In Korea, Kamita (1941, text-fig. 34) was the first to record *H. ducator* (as *H. pleione*) from a single male specimen from Jinsen (= Incheon). Kamita (1941) mentioned that Shen (1932) recorded the same species from North China. It also seems likely that the species recorded by the Japanese and Korean workers has been confused with *H. diacanthus* (De Haan, 1839) or *H. elongatus* Ortmann, 1893, which are also relatively large-size species and found in the same region.

Hyastenus ducator differs from *H. diacanthus* in having more granules on the carapace (figs. 4A, 5A) (versus with only one spine and one granule on the swollen gastric region and two lateral branchial spines on *H. diacanthus*; cf. Yamaguchi & Baba, 1993, fig. 115); and from *H. elongatus* in having relatively shorter pseudorostral spines of less than half the carapace length (figs. 4A, 5A) (versus *H. elongatus* having longer pseudorostral spines of slightly more than half the carapace length; cf. Sakai, 1938, pl. 36 fig. 6). The G1 of *H. ducator* is also slender, slightly curved, with the distal tip pointed (figs. 6A, B, 7A-D) (versus G1 is relatively stouter, curved medially, tapering to slender tip on distal two-thirds, with a lobe-like tip on *H. diacanthus*; cf. Griffin & Tranter, 1986, fig. 48b; and G1 slightly curved, distal tip pointed, with the distal top margin straight on *H. elongatus*; cf. Griffin & Tranter, 1986, fig. 47a, b).

Hyastenus ducator differs from *H. pleione* in having fewer large granules on the carapace (figs. 4A, 5A) (versus more distinct large granules on carapace on *H. pleione*; figs. 2A, 3A); a blunt distal angle of the basal antennal article with the outer margin slightly constricted medially (figs. 4B, 5B) (versus rounded distal angle of basal antennal article with the outer margin distinctly constricted medially on *H. pleione*; figs. 2B, 3B); the male thoracic sternites 3 and 4 are proportionately more narrow with the lateral margin constricted (fig. 5B) (versus male thoracic sternites 3 and 4 proportionately wider with the lateral margin less constricted on *H. pleione*; fig. 2C); and there is only one branchial granule along the lateral margin of the carapace (fig. 5C) (versus with three branchial granules along the lateral margin of the carapace of *H. pleione*; figs. 2D, 3C). The G1 of *H. ducator* also differs from *H. pleione* in being more elongate and in having the subdistal part before the tip is only weakly dilated (figs. 6A, B, 7A-D) (versus G1 proportionately shorter with the subdistal part prominently dilated in *H. pleione*; cf. Tirmizi & Serène, 1971, text-fig. 2; reproduced here as fig. 1B).

Distribution.— *Hyastenus ducator* n. sp. is found in its type locality, Qingdao (this study), Guangdong, Fujian, and Jiaozhou Bay, China (Dai & Yang, 1991), possibly Japan (Miyake et al., 1962) and Korea (Kamita, 1941; Kim, 1973; Kim & Kim, 1982; Ko, 1997; Kim & Hong, 1999).

ACKNOWLEDGEMENTS

The specimens from Qingdao were made available through the kindness of the late Prof Liu Riuyu who invited the second author as a visiting scholar in 2002 and who facilitated his study in Qingdao. During this and later visits, he was kindly hosted by the late Dr Chen Hui-Lian, the doyen of brachyuran crabs in China. The second author's visit to ZMB was kindly facilitated by Oliver Coleman. The first author would also like to thank the Lee Kong Chian Natural History Museum (J.C.E. Mendoza and Muhammad Dzaki Bin Safaruan), Institute of Zoology (Meng Kai), and Institute of Oceanology, Chinese Academy of Sciences (Li Xinzheng and Chu Yan Ling) for their help in making important specimens available for this study. The authors thank Shane Ahyong, Jiang Wei and Charles Fransen, as well as an anonymous referee, for their many helpful comments which have helped improve the paper.

REFERENCES

AL-KHAFAJI, K. K. S., I. M. ABDUL-SAAHIB & S. G. AJEEL, 2017. The epialtid crab: *Hyastenus hilgendorfi* (De Man, 1887) a new spider crab record from Iraqi coast, NW-Arabian Gulf. Int. J. Mar. Sci., **7**(49): 462-465.

ALCOCK, A., 1895. Materials for a carcinological fauna of India. No. 1. The Brachyura Oxyrhyncha. J. Asiat. Soc. Bengal, **44**(2): 157-291, pls. 3-5.

BALSS, H., 1935. Brachyura of the Hamburg Museum Expedition to South-Western Australia, 1905. J. R. Soc. West. Aust., **21**: 113-151, pl. 13.

BUITENDIJK, A. M., 1939. Biological results of the Snellius Expedition. V. The Dromiacea, Oxystomata, and Oxyrhyncha of the Snellius Expedition. Temminckia, **4**: 223-276, figs. 1-27, pls. 7-11.

BUITENDIJK, A. M., 1949. On a small collection of Decapoda Brachyura, chiefly Dromiidae and Oxyrhyncha, from the neighbourhood of Singapore. Bulletin of the Raffles Museum, **21**: 59-82.

DAI, A. Y., S. L. YANG, Y. Z. SONG & G. X. CHEN, 1986. [= Crabs of the China Seas: 1-17 [table of content], 1-642 [main text], pls. 1-74]. (China Ocean Press, Beijing). [In Chinese.]

DAI, A. Y. & S. L. YANG, 1991. Crabs of the China Seas: 1-21 [table of content], 1-682 [main text], pls. 1-74. (China Ocean Press, Beijing & Springer-Verlag, Berlin & Heidelberg).

DE MAN, J. G., 1887a. Bericht über die von Herrn Dr. J. Brock im indischen Archipel gesammelten Decapoden und Stomatopoden. Arch. Naturgesch., Berlin, **53**(2): 215-288, pls. 7-10.

DE MAN, J. G., 1887b. Report on the podophthalmous Crustacea of the Mergui Archipelago, collected for the Trustees of the Indian Museum, Calcutta, by Dr. John Anderson, F.R.S., Superintendent of the Museum. — Part I. J. Linn. Soc. London (Zoology), **22**(136): 1-64, pls. 1-3.

DEV ROY, M. K., 2008. An annotated checklist of mangrove and coral reef inhabiting brachyuran crabs of India. Rec. zool. Surv. India, Occasional Paper, **289**(136): 1-212.

DEV ROY, M. K. & N. C. NANDI, 2005. Brachyuran diversity of coral reef ecosystems in India. In: Proceedings of the National Seminar on Reef Ecosystem Remediation. SDMRI Research Publication, **9**: 220-231.

DEV ROY, M. K. & N. C. NANDI, 2007. Brachyuran diversity in coastal ecosystems of Tamil Nadu. Journal of Environment and Sociobiology, **4**(2): 169-192.

DEV ROY, M. K. & N. C. NANDI, 2012. Brachyuran crabs (Crustacea). In: DIRECTOR ZOOLOGICAL SURVEY OF INDIA, KOLKATA (ed.), Fauna of Andaman and Nicobar Islands. State Fauna Series, **19**(Part 1): 185-236. (Zoological Survey of India, Kolkata).

GERSTAECKER, A., 1856. Carcinologische Beiträge. Arch. Naturgesch., Berlin, **22**(1): 101-162, pls. 4-6. [In German.]

GOPALAKRISHNAN, A., P. R. DIVYA, V. S. BASHEER, T. RAJA SWAMINATHAN, A. KATHIRVELPANDIAN, K. K. BINEESH, R. G. KUMAR & J. K. JENA, 2012. Macro flora and fauna of the Gulf of Mannar — a checklist: 1-127. (National Bureau of Fish Genetic Resources, Lucknow, U.P., India).

GRAVELY, F. H., 1927. Decapoda and Stomatopoda. In: The littoral fauna of Krusadai Island in the Gulf of Manaar. Bulletin of the Madras Government Museum, New Series (Natural History), **1**: 141-155, figs. 1, 2, pls. 20-26.

GRIFFIN, D. J. G. & H. A. TRANTER, 1986. The Decapoda Brachyura of the Siboga expedition. Part VIII: Majidae. Siboga Expeditie Monografie, **39**: [i]-[vii], 1-335.

HENDERSON, J. R., 1893. A contribution to Indian carcinology. Trans. Linn. Soc. Lond., (2) Zoology, **5**(10): 325-458, pls. 36-40.

HERBST, J. F. W., 1803. Versuch einer Naturgeschichte der Krabben und Krebse nebst einer systematischen Beschreibung ihrer verschiedenen Arten, **3**(3): 1-54, pls. 55-58. (Gottlieb August Lange, Berlin und Stralsund).

HUANG, Z. G., C. X. ZHENG, S. LIN, C. Y. LI, J. J. WANG & S. K. YAN, 1993. Fouling organisms at Daya Bay nuclear power station, China. In: B. MORTON (ed.), The Marine Biology of the South China Sea. Proceedings of the First International Conference on the Marine Biology of Hong Kong and the South China Sea, Hong Kong, 28 October-3 October 1990: 121-130. (Hong Kong University Press, Hong Kong).

ICZN (INTERNATIONAL COMMISSION ON ZOOLOGICAL NOMENCLATURE), 1999. International Code of Zoological Nomenclature. Fourth Edition. The International Trust for Zoological Nomenclature, London: i-xxix, 1-306.

JEYABASKARAN, R., A. S. KHAN & V. RAMAIYAN, 2000. Brachyuran crabs of Gulf of Mannar, biodiversity project on Gulf of Mannar biosphere reserve. Parangipettai, India: 1-177, pls. 1-78. (Centre of Advanced Study in Marine Biology, Annamalai University, Chidambaram).

KAMITA, T., 1941. Studies on the Decapod Crustaceans of Chosen. Part 1. Crabs. The Fisheries Society of Chosen, Keijo: 1-289. [In Japanese and English.]

KAZMI, Q. B., 1997. A review of spider majid crabs of the northern Arabian Sea, with particular reference to Pakistani species, and with description of an undescribed *Pleistacantha*. Pakistan Journal of Marine Sciences, **6**(1-2): 79-91.

KIKUCHI, K., 1959. Decapod crustaceans of Sado Island and adjacent waters. Bull. Biol. Soc. Hiroshima Univ., **10**(26): 49-51. [In Japanese].

KIM, D. N. & S. Y. HONG, 1999. Larval development of *Scyra compressipes* (Decapoda: Brachyura: Majidae: Pisinae) reared in the laboratory. J. Crust. Biol., **19**(4): 782-791.

KIM, H. S., 1973. Anomura Brachyura. Ilustrated Encyclopedia of Fauna and Flora of Korea, **14**: 1-694. (Samhwa Publishing Company, Seoul, Korea). [In Korean.]

KIM, W. & H. S. KIM, 1982. Classification and geographical distribution of Korean crabs (Crustacea, Decapoda, Brachyura). Proceedings of the College of Natural Science, Seoul National University, **7**(1): 133-159.

KO, H. S., 1997. The first zoeal stage of *Hyastenus elongatus* (Ortmann, 1893) (Decapoda, Brachyura, Majidae). Korean J. Syst. Zool., **13**(1): 1-8.

LATREILLE, P. A., 1825-1828. Triangulaires, trigona. In: P. A. LATREILLE, A.-L.-M. LE PELETIER, J.-G.-A. SERVILLE & F.-E. GUÉRIN (eds.), Encyclopédie méthodique. Histoire naturelle. Entomologie, ou Histoire Naturelle des Crustacés, des Arachnides et des Insects, **10**(1-2), [1-6]: 1-832. (Mme Veuve Agasse, Paris). [Dates of publication follow EVENHUIS (2003a: 36) and SHERBORN & WOODWARD (1906b: 578): part 1 (pp. 1-344), published 1 October 1825; part 2 (pp. 345-832), published 13 December 1828.]

LAURIE, R. D., 1906. Report on the Brachyura collected by professor Herdman, at Ceylon, in 1902. In: W. A. HERDMAN (ed.), Report to the Government of Ceylon on the Pearl Oyster Fisheries of the Gulf of Manaar. With supplementary reports upon the Marine Biology of Ceylon by Other Naturalists. The Royal Society, London, part V, **40**(supplement): i-viii, 349-432, pls. 1, 2.

LEE, B. Y. & P. K. L. NG, 2019. On the identity of *Hyastenus inermis* Rathbun, 1911, and description of a new species from Sulawesi, Indonesia (Crustacea: Decapoda: Majoidea: Epialtidae). Raffles Bull. Zool., **67**: 490-497.

LOVETT, D. L., 1981. A guide to the shrimps, prawns, lobsters, and crabs of Malaysia and Singapore. Faculty of Fisheries and Marine Science, University Pertanian Malaysia, Serdang, Selangor, Malaysia, Occasional Publication, **2**: 1-156.

MACLEAY, W. S., 1838. On the brachyurous decapod Crustacea brought from the Cape by Dr. Smith. In: A. SMITH (ed.), Illustrations of the zoology of South Africa; consisting chiefly of figures and descriptions of the objects of natural history collected during an expedition into the interior of South Africa, in the years 1834, 1835, and 1836; fitted out by 'the cape of good hope association for exploring central Africa'. published under the authority of the lords commissioners of her majesty's treasury, London: i-iv, 53-71, 2 pls.

MIERS, E. J., 1886. Report on the Brachyura collected by H.M.S. Challenger during the years 1873–1876. In: C. WYVILLE THOMSON & J. MURRAY, (eds.), Report on the Scientific Results of the Voyage of H.M.S. Challenger during the years 1873–1876 under the command of Captain George S. Nares, R.N., F.R.S. and the late Captain Frank Tourle Thomson, R.N. prepared under the Superintendence of the late Sir C. Wyville Thomson, Knt., F.R.S. &c. Regius Professor of Natural History in the University of Edinburgh Director of the civilian scientific staff on board and now of John Murray, LL.D., Ph.D. &c. one of the naturalists of the Expedition. Zoology. Published by Order of Her Majesty's Government, London, Edinburgh and Dublin, volume **17**(49): 1 [=50], 1-362, pls. 1-29.

MILNE-EDWARDS, A., 1872. Recherches sur la faune carcinologique de la Nouvelle-Calédonie. Nouvelles Archs Mus. Hist. nat. Paris, **8**: 229-267, pls. 10-14.

MIYAKE, A., K. SAKAI & S. NISHIKAWA, 1962. A fauna-list of the decapod Crustacea from the coasts washed by the Tsushima warm current. Rec. oceanogr. Works in Japan, (Special No.) **6**: 121-131.

MIYAKE, S., 1983. Japanese crustacean decapods and stomatopods in color. Vol. **2**. Brachyuran (Crabs): i-vii, 1-277, 64 pls. (Hoikusha, Osaka, Japan). [In Japanese.]

NG, P. K. L., D. GUINOT & P. J. F. DAVIE, 2008. Systema Brachyurorum: Part I. An annotated checklist of extant brachyuran crabs of the world. Raffles Bull. Zool., (Suppl.) **17**: 1-286.

NG, P. K. L., T. KOMAI & T. SATO, 2017. On the trail of a Japanese "ghost species" –the identity of *Goniopugettia tanakae* Sakai, 1986, and the establishment of a new genus for *Pugettia sagamiensis* Gordon, 1930 (Decapoda, Brachyura, Epialtidae). Crustacean Research, **46**: 133-152.

ORTMANN, A. E., 1893. Die Decapoden-Krebse des Strassburger Museums, mit besonderer Berücksichtigung der von Herrn Dr. Döderlein bei Japan und bei den Liu-Kiu-Inseln gesammelten und z. Z. im Strassburger Museum aufbewahrten Formen. Theil VI. Abtheilung: Brachyura (Brachyura genuina Boas), 1. Unterabtheilung: Majoidea und Cancroidea, 1. Section Portuninea. Zoologische Jahrbücher, Abtheilung für Systematik, Geographie und Biologie der Thiere, **7**(1): 23-88, pl. 3.

ORTMANN, A. E., 1894. Crustaceen. In: R. SEMON (ed.), Zoologische Forschungreisen in Australien und dem Malayischen Archipel in den Jahren 1891–93. V. Denkschriften der Medizinisch-Naturwissenschaftlichen Geselschaft, Jena, **8**: 1-80, pls. 1-3. [In German.]

OW-YANG, C. K., 1963. Studies on the Systematics and Distribution of Marine Brachyura in Malaya with special references to the families Portunidae and Majidae: 1-268, 40 pls. (Unpublished M.Sc. thesis, Department of Zoology, University of Singapore).

PILLAI, N. K., 1951. Decapoda (Brachyura) from Travancore. Bull. cent. Res. Inst. Univ. Travancore, Trivandrum, (C), **2**(1): 1-46.

SAKAI, K., 1999. J. F. W. Herbst-collection of Decapod Crustacea of the Berlin Zoological Museum, with remarks on certain species. Naturalists, Publications of Tokushima Biological Laboratory, Shikoku University, **6**: 1-45, pls. 1-21.

SAKAI, T., 1938. Studies of the crabs of Japan III. Brachygnatha, Oxyrhyncha: 193-364, pls. 20-41. (Yokendo Co., Tokyo).

SAKAI, T., 1976. Crabs of Japan and the Adjacent Seas: volume **1** [English text]: i-xxxix, 1-773; volume **2** [Japanese text]: 1-461; volume **3** [plates]: 1-16, 251 pls. (Kodansha, Tokyo).

SAMOUELLE, G., 1819. The Entomologist's Useful Compendium, or an Introduction to the Knowledge of the British Insects, Comprising the Best Means of Obtaining and Preserving Them, and a Description of the Apparatus Generally Used; Together with the Genera of Linné, and the Modern Method of Arranging the Classes Crustacea, Myriapoda, Spiders, Mites and Insects, From Their Affinities and Structure, According to the Views of Dr. Leach. Also an Explanation of the Terms used in Entomology; a Calendar of the Times of Appearance and Usual Situations of 3,000 Species of British Insects; with Instructions for Collecting and Fitting up Objects for the Microscope: [v], 6-496, 12 pls. (Thomas Boys, London).

SHEN, C.-J., 1932. The brachyuran Crustacea of North China. Zoologia Sinica, (A, Invertebrates of China), **9**(1): i-x, 1-320, pls. 1-10.

SHEN, C.-J. & A. Y. DAI, 1964. Illustrations of animals in China. Crustacea part II: 1-142, 277 figs. (Science Press, Beijing). [In Chinese.]

TIRMIZI, N. M. & R. SERÈNE, 1971. The rediscovery of two species of crabs (Decapoda, Brachyura) with observations on three other species from Pakistan. Crustaceana, **21**(1): 21-32, pl. 1, 2.

TRIVEDI, J. N., D. J. TRIVEDI, K. D. VACHHRAJANI & P. K. L. NG, 2018. An annotated checklist of the marine brachyuran crabs (Crustacea: Decapoda: Brachyura) of India. Zootaxa, **4502**(1): 1-83.

VIDHYA, V., P. JAWAHAR & K. KARUPPASAMY, 2017. Annotated check list of the brachyuran crabs (Crustacea: Decapoda) from Gulf of Mannar region, south east coast of India. J. Entomol. Zool. Stud., **5**(6): 2331-2336.

WHITE, A., 1847a. List of Specimens of Crustacea in the Collection of the British Museum. By Order of the Trustees [of the British Museum], London: i-viii, 1-143.

WHITE, A., 1847b. Descriptions of new Crustacea from the Eastern Seas. In: April 27, 1847. William Tarrell, Esq., Vice-President, in the Chair. Proc. Zool. Soc. Lond., **15**(172): 56-58.

WHITE, A., 1847c. Descriptions of new Crustacea from the Eastern Seas. In: Proceedings of Learned Societies. Zoological Society. April 13, 1847. — Wm. Yarrell, Esq., Vice-President, in the Chair. Ann. Mag. nat. Hist., **20**(130): 61-63.

WINDSOR, A. M. & S. T. AHYONG, 2013. *Hyastenus baru*, a new species of spider crab from Indonesia (Brachyura, Majoidea, Epialtidae) with a key to the species of *Hyastenus*. Crustaceana, **86**(6): 718-727.

YAMAGUCHI, T. & K. BABA, 1993. Crustacean specimens collected in Japan by Ph. F. Von Siebold and H. Bürger and held by the Nationaal Natuurhistorisch Museum in Leiden and other museums. In: T. YAMAGUCHI (ed.), Ph. F. von Siebold and Natural History of Japan, Crustacea: 145-570. (The Carinological Society of Japan, Tokyo).

First received 9 October 2019.
Final version accepted 2 July 2020.

[When citing this volume, please refer to *Crustaceana* 93 (2020), 1233-1546]

Prof. Ruiyu Liu Memorial Issue

THE FRESHWATER CRAB GENUS *LACUNIPOTAMON* DAI, SONG, HE, CAO, XU & ZHONG, 1975 (DECAPODA, BRACHYURA, POTAMIDAE), WITH DESCRIPTIONS OF TWO NEW SPECIES FROM SOUTHWESTERN CHINA

BY

CHAO HUANG[2,3,5]), HSI-TE SHIH[4]) and SHANE T. AHYONG[1,2])

[1]) University of New South Wales, Kensington, NSW 2052, Australia

[2]) Australian Museum, 1 William St, Sydney NSW 2010, Australia

[3]) School of Life Sciences, Sun Yat-sen University, Guangzhou 510275, P.R. China

[4]) Department of Life Science and Research Center for Global Change Biology, National Chung Hsing University, 250, Kuo Kuang Road, Taichung 402, Taiwan, Republic of China

ABSTRACT

The poorly known genus *Lacunipotamon* Dai, Song, He, Cao, Xu & Zhong, 1975, is studied and reviewed. Two new species, *Lacunipotamon yuanshi* sp. nov. and *Lacunipotamon cymatile* sp. nov. from Yunnan Province and Guangxi Zhuang Autonomous Region, respectively, are described based on morphology and mitochondrial cytochrome c oxidase subunit I (COI) sequences. Both superficially resemble *Lacunipotamon albusorbitum* Dai, Song, He, Cao, Xu & Zhong, 1975, the type and previously only known species in the genus, but can be distinguished by the combination of live colour, size, maxilliped III, male thoracic sternum, male pleon, male gonopod, and female gonopore characters. Phylogenetic analysis of the molecular data corroborates the novelty of the two species.

Key words. — *Lacunipotamon*, COI, new species, Potamidae, systematics, China, freshwater crab

RÉSUMÉ

Le genre peu connu *Lacunipotamon* Dai, Song, He, Cao, Xu & Zhong, 1975, est étudié et revu. Deux nouvelles espèces *Lacunipotamon yuanshi* nov. sp. et *Lacunipotamon cymatile* nov. sp. respectivement de la province du Yunnan et de la région autonome Guangxi Zhuang, sont décrites sur la base de leur morphologie et des séquences de la sous-unité de l'oxydase du cytochrome c mitochondrial (COI). Les deux ressemblent approximativement à *Lacunipotamon albusorbitum* Dai, Song, He, Cao, Xu & Zhong, 1975, le type et la seule espèce du genre connue jusqu'alors, mais peuvent être distingués grâce à la combinaison de leur couleur sur le vivant, la taille, le maxillipède III, le sternum thoracique, le pléon, les gonopodes chez le mâle et les gonopores femelles. L'analyse phylogénétique des données moléculaires corrobore la nouveauté des deux espèces.

[5]) Corresponding author; e-mail: chaohuang1777@yahoo.com

 | DOI:10.1163/9789004464353_009

Mots clés. — *Lacunipotamon*, COI, nouvelles espèces, Potamidae, systématique, Chine, crabe d'eau douce

INTRODUCTION

Prior to the present study, the freshwater crab genus *Lacunipotamon* Dai, Song, He, Cao, Xu & Zhong, 1975, only contained the little-known type species, *L. albusorbitum* Dai, Song, He, Cao, Xu & Zhong, 1975 from Basa, Hekou County, Yunnan Province, China (Dai, 1999; Ng et al., 2008; Shih & Ng, 2011). Recently, the first author was informed by botanist Bing-Mou Wang (Panyu Central Hospital) of the occurrence of orange coloured, semi-terrestrial freshwater crabs from Malipo County, Yunnan, discovered by his local guide Ming-Feng Long, who on our request, kindly provided specimens to us. These crabs fitted well the description of *Lacunipotamon* externally but the male first gonopod differed from that of *L. albusorbitum*, and thus probably represented a species new to science. Subsequently, the first author visited Malipo to make further collections and observe the habitat. An attempt was also made to locate and collect *L. albusorbitum* from the type locality, but the original feeder stream from which it had been recorded had been turned into a concrete waterway and so none were found. At around the same time, the first author also noticed large, blue, semi-terrestrial freshwater crabs were being sold from Guangxi as pets in the aquarium trade. The crabs also fit the description of *Lacunipotamon* and had generally similar male gonopods as the Malipo new species but were obviously different in live colour and mature size. With the aid of the seller of these crabs, the first author was able to visit the original locality to photograph and collect these crabs. As of such, we herein describe two new species of *Lacunipotamon* from Yunnan and Guangxi, China.

MATERIAL AND METHODS

Specimens were collected by hand and preserved in 75% ethanol from Guangxi and Yunnan, China. They are deposited in the Sun Yat-sen Museum of Biology, Sun Yat-sen University, Guangzhou, China (SYSBM); the Australian Museum, Sydney, Australia (AM); the Zoological Reference Collection of the Lee Kong Chian Natural History Museum, National University of Singapore, Singapore (ZRC); the Zoology Collection of the National Chung Hsing University, Taichung, Taiwan (NCHUZOOL); and the Chinese Academy of Sciences, Beijing (CAS). Measurements, in millimetres, are of the carapace width and length, respectively. Other abbreviations are as follows: G1, male first gonopod; G2, male second gonopod; CW, carapace width; a.s.l., above sea level. The terminology used primarily follows that of Dai (1999) and Davie et al. (2015).

Genomic DNA was isolated from muscle tissue using kits (see Shih et al., 2009, for details). A portion of the COI gene was amplified with a polymerase chain reaction (PCR) using the primers LCO1490, HCO2198, and COL14 (Folmer et al., 1994; Roman & Palumbi, 2004; Schubart, 2009), after verification with the complementary strand. Sequences of different haplotypes have been deposited in Gen Bank (fig. 10). Because a preliminary analysis with a wide suite of potamids showed that the two new species of *Lacunipotamon* are closely related to members of the genus *Tiwaripotamon* Bott, 1970, the COI sequences of *Tiwaripotamon* and its closely related genera (see Shih & Do, 2014; Do et al., 2016) were included for phylogenetic analyses.

The best-fitting model of sequence evolution was determined by PartitionFinder ver. 2.1.1 (Lanfear et al., 2017) and selected by the Bayesian information criterion (BIC). The obtained best model (GTR + I + G) was subsequently used for a Bayesian inference (BI) analysis. BI analysis was performed with MrBayes ver. 3.2.3 (Ronquist et al., 2012). Phylogenetic analysis was performed with four chains for 10 million generations and four independent runs, with trees sampled every 1000 generations. The convergence of chains was determined by the average standard deviation of split frequency values below the recommended 0.01 (Ronquist et al., 2019), and the first 700 trees were discarded as the burnin accordingly. Maximum likelihood (ML) analysis was conducted using RAxML vers. 7.2.6 (Stamatakis, 2006). The model GTR + G (i.e., GTRGAMMA) was used with 100 runs and finding the best ML tree by comparing the likelihood scores. The robustness of the ML tree was evaluated by 1000 bootstrap pseudoreplicates under the model GTRGAMMA. Base pair (bp) differences and pairwise estimates of Kimura 2-parameter (K2P) distances (Kimura 1980) for genetic divergences between specimens were calculated with MEGA ver. 10.0.5 (Kumar et al., 2018).

SYSTEMATICS

Family POTAMIDAE Ortmann, 1896
Subfamily POTAMISCINAE Bott, 1970 (sensu Yeo & Ng, 2004)
Genus *Lacunipotamon* Dai, Song, He, Cao, Xu & Zhong, 1975

Type species.— *Lacunipotamon albusorbitum* Dai, Song, He, Cao, Xu & Zhong, 1975, by original designation.

Diagnosis.— Carapace broader than long, convex, regions indistinct, dorsal surface smooth (fig. 8B); epigastric cristae and postorbital cristae inconspicuous (fig. 8B); external orbital tooth blunt, separated from anterolateral margin by gap (fig. 8B). Epistome median lobe broadly rounded (fig. 8A). Maxilliped 3 exopod reaching to proximal one-fifth of merus height, flagellum short to medium length (figs. 3A, 6A). Cheliped palm surface smooth (figs. 3F–I, 6F–I). Ambulatory legs slender (fig. 8B). Male anterior thoracic sternum relatively narrow, width 1.4 × length (figs. 2B, 5B). Male pleon triangular (figs. 2C, 5C). Female pleon broadly ovate (figs. 2E, 5E). G1 generally slender, terminal segment with basal flap and dorsal projection (figs. 3C–E, 6C–E, 7). G2 with long flagellum-like terminal segment (figs. 3B, 6B). Female vulva ovate, medium-sized, reaching sternite 5 but not sternite 7; positioned closely to one another (figs. 2F, 5F).

Remarks.— This genus was briefly described in Dai et al. (1975), and later more details were documented by Dai (1999). Our diagnosis of *Lacunipotamon* builds upon that of Dai's (1999) with more detail and the third maxilliped and G1 characters updated. Dai (1999) describes the third maxilliped exopod flagellum as being short and degenerated and the G1 as sturdy with a short, key-shaped terminal segment. With the addition of the two new species described below, however, we update the third maxilliped exopod flagellum as being short to medium in length and the G1 as generally slender, with basal flap and dorsal projection in the terminal segment.

Key to the Species of *Lacunipotamon*

1. G1 terminal segment with basal flap and bent dorsal projection .... *Lacunipotamon albusorbitum*
– G1 terminal segment with basal flap and straight dorsal projection 2
2. Size large (CW < 53.6 mm), blue coloured, G1 terminal segment with relatively long dorsal projection .... *Lacunipotamon cymatile* sp. nov.
– Size relatively small (CW < 30.7 mm), bright orange to red coloured, G1 terminal segment with relatively short dorsal projection .... *Lacunipotamon yuanshi* sp. nov.

Lacunipotamon albusorbitum Dai, Song, He, Cao, Xu & Zhong, 1975 (fig. 8A–D)

Type material.— Holotype male: CAS CB 05108, Basa, Hekou County, Yunnan, China. August, 2018 [photographs examined; fig. 8A–D].

Diagnosis.— Carapace broader than long, convex, regions indistinct, dorsal surface smooth (fig. 8B); epigastric cristae and postorbital cristae inconspicuous (fig. 8B); external orbital tooth blunt, separated from anterolateral margin by gap (fig. 8B). Epistome median lobe broadly rounded (fig. 8A). Maxilliped 3 exopod reaching to proximal one-fifth of merus height, flagellum short (Dai, 1999: fig. 74(1)). Cheliped palm surface smooth (figs. 8A). Ambulatory legs slender (fig. 8B). Male anterior thoracic sternum relatively narrow, width 1.4 × length (fig. 8C). Male pleon triangular (fig. 8C). Female pleon broadly ovate (Dai, 1999:

fig. 74(6)). G1 generally slender, terminal segment with basal flap and bent dorsal projection (fig. 8D). Female vulva ovate, medium-sized, reaching sternite 5 but not sternite 7; positioned closely to one another (Dai, 1999: fig. 74(7)).

Remarks.— *Lacunipotamon albusorbitum*`remains known only from the two type specimens collected in 1963 from Basa, Hekou County, Yunnan. There is no figure or description available concerning the G2, therefore this character remains undescribed. About eight years ago, ichthyologist Zhuo-Cheng Zhou purchased a pair of large bright red semi-terrestrial freshwater crabs from the local wet market in Hekou, Yunnan. After photographing the live colouration (fig. 8A), the specimens were fixed and deposited in the Shanghai Ocean University Museum. The male specimen in his colour photo (fig. 8A) is considerably larger than the holotype of *L. albusorbitum*, but agrees well with its description in colour, external morphology and locality, suggesting that they are indeed the same species. A recent attempt to locate this pair of specimens was unsuccessful, and the crabs have not been seen from the markets since. Attempts to locate these crabs in the wild have also been unsuccessful so far. Further surveying attempts around Hekou are needed to collect new material and gain insight as to the conservation status of this species.

Lacunipotamon yuanshi sp. nov.
(figs. 1–3, 7A, B, 9A)

Type material.— Holotype: SYSBM 001816, male (30.7 × 23.3 mm), Malipo County, Wenshan Zhuang and Miao Autonomous Prefecture, Yunnan Province, China, 23.13°N, 104.70°E, mud hole next to small hillstream, 1000 m a.s.l., coll. C. Huang, July 2018. Paratypes: SYSBM 001818-001819, 2 males (25.5 × 20.1 mm, 22.7 × 17.7 mm), same data as holotype; SYSBM 001817, 001820, 2 females (25.0 × 19.5 mm, 24.4 × 19.4 mm), same data as holotype; AM P.103562, 1 male (29.1 × 22.4 mm), 1 female (25.3 × 20.6 mm), Malipo County, Wenshan Zhuang and Miao Autonomous Prefecture, Yunnan Province, China, coll. M.-F. Long, May 2018; ZRC 2018.1185, 1 male (29.1 × 22.4 mm), 1 female (25.3 × 20.6 mm), same data as above. NCHUZOOL 15083, 1 male, (26.2 × 21.0 mm), same data as above. Additional material: SYSBM 001821–001823, 3 males (27.2 × 20.2 mm, 26.4 × 20.0 mm, 23.4 × 17.7 mm), Malipo County, Wenshan Zhuang and Miao Autonomous Prefecture, Yunnan Province, China, coll. M.-F. Long, May 2018. SYSBM 001824–001825, 2 females (27.2 × 21.4 mm, 24.4 × 18.7 mm), same data as above.

Description.— Carapace width 1.2-1.3 × length (n = 14), regions indistinct. Dorsal surface smooth, finely pitted, convex (fig. 1). Front deflexed, margin almost straight in dorsal view (fig. 1). Epigastric cristae smooth, very low, inconspicuous, divided by a narrow gap (figs. 1, 2A). Postorbital cristae smooth, very low, almost indiscernible (fig. 1). Branchial regions swollen; cervical groove very shallow, inconspicuous; mesogastric region convex (fig. 1). External orbital tooth blunt, triangular with gently convex outer margins, separated from anterolateral margin by gap (figs. 1, 2A). Epibranchial tooth very small, inconspicuous (figs. 1, 2A). Anterolateral margin smooth, lined with fused granules. Posterolateral

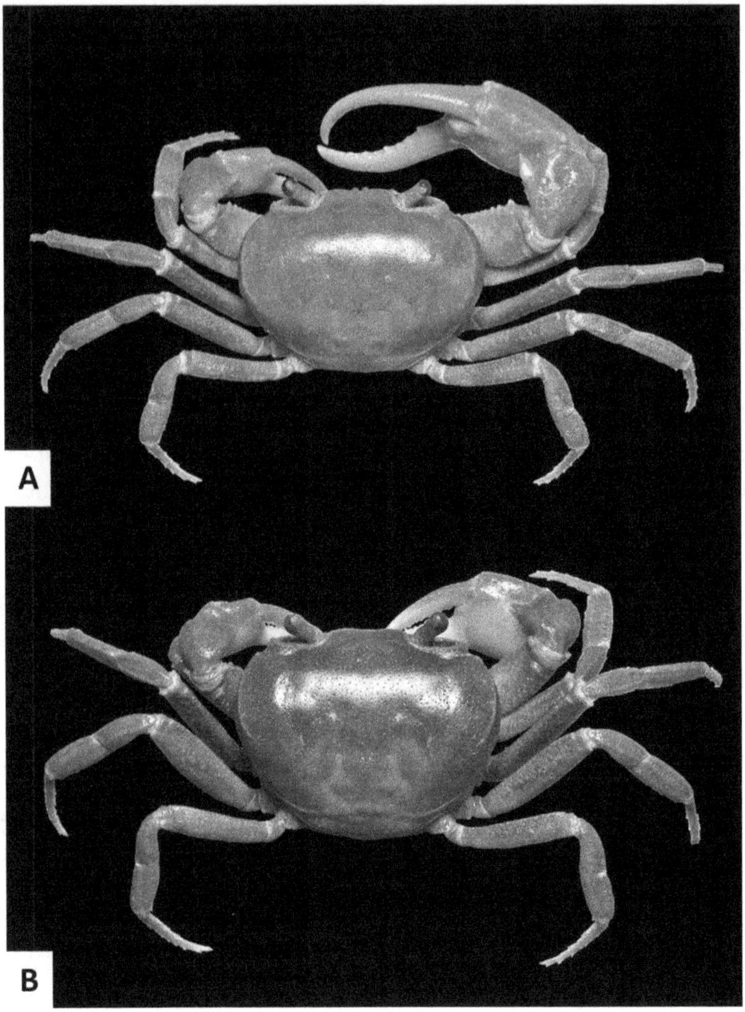

Fig. 1. *Lacunipotamon yuanshi* sp. nov. A, Male holotype (SYSBM 001816), dorsal overall view; B, female paratype (SYSBM 001817), dorsal overall view.

surface smooth (fig. 1). Orbits large, supraorbital and infraorbital margins ridged, smooth (figs. 1, 2A). Sub-orbital, sub-hepatic and pterygostomial regions divided by sutures; surfaces generally smooth (fig. 2A). Epistome median lobe broadly rounded, lateral margins curved downwards (fig. 2A).

Maxilliped 3 merus length and width equal; ischium width about 0.8 × length; merus subtrapezoidal with median depression; ischium subtrapezoidal, with distinct median sulcus, mesial margin rounded; exopod reaching to proximal one-fifth of merus height, flagellum as long as merus width (fig. 3A).

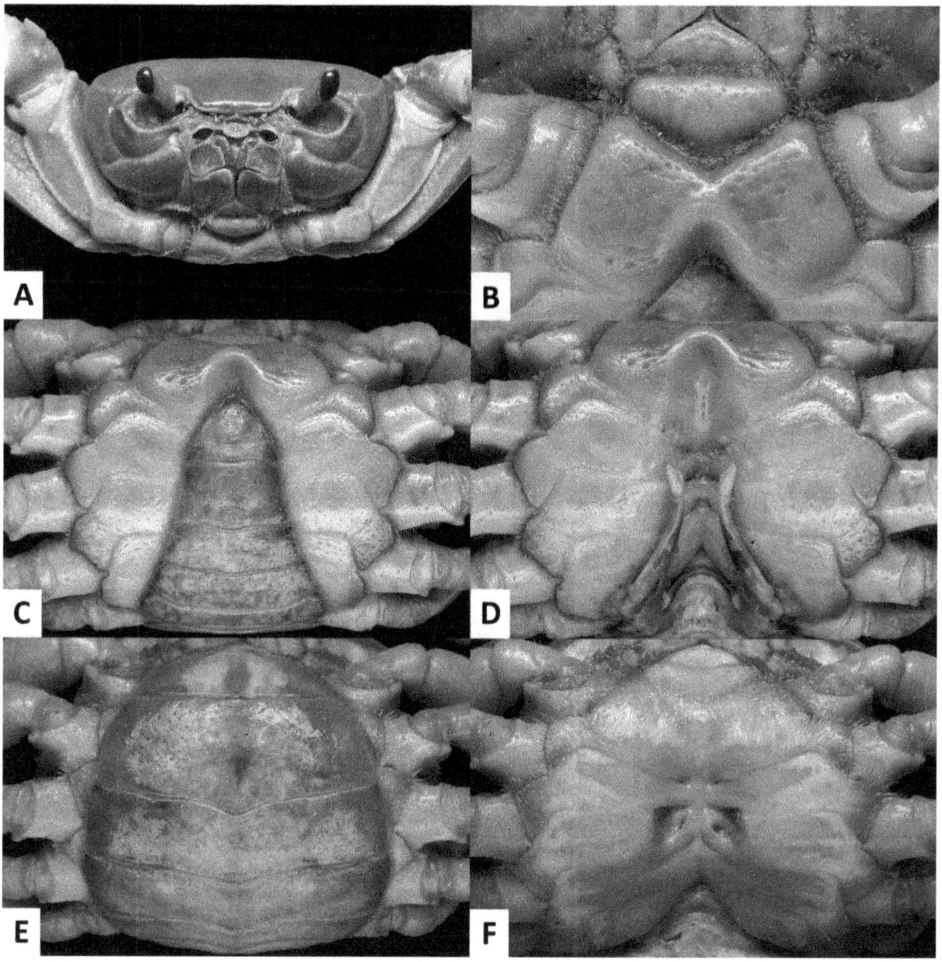

Fig. 2. *Lacunipotamon yuanshi* sp. nov. A-D, Male holotype (SYSBM 001816); E, F, female paratype (SYSBM 001817). A, Frontal view of cephalothorax; B, tip of the anterior thoracic sternum; C, ventral view of anterior thoracic sternum and pleon; D, ventral view showing sterno-pleonal cavity with right G1 in situ; E, ventral view of anterior thoracic sternum and pleon; F, ventral view showing female vulvae.

Chelipeds (pereiopod 1) unequal, less inflated in females (figs. 1, 3F–I). Merus trigonal in cross section, margins weakly crenulated, surfaces generally smooth (figs. 1, 2A). Carpus with sharp spine at inner-distal angle, spinule at base, surfaces generally smooth (fig. 1). Major cheliped palm length about 1.1–1.2× height in males ($n = 6$), 1.2–1.3× in females ($n = 4$); dactylus 1.1–1.5 × palm length in males ($n = 6$), 1.1 × in females ($n = 4$) (fig. 3F–I). Palm surface generally smooth, posterior region slightly granulated. Occlusal margin of fingers lined with blunt, round teeth; fingers of major chela of large males with significant gape

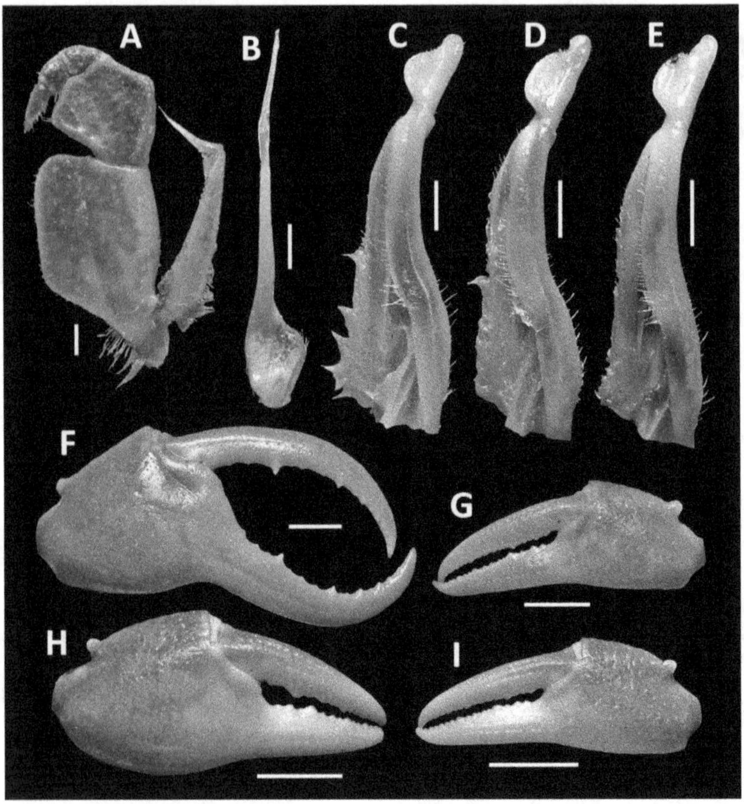

Fig. 3. *Lacunipotamon yuanshi* sp. nov. Male holotype (SYSBM 001816): A, left third maxilliped; B, left G2; C, male left G1; F, male right cheliped; G, male left cheliped. Male paratype (SYSBM 001818): E, male left G1. Male paratype (SYSBM 001819): F, male left G1. Female paratype (SYSBM 001817): H, female right cheliped; I, female left cheliped. Scale bars: A–E = 1.0 mm; F–I = 5.0 mm.

when closed, height up to 0.4 length of dactylus (fig. 3F); small gape in all others (fig. 3G–I).

Ambulatory legs (pereiopods 2–5) slender, with very sparse, short setae. Pereiopod 3 merus 0.7–0.8 × carapace length in males (n = 6) and females (n = 4). Pereiopod 5 propodus length 2.5–2.7 × width in males (n = 6), 2.5–2.7 × in females (n = 4), shorter than dactylus (fig. 1).

Male thoracic sternum generally smooth, pitted; sternites 1–4 relatively narrow, width 1.4 × length. Sternites 1 and 2 fused, forming a subtriangular structure; sternites 2 and 3 fused, separated by deep transverse sulcus; sternites 3 and 4 fused, with deep sulcus (fig. 2B). Male sterno-pleonal cavity reaching anteriorly to level of midlength of chelipeds coxae base (fig. 2B); median longitudinal groove separating sternites 7 and 8 deep (fig. 2D). Male pleonal locking tubercle positioned posterior to mid-length of sternites 5 (fig. 2D). Female vulva ovate, medium-sized, with relatively narrow outer rim, reaching sternite 5 but not sternite 7; positioned close together, orientated slightly oblique to midline (fig. 2F).

Pleon triangular in males (fig. 2C), broadly ovate in females (fig. 2E). Male pleonites 3–6 progressively narrower, lateral margins slightly concave; pleonite 6 width 1.7× length; telson width 1.3× length, apex blunt (fig. 2C).

G1 generally slender, in situ reaching to pleonal locking tubercle (fig. 2D). Subterminal segment 3.4–3.9× as long as terminal segment ($n = 3$), curved outwards and tapering anteriorly. Terminal segment with large rounded basal flap and stumpy dorsal projection, outer margin almost straight, apex blunt, directed obliquely anteriorly (figs. 3C–E, 7A, B). G2 subterminal segment 1.7–1.9× as long as the flagellum-like terminal segment ($n = 3$) (fig. 3B).

Etymology.— The species name is Hanyu Pinyin for "scholar" in Chinese, in honour of the late Prof. Rui-Yu Liu, for his major career-long contributions to Chinese marine sciences, especially in the field of carcinology.

Live colouration.— Carapace and chelipeds generally bright orange to red with purplish brown ambulatory legs in mature individuals (fig. 9A). Generally purplish brown all-over with bright orange tips of the chelipeds in smaller individuals.

Habitat.— This new semi-terrestrial species digs and inhabits mud burrows close to small hillstreams and seeps at a relatively high altitude (approx. 1000 m a.s.l.). Freshwater crabs from high altitudes tend to be smaller in size (Okano et al., 2000; Shih et al., 2004), as is also the case for *L. yuanshi* sp. nov., being the smallest and highest dwelling species in the genus. No other potamids where found at the type locality.

Remarks.— *Lacunipotamon yuanshi* sp. nov. externally resembles *L. albusorbitum*, but can be distinguished by its less inflated carapace (compare fig. 9A and 8A), third maxilliped exopod with flagellum as long as merus width (fig. 3A) (versus short flagellum reaching one third width of merus in *L. albusorbitum*; Dai, 1999: fig. 74 (1)), deep suture between male thoracic sternites 3/4 (fig. 2B) (versus suture shallow in *L. albusorbitum*; fig. 8C) and G1 terminal segment with large rounded basal flap and stumpy dorsal projection (figs. 3C–E, 7A, B) (versus G1 terminal segment with narrowly rounded basal flap and dorsal projection bent 90° inwards in *L. albusorbitum*; fig. 8D). Features distinguishing *L. yuanshi* sp. nov. from *L. cymatile* sp. nov. are discussed under the account of the latter.

Lacunipotamon cymatile sp. nov.
(figs. 4–6, 7C, D, 9B)

Type material.— Holotype: SYSBM 001826, male (53.6 × 40.7 mm), Leiping Town Daxin County, Chongzuo City, Guangxi Province, China, 22.65°N, 107.10°E, small hillstream, forest floor, 300 m a.s.l., coll. C. Huang, August 2018. Paratypes: SYSBM 001828–001831, 4 males (41.5 × 31.5 mm, 39.5 × 30.1 mm, 38.5 × 29.4 mm, 30.1 × 23.0 mm), same data as holotype. SYSBM

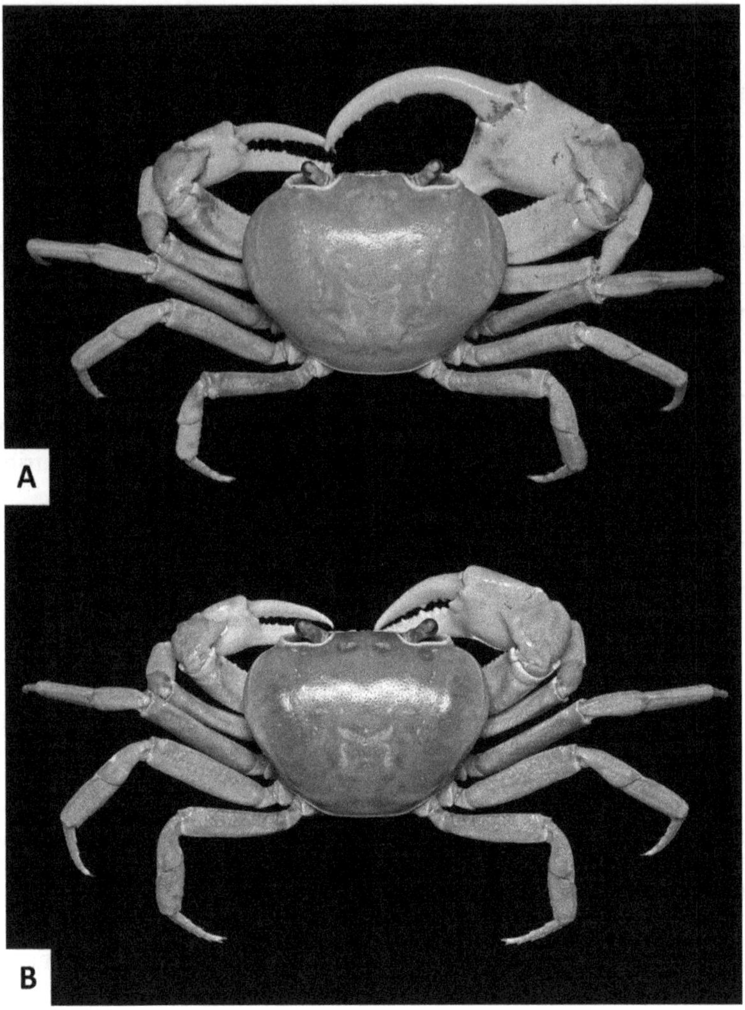

Fig. 4. *Lacunipotamon cymatile* sp. nov. Male holotype (SYSBM 001826): A, dorsal overall view; B, female paratype (SYSBM 001827), dorsal overall view.

001827, 001832–001833, 3 females (39.0 × 29.7 mm, 35.3 × 27.5 mm, 29.4 × 22.7 mm), same data as holotype. AM P.103563, 1 male (41.4 × 31.2 mm), same data as holotype. ZRC 2018.1182, 1 male (42.6 × 32.2 mm), 1 female (40.5 × 31.2 mm), same data as holotype. NCHUZOOL 15084, 1 male, (36.0 × 27.1 mm), same data as above.

Description.— Carapace width 1.3× length ($n = 11$), regions indistinct. Dorsal surface smooth, finely pitted, convex (fig. 4). Front deflexed, margin almost straight in dorsal view (fig. 4). Epigastric cristae smooth, very low, inconspicuous, divided by a narrow gap (figs. 4, 5A). Postorbital cristae smooth, very low, almost indiscernible (fig. 4). Branchial regions very swollen; cervical groove very shallow, inconspicuous; mesogastric region convex (fig. 4). External orbital tooth

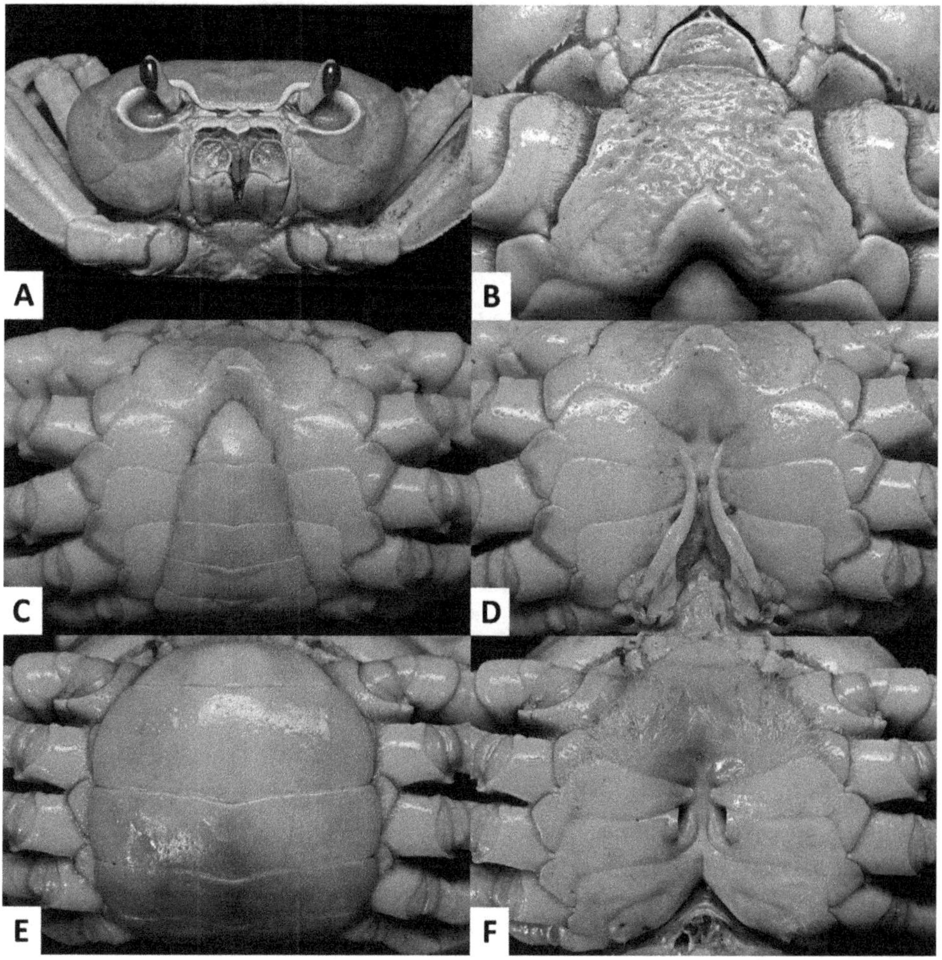

Fig. 5. *Lacunipotamon cymatile* sp. nov., A–D, male holotype (SYSBM 001826); E, F, female paratype (SYSBM 001827). A, Frontal view of cephalothorax; B, tip of the anterior thoracic sternum; C, ventral view of anterior thoracic sternum and pleon; D, ventral view showing sterno-pleonal cavity with right G1 in situ; E, ventral view of anterior thoracic sternum and pleon; F, ventral view showing female vulvae.

blunt, triangular with gently convex outer margins, separated from anterolateral margin by gap (figs. 4, 5A). Epibranchial tooth very small, inconspicuous (figs. 4, 5A). Anterolateral margin smooth, lined with fused granules. Posterolateral surface smooth (fig. 4). Orbits large, supraorbital and infraorbital margins ridged, smooth (figs. 4, 5A). Sub-orbital, sub-hepatic and pterygostomial regions divided by sutures; surfaces generally smooth (fig. 5A). Epistome median lobe broadly rounded, lateral margins curved downwards (fig. 5A).

Maxilliped 3 merus width about 0.9 × length; ischium width about 0.7× length; merus subtrapezoidal with median depression; ischium subtrapezoidal,

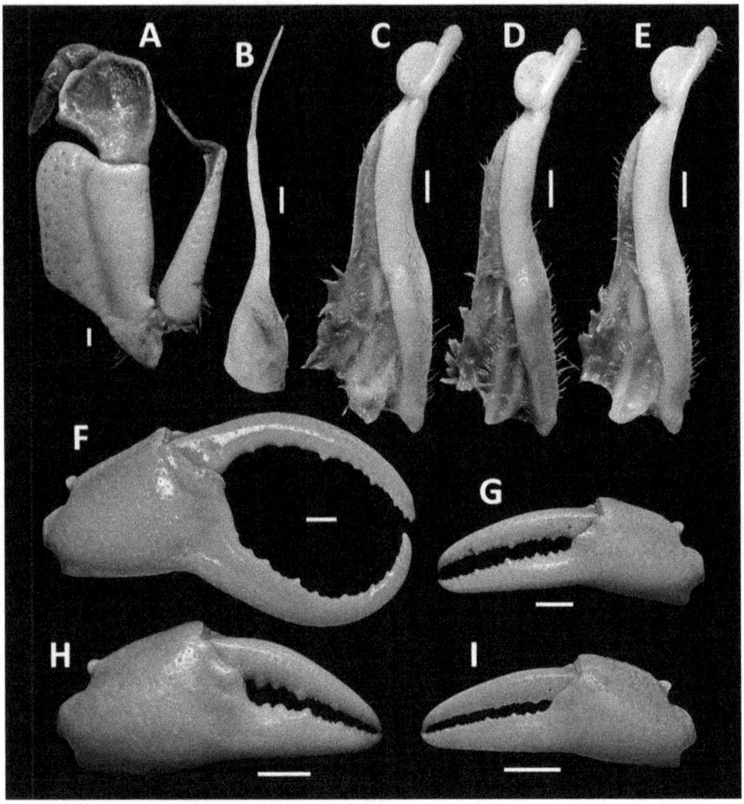

Fig. 6. *Lacunipotamon cymatile* sp. nov. Male holotype (SYSBM 001826): A, left third maxilliped; B, left G2; C, male left G1; F, male right cheliped; G, male left cheliped. Male paratype (AM P.103563): E, male left G1. Male paratype (ZRC): F, male left G1. Female paratype (SYSBM 001827): H, female right cheliped; I, female left cheliped. Scale bars: A–E = 1.0 mm; F–I = 5.0 mm.

with distinct median sulcus, mesial margin rounded; exopod reaching to proximal one-fifth of merus height, flagellum as long as merus width (fig. 6A).

Chelipeds (pereiopod 1) unequal, less inflated in females (figs. 4, 6F–I). Merus trigonal in cross section, margins weakly crenulated, surfaces generally smooth (figs. 4, 5A). Carpus with sharp spine at inner-distal angle, spinule at base, surfaces generally smooth (fig. 4). Major cheliped palm length about 1.2× height in males (n = 7), 1.2–1.3 × in females (n = 4); dactylus 1.1–1.5× palm length in males (n = 7), 1.1× in females (n = 4) (fig. 6F–I). Palm surface generally smooth, posterior region slightly granulated. Occlusal margin of fingers lined with blunt, round teeth; fingers of major chela of large males with significant gape when closed, height up to 0.5 length of dactylus (fig. 6F); small gape in all others (fig. 6G–I).

Ambulatory legs (pereiopods 2–5) slender, with very sparse and short setae. Pereiopod 3 merus 0.7–0.8× carapace length in males (n = 7) and females

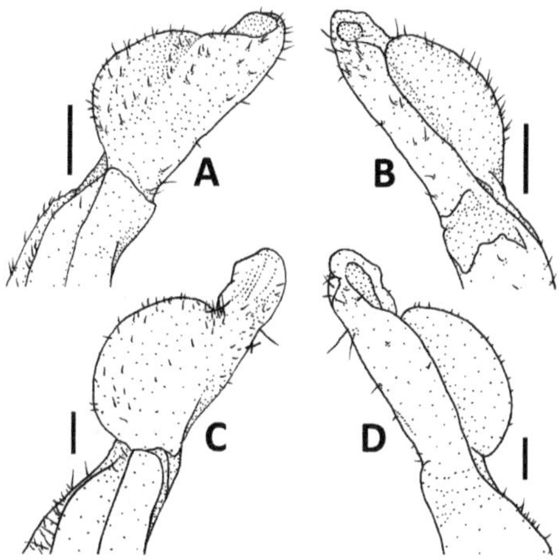

Fig. 7. *Lacunipotamon yuanshi* sp. nov., male holotype (SYSBM 001816), G1 terminal segment: A, ventral view; B, dorsal view. *Lacunipotamon cymatile* sp. nov., male holotype (SYSBM 001826), G1 terminal segment: C, ventral view; D, dorsal view. Scale bars 1 mm.

(n = 4). Pereiopods 5 propodus length 2.4–2.7× width in males (n = 7), 2.4–2.5 in females (n = 4), shorter than dactylus (fig. 4).

Male thoracic sternum generally smooth, wrinkled in large specimens; sternites 1–4 relatively narrow, width 1.4× length. Sternites 1, 2 fused, forming subtriangular structure; sternites 2 and 3 fused, separated by a deep transverse sulcus; sternites 3 and 4 fused, with shallow demarcation (fig. 5B). Male sterno-pleonal cavity reaching anteriorly to level of midlength of chelipeds coxae base (fig. 5B); median longitudinal groove separating sternites 7 and 8 deep (fig. 5D). Male pleonal locking tubercle positioned slightly posterior to mid-length of sternites 5 (fig. 5D). Female vulva ovate, medium-sized, with relatively wide outer rim, reaching sternite 5 but not sternite 7; positioned closely to one another, orientated slightly oblique to midline (fig. 5F).

Pleon triangular in males (fig. 5C) and broadly ovate in females (fig. 5E). Male pleonites 3–6 progressively narrower, lateral margins nearly straight; pleonite 6 width 1.7× length; telson width 1.4× length, apex blunt (fig. 5C).

G1 generally slender, in situ reaching beyond the pleonal locking tubercle (fig. 5D). Subterminal segment 3.5–3.8× as long as terminal segment (n = 3), curved outwards and tapering anteriorly. Terminal segment with large rounded basal flap and relatively long dorsal projection, outer margin almost straight, apex blunt, directed obliquely anteriorly (figs. 6C-E, 7C, D). G2 subterminal segment 2.0-2.2× as long as the flagellum-like terminal segment (n = 3) (fig. 6B).

Etymology.— The species name, derived from Greek, means "blue-coloured", pertaining to the colouration of the new species.

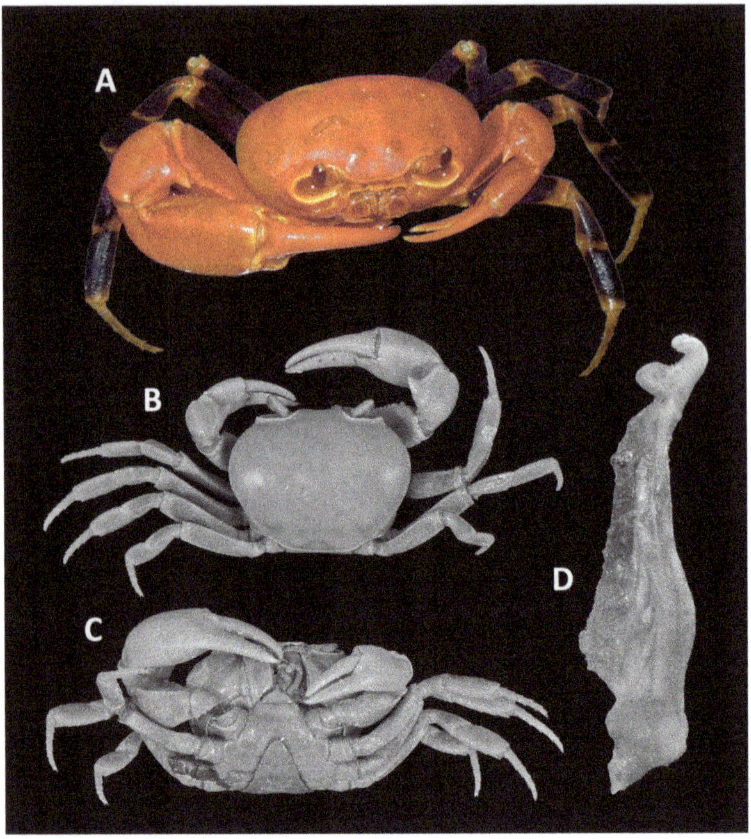

Fig. 8. *Lacunipotamon albusorbitum* Dai, Song, He, Cao, Xu & Zhong, 1975, specimen lost: A, colour in life. *Lacunipotamon albusorbitum*, male holotype (CAS CB 05108), courtesy of XM Zhou: B, dorsal overall view; C, ventral overall view; D, ventral view of G1.

Live colouration.— Carapace dorsal surface and merus of third maxilliped sky blue; orbits, epistome and chelipeds ivory white; ambulatory legs light olive brown in mature individuals (fig. 9B). Carapace dorsal surface and merus of third maxilliped dark brown with light olive brown in all other areas in smaller individuals.

Habitat.— This new semi-terrestrial species digs and inhabits mud burrows close to seeps and springs in a karstic mountain (approx. 300 m a.s.l.). According to the collector, *L. cymatile* sp. nov. can be sympatric with an undescribed blind cave potamid that has been found in the springs, currently under study.

Remarks.— The external morphology of *Lacunipotamon cymatile* sp. nov. is similar to that of *L. albusorbitum*, but can be distinguished by its blue colouration (fig. 9B) (versus bright red in *L. albusorbitum*; fig. 8A), third maxilliped exopod with flagellum as long as merus width (fig. 6A) (versus short flagellum reaching

Fig. 9. A, *Lacunipotamon yuanshi* sp. nov. in natural habitat; B, *Lacunipotamon cymatile* sp. nov. in natural habitat.

one third width of merus in *L. albusorbitum*; Dai, 1999: fig. 74 (1)), and G1 terminal segment with a large rounded basal flap and relatively long dorsal projection (figs. 6C–E, 7C, D) (versus G1 terminal segment with narrowly rounded basal flap and dorsal projection bent 90° inwards in *L. albusorbitum*; fig. 8D). *Lacunipotamon cymatile* sp. nov. can be separated from *L. yuanshi* sp. nov. by its larger size (CW < 53.6 mm) (versus smaller size in *L. yuanshi* sp. nov.; CW < 30.7 mm), shallow suture between male thoracic sternites 3/4 (fig. 5B) (versus suture deep in *L. yuanshi* sp. nov.; fig. 2B) and G1 terminal segment

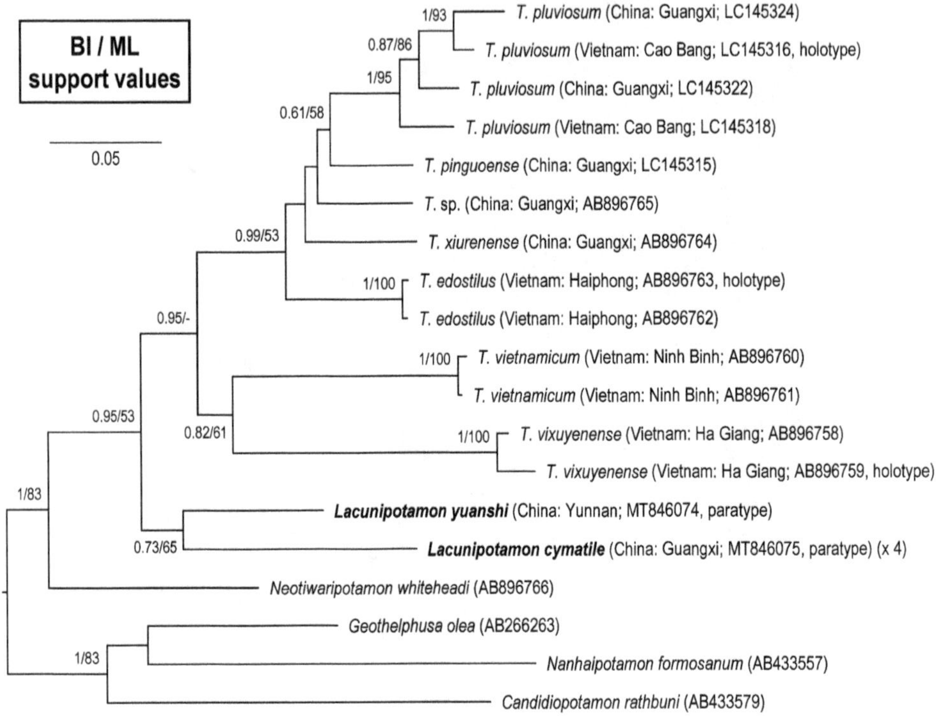

Fig. 10. Bayesian inference (BI) tree for species of *Lacunipotamon* and *Tiwaripotamon*, as well as the outgroups, based on the cytochrome c oxidase subunit I (COI) gene. Probability values at the nodes represent support values for BI and maximum likelihood (ML).

with a relatively long dorsal projection (figs. 6C–E, 7C, D) (versus stumpy dorsal projection in *L. yuanshi* sp. nov.; figs. 3C–E, 7A, B).

PHYLOGENETIC ANALYSIS AND DISCUSSION

The molecular analyses included eight specimens of the genus *Lacunipotamon*. Among them, shorter (incomplete) COI sequences (616 bp) were obtained for two specimens of *L. yuanshi* (SYSBM 001817, 001820) and one of *L. cymatile* (SYSBM 001831), but were otherwise identical to the longer sequences (658 bp) of *L. yuanshi* (SYSBM 001818) and *L. cymatile* (SYSBM 001832, 001833, 001827, 001830), respectively. Therefore, the shorter sequences were omitted from further analyses and only a single sequence was used for each species. The reconstructed COI phylogenetic tree (fig. 10) shows the genus *Lacunipotamon* to be sister to the genus *Tiwaripotamon*, together forming a highly-supported clade. Further analyses of their phylogenetic relationships are necessary by including more species of each genus, as well as more genetic markers. The pairwise nucleotide divergence with K2P distance (and bp difference) between the two species of *Lacunipotamon* is 9.97% (61 bp), which is greater than the distances for most species of Potamidae, including *Geothelphusa* Stimpson, 1858, *Nanhaipotamon* Bott, 1968, *Sinopotamon* Bott, 1968 and *Tiwaripotamon* Bott, 1970 (see Chu et al., 2015). Thus, the two species of *Lacunipotamon* are also well corroborated by genetic data. Preliminary analyses (not shown) that include more species of *Tiwaripotamon* using the 16S and 28S rRNA markers show *Lacunipotamon* to be clustered within the former. This either suggests that *Lacunipotamon* should be merged with *Tiwaripotamon*, or that *Tiwaripotamon* should be split. Externally, the two genera can easily be separated. The most easily recognizable character of *Tiwaripotamon* is the long and slender ambulatory legs, which appear to be an adaptation to living in karst environments (Chapman, 1982; Guinot, 1988, 1994; Shih & Ng, 2011). On the other hand, *Lacunipotamon*, possesses the inflated carapace, typical of semi-terrestrial burrowing species, and have shorter but stronger legs. These external characters are strongly subject to selective pressure from the habitat and may not always be a good indicator of relationship (Huang et al., 2017). The more reliable G1 characters of the two genera, nonetheless, are also distinctly different, especially the presence of a large flap on the terminal segment in *Lacunipotamon* (versus with no flap or very minor flap in *Tiwaripotamon*). Further research will have to be conducted using more specimens and markers to solve this problem.

ACKNOWLEDGEMENTS

This study was supported by grants from the Student Research Project Funding and PANGEA Small Research Grants (School of Biological, Earth and Environmental Sciences, University of New South Wales) to CH and STA; and grants from the Ministry of Science and Technology (MOST 108-2621-B-005-002-MY3; 108-2313-B-005-007-MY3), Executive Yuan, Taiwan, to HTS. We thank Xian-Min Zhou for his generosity in allowing us to use his photographs of the holotype of *L. albusorbitum* for use in this paper. We would also like to thank the reviewers for improving our manuscript.

REFERENCES

CHAPMAN, P., 1982. The origin of troglobites. Proceedings of the University of Bristol Speleological Society, **16**: 133-141.

CHU, K. H., C. D. SCHUBART, H.-T. SHIH & L. M. TSANG, 2015. Genetic diversity and evolution of Brachyura. In: P. CASTRO, P. J. F. DAVIE, D. GUINOT, F. R. SCHRAM & J. C. VON VAUPEL KLEIN (eds.), Treatise on zoology – anatomy, taxonomy, biology – the Crustacea, complementary to the volumes translated from the French of the Traité de Zoologie, **9**(C) (II), Decapoda: Brachyura (Part 2): 775-820. (Brill, Leiden).

DAI, A. Y., 1999. Fauna sinica (Arthropoda. Crustacea. Malacostraca. Decapoda. Parathelphusidae. Potamidae): 1-501, figs. 1-238, pls. 1-30. (Science Press, Beijing). [In Chinese with English summary.]

DAI, A. Y., Y. Z. SONG, L. Y. HE, W. J. CAO, Z. B. XU & W. L. ZHONG, 1975. Description of several new species of freshwater crabs from China that are intermediate hosts to lung flukes and other flukes. Acta Zoologica Sinica, **21**: 257-264, pls. 1-3. [In Chinese.]

DO, V. T., H.-T. SHIH & C. HUANG, 2016. A new species of freshwater crab of the genus *Tiwaripotamon* Bott, 1970 (Crustacea, Brachyura, Potamidae) from northern Vietnam and southern China. Raffles Bulletin of Zoology, **64**: 213-219.

GUINOT, D., 1988. Les crabes cavernicoles du monde. Mémoires de Biospéologie, **15**: 3-40.

GUINOT, D., 1994. Decapoda Brachyura. In: C. JUBERTHIE & V. DECOU (eds.), Encyclopaedia biospeologica, **1**: 165–179, pl. 3. (Société de Biospéologie, Moulis (CRNS) & Académie Roumaine, Bucharest).

HUANG, C., H.-T. SHIH & P. K. L. NG, 2017. A new genus and new species of Potamidae (Crustacea: Decapoda: Brachyura: Potamoidea), the first stygomorphic cave crab known from China and east Asia. Zootaxa, **4232**: 71-84.

KUMAR, S., G. STECHER, `M. LI, `C. KNYAZ & K. TAMURA, 2018. MEGA X: molecular evolutionary genetics analysis across computing platforms. Molecular Biology and Evolution, **35**: 1547-1549.

LANFEAR, R., P. B. FRANDSEN, A. M. WRIGHT, T. SENFELD & B. CALCOTT, 2017. PartitionFinder 2: new methods for selecting partitioned models of evolution for molecular and morphological phylogenetic analyses. Molecular Biology and Evolution, **34**: 772-773.

NG, P. K. L., D. GUINOT & P. J. F. DAVIE, 2008. Systema Brachyurorum: Part I. An annotated checklist of extant brachyuran crabs of the world. Raffles Bulletin of Zoology, (Supplement) **17**: 1-296.

OKANO, T., H. SUZUKI & T. MIURA, 2000. Comparative biology of two Japanese freshwater crabs *Geothelphusa exigua* and *G. dehaani* (Decapoda, Brachyura, Potamidae). Journal of Crustacean Biology, **20**: 299-308.

RONQUIST, F., J. P. HUELSENBECK, M. TESLENKO & A. A. NYLANDER, 2019. MrBayes 3.2 Manual: tutorials and Model Summaries. http://mrbayes.csit.fsu.edu/manual.php.

RONQUIST, F., M. TESLENKO, P. VAN DER MARK, D. L. AYRES, A. DARLING, `S. HÖHNA, B. LARGET, L. LIU, M. A. SUCHARD & J. P. HUELSENBECK, 2012. MRBAYES 3.2: efficient Bayesian phylogenetic inference and model choice across a large model space. Systematic Biology, **61**: 539-542.

SCHUBART, C. D., 2009. Mitochondrial DNA and decapod phylogenies: the importance of pseudogenes and primer optimization. In: J. W. MARTIN, K. A. CRANDALL & D. L. FELDER (eds.), Decapod crustacean phylogenetics. Crustacean issues, **18**: 47-65. (CRC Press, Taylor & Francis Group, Boca Raton, London, New York).

SHIH, H.-T. & V. T. DO, 2014. A new species of *Tiwaripotamon* Bott, 1970, from northern Vietnam, with notes on *T. vietnamicum* (Dang & Ho, 2002) and *T. edostilus* Ng & Yeo, 2001 (Crustacea, Brachyura, Potamidae). Zootaxa, **3764**: 26-38.

SHIH, H.-T. & P. K. L. NG, 2011. Diversity and biogeography of freshwater crabs (Crustacea: Brachyura: Potamidae, Gecarcinucidae) from east Asia. Systematics and Biodiversity, **9**: 1-16.

SHIH, H.-T., P. K. L. NG & H.-W. CHANG, 2004. Systematics of the genus *Geothelphusa* (Crustacea, Decapoda, Brachyura, Potamidae) from southern Taiwan: a molecular appraisal. Zoological Studies, **43**: 561-570.

SHIH, H.-T., D. C. J. YEO & P. K. L. NG, 2009. The collision of the Indian plate with Asia: molecular evidence for its impact on the phylogeny of freshwater crabs (Brachyura: Potamidae). Journal of Biogeography, **36**: 703-719.

STAMATAKIS, A., 2006. RAxML-VI-HPC: maximum likelihood-based phylogenetic analyses with thousands of taxa and mixed models. Bioinformatics, **22**: 2688-2690.

YEO, D. C. J. & P. K. L. NG, 2003. Recognition of two subfamilies in the Potamidae Ortmann, 1896 (Brachyura, Potamidae) with a note on the genus *Potamon* Savigny, 1816. Crustaceana, **76**: 1219-1235.

First received 26 October 2019.
Final version accepted 7 July 2020.

Decapoda, Dendrobranchiata

[When citing this volume, please refer to *Crustaceana* 93 (2020), 1233-1546]

Prof. Ruiyu Liu Memorial Issue

SICYONELLA LIUI SP. NOV., A NEW SERGESTID SHRIMP (DECAPODA, DENDROBRANCHIATA) DISCOVERED FROM MADAGASCAR

BY

TIN-YAM CHAN[1])
Institute of Marine Biology and Center of Excellence for the Oceans, National Taiwan Ocean University, Keelung 20224, Taiwan, R.O.C.

ABSTRACT

A new species of the sergestid shrimp genus *Sicyonella* Borradaile, 1910 was discovered in a recent expedition to southern Madagascar. The presence of *Sicyonella* in Madagascar is also a new genus record for the country. The new species has the petasma and thelycum very different from the three known species of the genus, and is also unique in the basal part of the mesial antennular flagellum not being modified in males.

Key words. — *Sicyonella*, new species, Madagascar

RÉSUMÉ

Une nouvelle espèce du genre de crevettes Sergestidae, *Sicyonella* Borradaile, 1910 a été découverte lors d'une expédition récente au sud de Madagascar. La présence du genre *Sicyonella* à Madagascar est aussi un nouveau signalement pour le pays. La nouvelle espèce présente un pétasma et un thélycum très différents de ceux des trois espèces connues du genre, et est aussi unique par la partie basale non modifiée du flagelle antennulaire mésial chez les mâles.

Mots clés. — *Sicyonella*, espèce nouvelle, Madagascar

INTRODUCTION

The sergestid genus *Sicyonella* Borradaile, 1910 is unique in the family by having a benthic lifestyle instead of a pelagic one. Although *Sicyonella* is widely distributed in the Indo-West Pacific, there are few reports on them (see Fukuoka et al., 2005) and currently only three species are recognized (Fukuoka et al., 2005; De Grave & Fransen, 2011; Vereshchaka et al., 2016). Burkenroad (1983) proposed a subfamily Sicyonellinae for this genus, which has not been recognized even though

[1]) e-mail: tychan@mail.ntou.edu.tw

 | DOI:10.1163/9789004464353_010

Vereshchaka (2017) treated this group as a family. The genus has been relatively recently revised by Fukuoka et al. (2005) with the known westernmost distribution being the Red Sea and Mauritius. During the ATIMO VATAE expedition organized by the Muséum national d'Histoire naturelle, Paris along southern Madagascar in 2010, several specimens of *Sicyonella* were collected and present the first record of the genus in Madagascar. Detailed examination of the Madagascan material revealed that it even represents a species new to science. The present work reports this finding and describes the new species.

MATERIAL AND METHODS

The types are deposited in the Muséum national d'Histoire naturelle, Paris (MNHN). The abbreviations "cl" represents carapace length which was measured dorsally from the posterior orbital margin to the posterior margin of the carapace; "DW" and "BP" represent the collecting gear Warén Dredge and brushing, respectively. The measurements given are the maximum length of the part measured. Morphological terminology follows Fukuoka et al. (2005) and Vereshchaka et al. (2016).

SYSTEMATICS

Genus *Sicyonella* Borradaile, 1910
Sicyonella liui sp. nov.
(figs. 1, 2)

Sicyonella sp — Robalino et al., 2016: fig. 1.

Material examined.— Holotype, Southern Madagascar, ATIMO VATAE, stn DW3609, 25°34′17.4036″S, 44°55′15.0024″E, 32 m, 13 May 2010, male 6.9 mm cl (MNHN IU-2014-8785). Paratypes: Southern Madagascar, ATIMO VATAE, stn DW3607, 25°45′43.8012″S, 44°51′58.7988″E, 40-41 m, 13 May 2010, 1 male 7.0 mm cl (MNHN IU-2010-4046, as *Sicyonella sp* in Robalino et al., 2016); stn DW3609, 25°34′17.4036″S, 44°55′15.0024″E, 32 m, 13 May 2010, 1 male 7.5 mm cl (MNHN IU-2010-4041), 1 male 7.7 mm cl (MNHN IU-2014-8784), 1 female 6.9 mm cl (MNHN IU-2014-8783); stn BP08, 25°27′18.18″S, 44°55 9.66″E, 25-26 m, 29 May 2010, 1 female 9.5 mm cl (MNHN IU-2010-4990, as *Sicyonella sp* in Robalino et al., 2016).

Description.— Integument moderately firm. Carapace (fig. 1A) with large postorbital and hepatic teeth, pterygostomial tooth distinct. Rostrum short, extending to around base of eye stalk; bearing 1 dorsal tooth above orbital margin and 1 postrostal tooth. Eye (fig. 1B) elongated and gradually wider distally, 2.4-2.8 (males 2.4-2.6, females 2.7-2.8) times as long as width of cornea; cornea large and with distinct ocellus (nebenauge), 1.9-2.2 (males 2.0-2.2, females 1.9-2.1) times as long as basal width of eye stalk. Antennular peduncle elongated; stylocerite

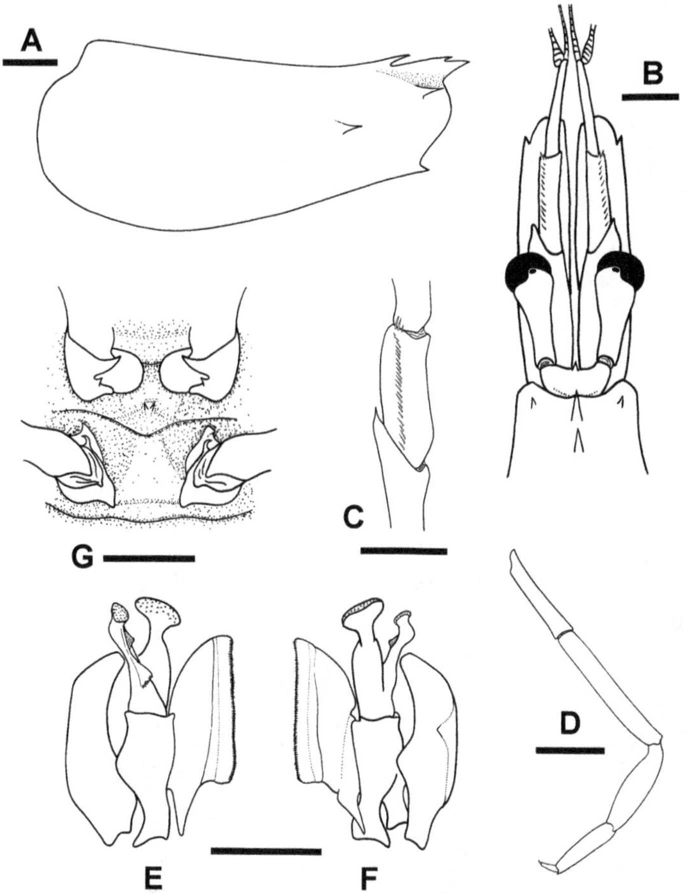

Fig. 1. *Sicyonella liui* sp. nov., southern Madagascar, ATIMO VATAE, stn DW3609, setae mostly omitted: A, C, E, F, holotype male 6.9 mm cl (MNHN IU-2014-8785); B, D, paratype male 7.7 mm cl (MNHN IU-2014-8784); G, paratype female 6.9 mm cl (MNHN IU-2014-8783). A, Carapace, lateral; B, anterior carapace and cephalic appendages, dorsal; C, left second antennular peduncle segment, dorsal; D, right pereiopod IV, lateral; E, left petasma, anterior; F, left petasma, posterior; G, thelycum, thoracic sternites VI and VII (pereiopods III and IV), ventral. Scales = 1 mm.

prominent but short, not extending to middle of first (proximal) peduncle segment; first peduncle segment extending to distal 1/5-2/5 of scaphocerite, anterolateral angle usually with distinct spine but occasionally blunt; second peduncle segment (fig. 1C) more or less reaching distal 1/10 (in both males and females) of scaphocerite, 0.4-0.6 (males 0.4-0.6, females 0.4-0.5) times as long as first peduncle segment and 2.2-3.1 (males 2.6-3.1, females 2.2-2.7) times as long as broad, armed with 2-3 spines at distodorsal margin; third peduncle segment 0.8-1.2 (males 0.9-1.2, females 0.8-0.9) times as long as second peduncle segment; distal 2 peduncle segments combined 0.7-1.1 (males 0.9-1.1, females 0.7-0.8) times as long

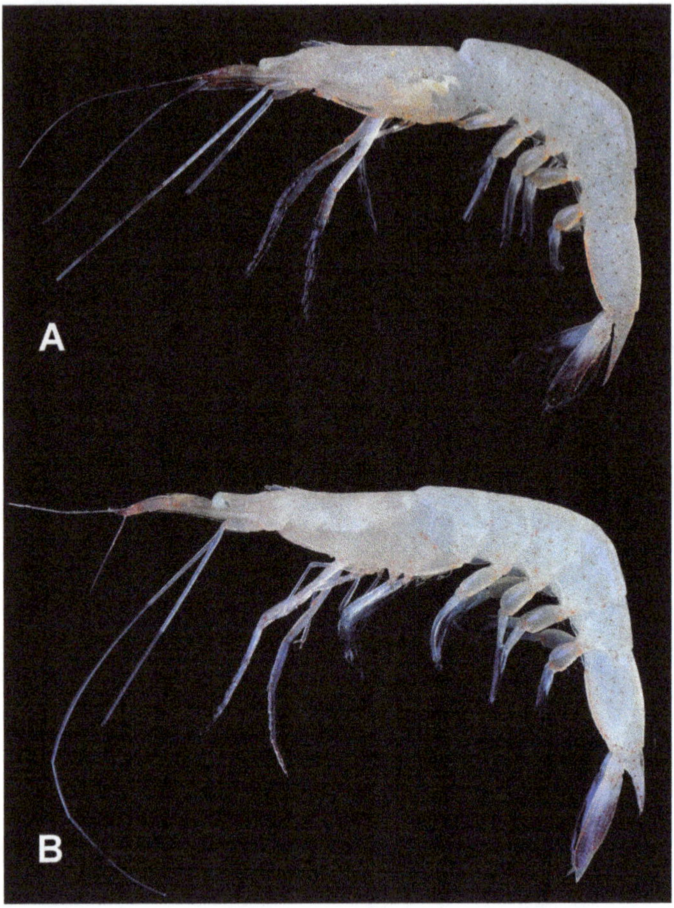

Fig. 2. *Sicyonella liui* sp. nov., southern Madagascar, ATIMO VATAE: A, stn BP08, female 9.5 mm cl (MNHN IU-2010-4990); B, stn BP16, male (MNHN, specimen not located).

as first peduncle segment. Basal part of mesial antennular flagellum similar in both sexes and simple, without any modification in males. Scaphocerite 3.7-4.8 times as long as wide, distolateral tooth distinctly falling short of distal margin of lamella.

Maxilliped III robust, much longer than all pereiopods and exceeding antennular peduncle by about full length of distal segment, distal segments spinous. Pereiopod I more or less extending to level of anterior end of carapace, chela similar length as carpus. Pereiopod II overreaching carapace by 1/3-1 chela and with carpus 1.2-1.4 times as long as chela. Pereiopod III reaching proximal 1/3 to 2/3 of scaphocerite, carpus 1.5-1.6 times as long as chela. Pereiopod IV (fig. 1D) extending to about level of anterior end of carapace, dactylus 0.3 times as long as propodus. Pereiopod V shortest, extending to level of base of hepatic spine, dactylus 0.2 times as long as propodus.

Abdomen generally smooth except posterior margin of tergite III broadly triangular with tip acute or obtuse, and posterior margin of tergite VI bearing small median tooth; pleura without any teeth; somite VI 1.8-1.9 times as long as somite V. Petasma (fig. 1E, F) with processus uncifer, processus ventralis, lobus terminalis, lobus accessories and lobus armatus, all rather simple; processus uncifer not differentiated from lamina externa (pars externa) and with rounded tip; lamina externa with long straight lateromedian margin accompanied by sublateral longitudinal ridge on posterior surface, sublateral longitudinal ridge bearing wide triangular protrusion directed mesially at about distal 2/5; processus ventralis not subdivided, with subdistal part narrowing and slightly twisted, terminal part broadened and slightly folded posteriorly, anterior surface of terminal part densely covered with minute spinules; lobus terminalis not subdivided, anterior surface densely covered with minute spinules and strongly folded posteriorly; lobus accessories small and bearing minute spinules; lobus armatus rudimentary. Thelycum (fig. 1G) with pair of small sharp to blunt median protrusions between thoracic sternites VI and VII; coxa of pereiopod III with distinct distomesial and mediomesial spines; coxa of pereiopod IV bearing hooked (directed ventrally) anterior spine and with base anteriorly expanded into wide flat triangular projection with tip recurved upwards. Anterodistal part of peduncle of pleopods I and II without seta, that of pleopod III bearing 1 short seta in males but without seta in females, that of pleopod IV generally with 1 (occasionally none) very long seta, that of pleopod V generally bearing 2 long setae, but sometimes with 1 long seta or 1 long and 3 short setae. Telson with distinct median longitudinal groove almost along entire dorsal surface, about 0.8 times as long as somite VI, bearing 4 pairs of lateral spines.

Coloration (fig. 2).— Body pinkish translucent and entirely covered with red dots. Cornea light grey. Antennal flagella covered with some red bands in females but without red bands in males.

Etymology.— The species is named after Prof. J. Y Liu for his considerable contribution to the shrimp taxonomy of China.

Distribution.— Currently only known from southern Madagascar, at depths of 25-41 m.

Remarks.— The holotype male has the maxilliped III, pereiopods II to V, and most of the antennular flagella missing. Moreover, its petasma has the right side and the pars astringens accidentally lost during the preparation of line drawings. Nevertheless, the six specimens examined are generally very similar except for a male (MNHN IU-2014-8784) with the anterolateral angle of the first antennular peduncle segment not produced into a large tooth (fig. 1B versus fig. 1C). Two paratypes (MNHN IU-2010-4046, IU-2010-4990) were used for the molecular work in Robalino et al. (2016) as "*Sicyonella sp*".

The present material from southern Madagascar has the petasma and thelycum very different from those of the three known species of *Sicyonella*. The petasma of the Madagascan species (fig. 1E, F) is rather simple for the genus and similar to that of *S. inermis* (Paulson, 1875) (see Fukuoka et al., 2005: fig. 9k). However, the petasma of *S. inermis* has the pars media not divided and with a distinct processus uncifer (versus pars media divided into two long branches and processus uncifer indistinct). The petasma of the other two known species, *S. antennata* Hansen, 1919 and *S. maldivensis* Borradaile, 1910, are quite intricate with the processus ventralis subdivided into many branches and bearing a well-developed lobus armatus (see Fukuoka et al., 2005: figs. 3J, 4F, 14J, K, 15E). The petasma of *S. liui* sp. nov. has the processus ventralis not subdivided and the lateral margin of the branch bearing the lobus terminalis slightly protruded at a subdistal position, probably representing a rudimentary lobus armatus. Furthermore, the posterior surface of the lamina externa is unique in *S. liui* sp. nov. in having a long sublateral longitudinal ridge bearing a broad triangular protrusion (vs. completely absent in the three other species). The thelycum of *S. liui* sp. nov. (fig. 1G), however, is the most intricate in the genus and similar to *S. antennata* in having many spines and projections (see Fukuoka et al., 2005: fig. 5E-H). Furthermore, the positions of the spines on the thelycum of *S. liui* sp. nov. are unique in the genus. No other species of *Sicyonella* (Fukuoka et al., 2005: figs. 5E, 10F, 16D) has a pair of small median protrusions between the thoracic sternites VI (pereiopod III) and VII (pereiopod IV), and bearing spines on coxae of pereiopods III and IV (no spine in *S. inermis* and *S. maldivensis*, see Fukuoka et al., 2005: figs. 10F, G, 16E-G; only 1 spine at coxa of pereiopod III in *S. antennata*, see Fukuoka et al., 2005: fig. 5E-H). Nevertheless, the flat triangular projection at the base of the coxa of the pereiopod IV in the new species is similar to *S. antennata* (see Fukuoka et al., 2005: fig. 5E, H), and there is a posterior elongated projection at the base of the pereiopod III in *S. maldivensis* (see Fukuoka et al., 2005: fig. 16E, F).

Other than the shape of the petasma and thelycum, the new species differs from the congeneric species in the absence of sexual dimorphism at the basal part of the mesial antennular flagellum and with that of the male without any other modification (fig. 1B). The basal part of the mesial antennular flagellum in males is modified into a clasping organ in the other species of the genus and rather intricate in structure in *S. antennata* and *S. maldivensis* with many stout setae (Fukuoka et al., 2005: figs. 4A, 15A). The clasping organ of *S. inermis* is rather simple but has the fourth segment greatly elongated and densely covered with fine scales (Fukuoka et al., 2005: fig. 10A-C). Furthermore, there is only a minor sexual dimorphism in the distal two antennular peduncle segments in *S. liui* sp. nov., with those of males slightly longer than females. Therefore, this character alone cannot be used to separate the new species from *S. inermis* and *S. maldivensis* (see Fukuoka et al., 2005). The presence of two or three distodorsal spines on the second antennular peduncle segment in *S. liui* sp. nov. (fig. 1B, C) differs from that of *S. antennata* (no distodorsal spines, Fukuoka et al., 2005: fig. 1A, B) but is also similar to *S. inermis* and *S. maldivensis* (Fukuoka et al., 2005: figs. 7D, 12D). The eyes of the *S. liui* sp. nov. (2.4-2.8 times as long as cornea, cornea 1.9-2.2 times wider than base of eye stalk, fig. 1B) is more or less as large as those of *S. inermis* and *S. maldivensis* (2.0-2.9 times as long as cornea, cornea 1.8-2.6 times wider than base of eye stalk, Fukuoka et al., 2005: figs. 7A, B, 12A-C), but smaller than *S. antennata* (1.9-2.3 times as long as cornea, cornea 2.0-2.9 times wider than base of eye stalk, Fukuoka et al., 2005: fig. 1A-C).

The other two characters proposed by Fukuoka et al. (2005) for separating the species of *Sicyonella*, such as the color of antennal flagella and the setae at the anterodistal part of pleopod peduncles are found to be rather variable in the new species. For example, the antennal flagella have red bands in a female (fig. 2A) but are not banded in a male (fig. 2B) for *S. liui* sp. nov.

ACKNOWLEDGEMENTS

The ATIMO VATAE expedition was part of a cluster of Mozambique-Madagascar expeditions (La Planète Revistée/Our Planet Reviewed) funded by the Total, Prince Albert II of Monaco, and Stavros Niarchos foundations, and conducted by the Muséum national d'Histoire naturelle, Paris (MNHN, P. Bouchet principal investigator) and Pro-Natura International (PNI). The expedition was undertaken in partnership with Institut d'Halieutique et des Sciences Marines, University of Toliara, and the Madagascar bureau of the Wildlife Conservation Society. Sincere thanks are extended to L. Corbari and P. Martin-Lefevre (MNHN) for loaning the Madagascan material for this study and to C.C. Lin of National Taiwan Ocean University for preparing the line drawings. This work was supported by grants from the Ministry of Science and Technology, Taiwan, R.O.C. and the Center of Excellence for the Oceans (National Taiwan Ocean University), which is financially supported from The Featured Areas Research Center Program within the framework of the Higher Education Sprout Project by the Ministry of Education in Taiwan, R.O.C.

REFERENCES

BORRADAILE, L. A., 1910. The Percy Sladen Trust Expedition to the Indian Ocean in 1905, under the leadership of Mr. J. Stanley Gardiner. No. X. Penaeidea, Stenopodidea, and Reptantia from the western Indian Ocean. Transactions of the Linnean Society of London (2) Zoology, **13**: 257-264, pl. 16.

BURKENROAD, M. D., 1983. Natural classification of Dendrobranchiata, with a key to recent genera. In: F. R. SCHRAM (ed.), Crustacean issues I: crustacean phylogeny: 279-290. (A.A. Balkema, Rotterdam).

DE GRAVE, S. & C. H. J. M. FRANSEN, 2011. Carideorum catalogus: the recent species of the dendrobranchiate, stenopodidean, procarididean and caridean shrimps (Crustacea: Decapoda). Zoologische Mededelingen, **85**: 195-589.

FUKUOKA, K., M. TAMAKI & T. KIKUCHII, 2005. The redescription of three species of *Sicyonella* (Crustacea: Decapoda: Dendrobranchiata: Sergestidae). Zootaxa, **833**: 1-31.

HANSEN, H. J., 1919. The Sergestidae of the Siboga Expedition. Siboga-Expeditie Monograph, **38**: 1-65, pls 1-7.

PAULSON, O., 1875. Studies on the Crustacea of the Red Sea with notes regarding other seas. Part 1 Podophthalmata and Edriophthalmata (Cumacea): i-xiv, 1-144, pls. 1-22. (Kiev). [in Russian.]

ROBALINO, J., B. WIKINS, H. D. BRACKEN-GRISSOM, T. Y. CHAN & M. A. O'LEARY, 2016. The origin of large-bodied shrimp that dominate modern global aquaculture. PLoS ONE, **11**(7): 1-24.

VERESHCHAKA, A. L., A. A. LUNINA & J. OLESEN, 2016. Phylogeny and classification of the shrimp genera *Acetes*, *Peisos*, and *Sicyonella* (Sergestidae: Crustacea: Decapoda). Zoological Journal of the Linnean Society, **177**: 353-377.

VERESHCHAKA, A. L., 2017. The shrimp superfamily Sergestoidea: a global phylogeny with definition of new families and an assessment of the pathways into principal biotopes. Royal Society Open Science, **4**: e170221 (11 pp.).

First received 9 July 2019.
Final version accepted 25 November 2019.

Decapoda, Caridea

[When citing this volume, please refer to *Crustaceana* 93 (2020), 1233-1546]

Prof. Ruiyu Liu Memorial Issue

DESCRIPTION OF A NEW SPECIES OF THE GENUS *LEBBEUS* WHITE, 1847 (DECAPODA, THORIDAE) FROM THE BOHAI SEA, CHINA

BY

YANRONG WANG[1,2,5]), CHAODONG ZHU[1,5]) and ZHONGLI SHA[2,3,4,5,6])

[1]) Institute of Institute of Zoology, Chinese Academy of Sciences, Beijing 100101, China
[2]) Institute of Oceanology, Chinese Academy of Sciences, Qingdao 266071, China
[3]) Laboratory for Marine Biology and Biotechnology, Qingdao National Laboratory for Marine Science and Technology, Qingdao, China
[4]) Center for Ocean Mega-Science, Chinese Academy of Sciences, Qingdao 266071, China
[5]) University of Chinese Academy of Sciences, Beijing 100049, China

ABSTRACT

A new species of the caridean family Thoridae, *Lebbeus liui* sp. nov., is described from the Bohai Sea, China. It belongs to the group characterized by the presence of epipods on the third maxilliped to third pereiopod. The new species is most similar to *L. fasciatus* and *L. speciosus*. It differs from the above two species by the number of the dorsal teeth and the spacing of those on the rostrum, and by the posterior median margin of the telson not produced into a triangular tooth.

Key words. — Decapoda, Thoridae, *Lebbeus*, new species, Bohai Sea

RÉSUMÉ

Une nouvelle espèce de la famille de Caridea Thoridae, *Lebbeus liui* sp. nov., est décrite de la mer de Bohai, Chine. Elle appartient au groupe caractérisé par la présence d'épipodites sur le troisième maxillipède et jusqu'au troisième péréiopode. La nouvelle espèce est la plus proche de *L. fasciatus* et de *L. speciosus*. Elle diffère de ces deux espèces par le nombre de dents dorsales et l'espacement de celles situées sur le rostre, et par le bord médian postérieur du telson non prolongé en une dent triangulaire.

Mots clés. — Decapoda, Thoridae, *Lebbeus*, espèce nouvelle, mer de Bohai

INTRODUCTION

Lebbeus White, 1847 is the most species rich genus within the Thoridae, and currently contains 67 species (De Grave & Fransen, 2011; Komai et al., 2012; Komai, 2013; 2015; Nye, 2013; Nye et al., 2013; Matsuzaki et al., 2015;

[6]) Corresponding author; e-mail: shazl@qdio.ac.cn

 | DOI:10.1163/9789004464353_011

Schiaparelli et al., 2015; Chan & Komai, 2017). The genus is widely distributed from shallow to deeper waters (exceeding 2000 m), and from the tropics to polar seas (e.g., Crosnier, 1999; Chang et al., 2010; Nye, 2013). Furthermore, several species occur in hydrothermal vents (Komai et al., 2012; Nye et al., 2012). The genus has traditionally been divided into four species groups based on the number of the epipods (e.g., Butler, 1980; Wicksten, 1990; Hayashi, 1992a; Komai et al., 2004). Liu (2008) listed only two species of this genus (*L. balssi* Hayashi, 1992 and *L. speciosus* (Urita, 1942) from Chinese waters, but no material of the genus was available to Xu & Li (2015), who recently reviewed the Hippolytidae from the area. The present study describes a new species of *Lebbeus* on the basis of material from shallow water in the Bohai Sea, northern China. The new species, *L. liui* sp. nov., is referred to the species group characterized by the possession of strap-like epipods on the third maxilliped through to the third pereiopod. Characters differentiating the new species from the allied species are discussed.

MATERIAL AND METHODS

Specimens were collected from the Bohai Sea, northern China, during the National Comprehensive Oceanographic Survey (1958-1960), by Agassiz trawl at a depth of about 50 m. The specimens were immediately fixed in 10% formalin and then deposited in the Marine Biological Museum, Chinese Academy of Sciences (MBMCAS), Qingdao, China, and subsequently preserved in 75% ethanol. Specimens were examined and dissected under a stereomicroscope (Nikon SMZ1500, Japan). Line drawings were completed in the Adobe Photoshop CS6, with a graphics tablet. Post-orbital carapace length (POCL) was measured from the posterior margin of the orbit to the posterior margin of the carapace and is used herein as an indication of specimen size. All measurements are in millimeters (mm).

SYSTEMATICS

Family THORIDAE Kingsley, 1879
Genus *Lebbeus* White, 1847
Lebbeus liui sp. nov.
(figs. 1-2)

Material examined.— Holotype: MBM 081122, female, POCL 7 mm, Sta. 1098 (121.175°E, 38.66°N), Bohai Sea, depth 53 m, 28 January 1959. Paratype: MBM 270831, female, POCL 6.5 mm, Sta. 1098 (121.175°E, 38.66°N), Bohai Sea, depth 51.5 m, 1 July 1959.

Description.— Rostrum (fig. 1A, C, D) styliform, with well-developed ventral blade, distinctly overreaching distal end of antennular peduncle, nearly as long

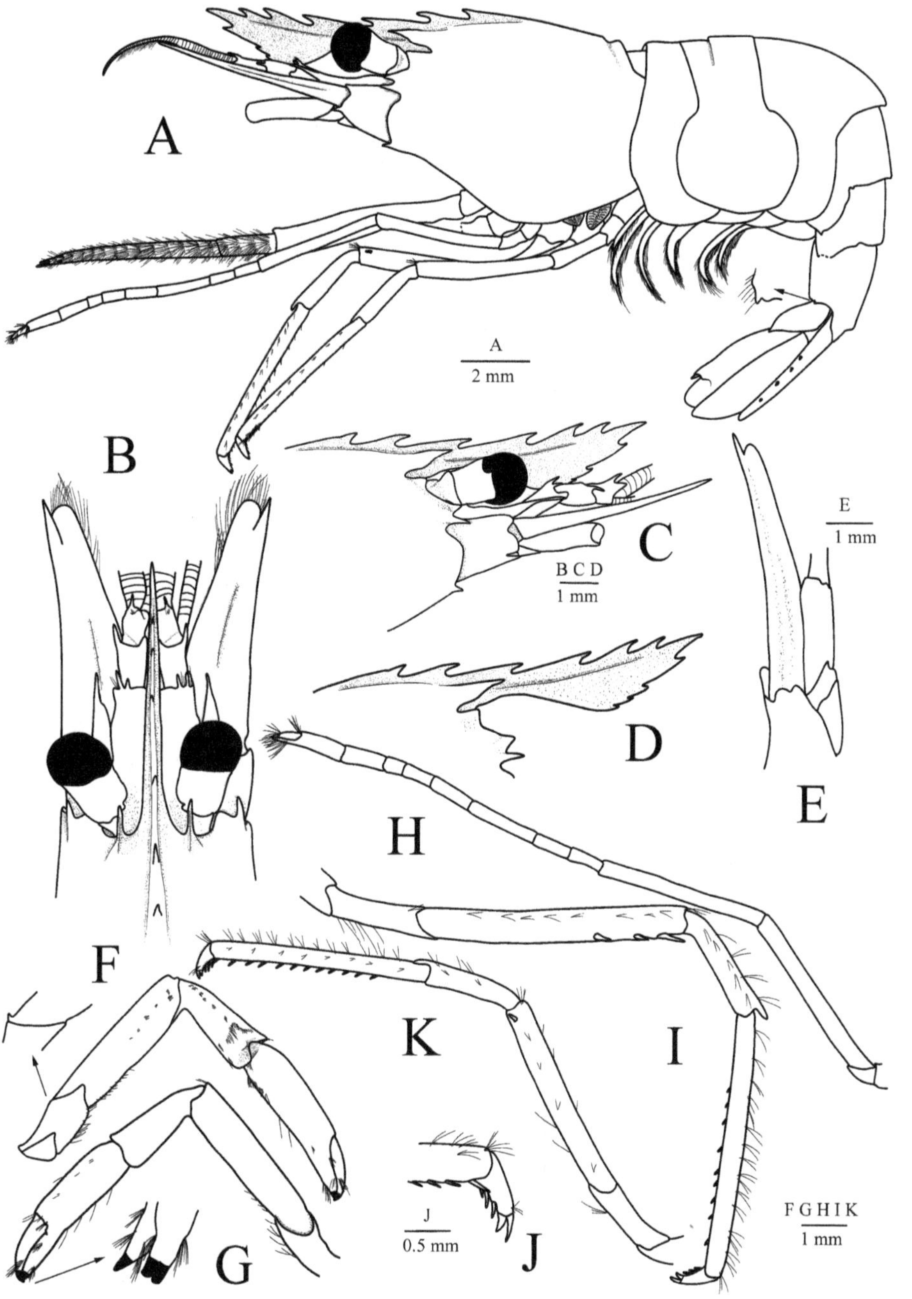

Fig. 1. *Lebbeus liui* sp. nov. Holotype, MBM 081122, female, CL 7 mm. A, Entire animal, lateral view; B, anterior part of carapace, dorsal view; C, same, lateral view; D, rostrum, lateral view; E, antenna; F, first pereiopod, mesial side (base of merus enlarged); G, same, lateral side (tip of fingers in dorsal view enlarged); H, left second pereiopod; I, right third pereiopod; J, dactylus of third pereiopod; K, left fourth pereiopod.

as carapace; 6 dorsal teeth, including 4 teeth on rostrum proper; ventral margin armed with 4 teeth in distal 0.4. Carapace (fig. 1A, C) with low, blunt post-rostral carina extending nearly to mid-length of carapace; posteriormost post-rostral tooth arising at about 0.2 of carapace length; supraorbital tooth moderately small, arising at level of rostral base, reaching tip of suborbital lobe, but not reaching tip of antennal tooth; orbital margin generally concave, with deep U-shaped notch below base of supraorbital tooth; suborbital lobe well developed, triangular, not reaching distal end of antennal tooth; anterolateral margin between antennal and pterygostomial teeth strongly sinuous with deep excavation just inferior of antennal tooth; pterygostomial tooth small.

Pleon (figs. 1A, 2B) dorsally rounded. Second somite with shallow transverse groove on tergum. Pleura of anterior three somites broadly rounded; fourth pleuron with small posteroventral tooth; fifth pleuron with strong posteroventral tooth. Sixth somite about 1.7 times as long as fifth somite, bearing posteroventral tooth; posterolateral process terminating in small sharp teeth. Telson (fig. 2C, J) about 1.4 times longer than sixth somite; tapering posteriorly to convex posterior margin (tip of telson damaged in holotype), bearing 4 dorsolateral spines on each side; posterior margin narrow, with 2 pairs of lateral spines (mesial pair about 5 times longer than lateral pair) and 2 median spiniform plumose setae, setae slightly shorter than lateral spines.

Eye sub-pyriform with stalk slightly narrowing proximally; cornea not particularly dilated, its maximum width 0.19 of carapace length; ocellus absent.

Antennular peduncle (fig. 1B, C) slightly overreaching midlength of antennal scale. First segment distinctly longer than distal two segments combined, reaching about 1/3 length of antennal scale, dorso-distal margin armed with 2 small teeth; stylocerite slightly overreaching distolateral angle of first peduncular segment (not including teeth), sharply pointed, mesial margin sinuous. Second segment 0.4 length of first segment, with 1 dorsolateral distal tooth. Third segment more than half as long as second segment, bearing 1 dorso-distal tooth.

Antenna (fig. 1B, C, E) with basicerite bearing moderately large ventrolateral tooth. Carpocerite reaching about midlength of antennal scale. Antennal scale about as long as carapace and 3.8 times longer than wide; lateral margin slightly concave; distolateral tooth distinctly overreaching rounded distal margin of narrow lamella.

Mouthparts typical for genus. Mandibles (fig. 2D) with two segmented, slender palp and slender incisor process with 4 distal teeth; molar process large, setose. Maxillule (fig. 2E) with shallowly bilobed palp, lateral lobe tipped with plumose seta; mesial lobe with slender, spiniform seta; distal endite oval, densely armed with spinules and setae on mesial margin; basal endite lost during dissection. Maxilla (fig. 2F) with palp tipped by a long spiniform seta and a short plumose

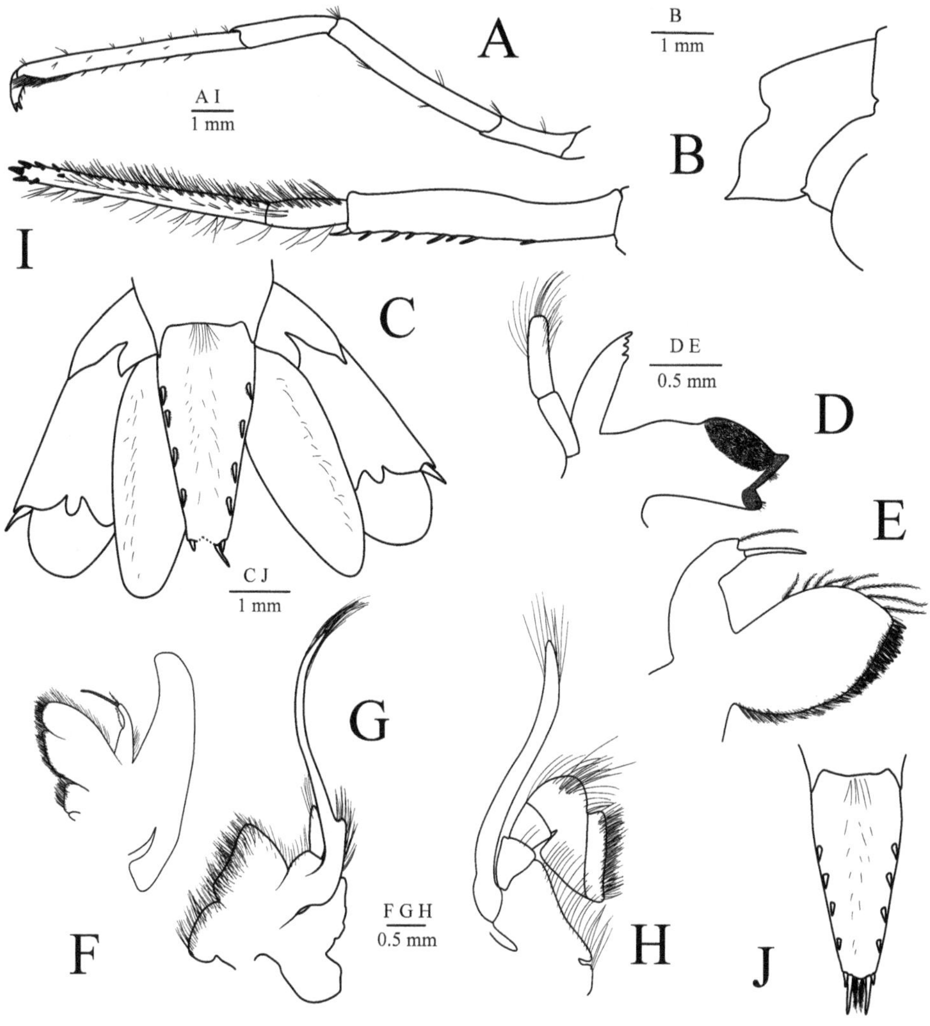

Fig. 2. *Lebbeus liui* sp. nov. Holotype, MBM 081122, female, CL 7 mm. A, Left fifth pereiopod; B, fourth to fifth pleon, lateral view; C, telson and uropods, dorsal view; D, left mandible; E, left maxillule; F, left maxilla; G, left first maxilliped; H, left second maxilliped; I, left third maxilliped. Paratype, MBM 270831, female, CL 6.5 mm. J, Telson, dorsal view.

seta; basal endite bilobed, mesial margin densely setose. Coxal endite small, with densely setose mesial margin; scaphognathite fringed with setae. First maxilliped (fig. 2G) with two segmented palp, distal article subequal to proximal article; caridean lobe well developed and fringed with long setae; exopod long, tipped with setae; endites with densely setae on mesial margin; epipod large, bilobed. Second maxilliped (fig. 2H) with ultimate article densely setose on mesial margin; exopod well developed, tipped with long setae. Third maxilliped (fig. 2I) slender, overreaching distal end of antennal scale by half length of ultimate segment;

ultimate segment with 7 darkly pigmented corneous spinules on distal margin; penultimate segment slightly shorter than 1/3 length of ultimate segment, only with simple long setae; antepenultimate segment almost as long as ultimate segment, armed with 1 and 7 slender, spine-like setae on distolateral margin and ventrolateral margin respectively.

Strap-like, terminally hooked epipods present on third maxilliped to third pereiopod, corresponding setobranchs present on first to fourth pereiopod.

First pereiopod (fig. 1F, G) moderately stout; fingers about 0.6 times as long as palm, dactylus terminating in 2 darkly pigmented corneous claws and fixed finger terminating in single darkly pigmented corneous claw; carpus slightly shorter than palm, inner margin bearing six tufts of setae and one transverse row of setae; merus bearing six tufts of setae on mesial surface, proximolateral with a small tubercle.

Second pereiopod (fig. 1H) slender, far overreaching antennal scale; fingers distinctly shorter than palm; carpus divided into 7 articles. Third to fifth pereiopods moderately long and slender, similar and slightly decreasing in length posteriorly.

Third pereiopod (fig. 1I, J) with dactylus moderately stout, about 0.16 times as long as propodus, terminating in acute, darkly pigmented unguis, armed with 4 darkly pigmented accessory spinules on flexor margin over entire length, distalmost accessory spinule distinctly larger than others, making tip of dactylus appearing biunguiculate; propodus subequal in length to merus, ventral margin bearing 9 small spines; carpus less than half length of propodus; merus armed with 3 ventrolateral spines. Fourth pereiopod (fig. 1K) nearly reaching distal margin of antennal scale; merus armed with 1 ventrolateral spine. Fifth pereiopod (fig. 2A) not overreaching antennal scale; merus unarmed.

Uropod (fig. 2C) distinctly longer than telson. Exopod slightly shorter than endopod, outer margin almost straight, diaeresis distinct, armed with one movable spine. Second pleopod of female only bearing appendix interna.

Etymology.— The species is named in honour of the late Professor Ruiyu Liu (J. Y. Liu) of the Institute of Oceanology, Chinese Academy of Sciences, for his great contributions to Chinese carcinology. The name is a noun in the genitive singular.

Remarks.— *Lebbeus liui* sp. nov. is characteristic in the species group in combination of the following female characters: the relatively long, styliform rostrum distinctly overreaching the distal end of the antennular peduncle, with a well-developed, deep ventral blade; and the relatively narrow antennal scale with the distolateral spine distinctly overreaching the distal lamella. There are 40 species belonging to this species group, including the new species described herein (shown in table I). Except the new species, the following 10 species are also known from shallow waters (<100 m): *L. catalepsis* Jensen, 1987, *L. comanthi*

TABLE I

Distribution and depth range the species belonging to the species group characterized by the possession of epipods from the third maxilliped to the third pereiopod (based on Komai et al., 2004)

Species	Distribution	Depth range
L. acudactylus Jensen, 2006	Northeastern Pacific	155-166 m
L. africanus Fransen, 1997	Southwestern Africa	1500 m
L. antarcticus (Hale, 1941)	Antarctica	640 m
L. armatus (Owen, 1839)	Japan	138-338 m
L. bidentatus Zarenkov, 1976	Eastern Pacific off Chile	1680 m
L. liui sp. nov.	Bohai Sea, China	32 m
L. carinatus Zarenkov, 1976	Eastern Pacific off Peru	1680-1860 m
L. catalepsis Jensen, 1987	Washington State	low intertidal
L. comanthi Hayashi & Okuno, 1997	Pacific coast of Japan	60 m
L. cristatus Ahyong, 2010	New Zealand	1231-1226 m
L. eludus Jensen, 2006	Northeastern Pacific	14 m
L. fasciatus (Kobyakova, 1936)	Northwestern Pacific	3-32 m
L. formosus Chang, Komai & Chan, 2010	Taiwan	1260-1275 m
L. groenlandicus (Fabricius, 1798)	North Atlantic, North Pacific	90-300 m
L. indicus Holthuis, 1947	Indonesia	1018 m
L. ketophilos Nye, 2013	Southwest Indian	732 m
L. lagunae (Schmitt, 1921)	California	0-55 m
L. lamina Komai, 2013	Japan	1988-2007 m
L. magnificus Komai, 2015	Japan	156 m
L. microceros (Krøyer, 1841)	Northwestern Atlantic	10-140 m
L. miyakei Hayashi, 1992	Japan	30-40 m
L. mundus Jensen, 2006	Northeastern Pacific	9-23 m
L. nudirostris Komai & Takeda, 2004	Japan	250 m
L. pacmanus Komai, Tsuchida & Segonzac, 2012	Pacific	1662 m
L. polyacanthus Komai, Hayashi & Kohtsuka, 2004	Japan	400 m
L. profundus (Rathbun, 1906)	Hawaii	762-1000 fms
L. rubrodentatus Bruce, 2010	Australian	360-396 m
L. saldanhae (Barnard, 1947)	South Africa	145 fms
L. schrencki (Bražnikov, 1907)	North Pacific	9-134 m
L. shinkaiae Komai, Tsuchida & Segonzac, 2012	Pacific	691-1474 m
L. similior Komai & Komatsu, 2009	Northern Japan	1196 m
L. speciosus (Urita, 1942)	Northern North Pacific	0-57 m
L. spinirostris (Kobyakova, 1936)	Sea of Okhotsk	182 m
L. spongiaris Komai, 2001	Pacific coast of Japan	257-264 m
L. thermophiles Komai, Tsuchida & Segonzac, 2012	Pacific	1512-1842 m
L. tosaensis Hanamura & Abe, 2003	Pacific coast of Japan	250 m
L. uschakovi (Kobjakova, 1936)	Sea of Okhotsk	185 m
L. virentova Nye, Copley, Plouvier & Van Dover, 2013	Caribbean Sea	2294 m
L. washingtonianus (Rathbun, 1902)	Northeastern Pacific	820-1808 m
L. wear Ahyong, 2009	New Zealand	1208-1324 m

Hayashi & Okuno, 1997, *L. eludus* Jensen, 2006, *L. fasciatus* (Kobyakova, 1936), *L. lagunae* (Schmitt, 1921), *L. microceras* (Krøyer, 1841), *L. miyakei* Hayashi, 1992, *L. mundus* Jensen, 2006, *L. schrencki* (Bražnikov, 1907) and *L. speciosus* (Urita, 1942). The new species differs from *L. catalepsis*, *L. comanthi*, *L. eludus*, *L. lagunae*, *L. mundus* and *L. schrencki* by the rostrum being styliform rather than spiniform. It is distinguished from *L. microceras* and *L. miyakei* by the rostrum overreaching the distal end of the antennular peduncle rather than at most reaching the end of the first segment. Moreover, the new species has the fourth abdominal segment with a marginal posterior tooth on the pleura which is absent in *L. miyakei*.

L. liui sp. nov. is also morphologically close to *L. ushakovi* (Kobyakova, 1936). Nevertheless, the rostrum is barely reaching the distal end of the second segment of the antennular peduncle, and it is armed with 9 dorsal teeth in *L. uschakovi*. In the general shape of the rostrum, the new species is most similar to *L. fasciatus* and *L. speciosus* which both have a longer rostrum.

The antennal scale of new species is moderately narrow and the distolateral tooth distinctly reaches beyond the distal margin of the lamella, while in *L. speciosus* the antennal scale is relatively wider (3.0 times as long as wide) with the distolateral tooth just reaching the distal margin of the lamella. In the new species, the rostrum is armed with 6 dorsal teeth rather than at least 7 teeth, and the space between the dorsal teeth it is larger than in of *L. speciosus*. The ultimate segment of the third maxilliped bears 7 apical and subapical, darkly pigmented corneous spinules in the new species, while in *L. speciosus* it bears 12 (6 movable, 6 immovable) spinules. The posterior margin of the telson is slightly convex medially in the new species, rather than produced to be a triangular tooth in *L. speciosus* (see Urita, 1942; Hayashi, 1992a, 1993; Fransen, 1997; Wicksten, 2012; Lee et al., 2013).

The rostrum distinctly overreaches the distal end of the antennular peduncle in the new species, while in *L. fasciatus* it does not reach that end. The rostrum bears 6 dorsal teeth rather than at least 7 teeth in *L. fasciatus*. The posterior margin of the telson is slightly convex medially in the new species, rather than produced in an acutely pointed triangular tooth in *L. fasciatus*.

Distribution.— Presently only known from the type location, Bohai Sea, Northern China, at a depth of 51-53 m.

ACKNOWLEDGEMENTS

This work was supported by the National Key R&D Program of China (2018YFD0900804), the Key Research Program of Frontier Sciences, Chinese Academy of Sciences (QYZDB-SSWDQC036), and partly supported by the project of S & T basic work from the Ministry of Science and Technology of China (2015FY210300).

REFERENCES

AHYONG, S. T., 2010. New species and new records of Caridea (Hippolytidae: Pasiphaeidae) from New Zealand. Zootaxa, **2372**: 341-357.

BRAŽNIKOV, V., 1907. Matériaux pour servir à la connaissance de la faune des mers russes de l'Est rassemblés par le schooner "Storož" en 1899-1902. Mémoires de l'Académie Impériale des Sciences de St-Pétersbourg. Classe des Sciences Physiques et Mathématiques, **20**(8): i-ii, 1-185, unumbered plate. [In Russian.]

BRUCE, A. J., 2010. *Lebbeus rubrodentatus* sp. nov. (Crustacea: Caridea: Hippolytidae) from the Australian North West Shelf. The Beagle, Records of the Museums and Art Galleries of the Northern Territory, **26**: 75-77.

BUTLER, T. H., 1980. Shrimps of the Pacific coast of Canada. Canadian Bulletin of Fisheries and Aquatic Sciences, **202**: 1-280.

CHAN, T. Y. & T. KOMAI, 2017. A new shrimp species of the genus *Lebbeus* White, 1847 (Crustacea: Decapoda: Caridea: Thoridae) from a deep-sea cold seep site off southwestern Taiwan. Zootaxa, **4238**(3): 426-432.

CHANG, S. C., T. KOMAI & T. Y. CHAN, 2010. First record of the hippolytid shrimp genus *Lebbeus* White, 1847 (Decapoda: Caridea) from Taiwan, with the description of three new species. Journal of Crustacean Biology, **30**: 727-744.

CROSNIER, A., 1999. Une nouvelle espèces de *Lebbeus* d'Indonésie (Crustacea, Decapoda, Caridea, Hippolytidae). Zoosystema, **21**: 453-460.

DE GRAVE, S. & C. H. J. M. FRANSEN, 2011. Carideorum catalogus: the recent species of the dendrobranchiate, stenopodidean, procarididean and caridean shrimps (Crustacea: Decapoda). Zoologische Mededelingen, Leiden, **89**(5): 195-589.

FABRICIUS, J. C., 1775. Systema Entomologiae, sistens Insectorum Classes, Ordines, Genera, Species, adjectis Sysnonymis, Locis, Descriptionibus, Observationibus: 832 pp. (Officina Libraria Kortii, Flensburgi et Lipsiae).

FRANSEN, C. H. J. M., 1997. *Lebbeus africanus* spec. nov., a new shrimp (Crustacea, Decapoda, Caridea, Hippolytidae) from Mauritanian waters, with redescriptions of four other species in the genus. Zoologische Mededelingen, Leiden, **71**(20): 231-260.

HALE, H. M., 1941. Decapod Crustacea. Report of the British Australia and New Zealand Antarctic Research Expedition, 1929-1931. Reports — Series B (Zoology and Botany), **4**: 257-286.

HANAMURA, Y. & Y. ABE, 2003. *Lebbeus tosaensis*, a new hippolytid shrimp (Decapoda, Caridea, Hippolytidae) from southwestern Japan. Biogeography, **5**: 17-24.

HAYASHI, K. I., 1992a. Studies on the hippolytid shrimps from Japan VIII. The genus *Lebbeus* White. The Journal of Shimonoseki University of Fisheries, **40**(3): 107-138.

HAYASHI, K. I., 1992b. Prawns, shrimps and lobsters from Japan (68). Family Hippolytidae-Genus *Lebbeus* 2. Aquabiology, **14**(6): 436-439.

HAYASHI, K. I., 1993. Prawns, shrimps and lobsters from Japan (69). Family Hippolytidae-Genus *Lebbeus* 3. Aquabiology, **15**(1): 6-9.

HAYASHI, K. I. & J. OKUNO, 1997. Two associated hippolytids, *Lebbeus comanthi* sp. nov. and *Lebbeus balssi* Hayashi (Decapoda, Caridea, Hippolytidae) from Japan. Journal of National Fisheries University, **46**(1): 47-56.

HOLTHUIS, L. B., 1947. The Decapoda of the Siboga Expedition. Part IX. The Hippolytidae and Rhynchocinetidae collected by the Siboga and Snellius expeditions with remarks on other species. Siboga Expeditie, **39a^{8}**: 1-100.

JENSEN, G. C., 1987. A new species of the genus *Lebbeus* (Caridea: Hippolytidae) from the northeastern Pacific. Bulletin of the Southern California Academy of Sciences, **86**(2): 89-94.

JENSEN, G. C., 2006. Three new species of *Lebbeus* (Crustacea: Decapoda: Hippolytidae) from the Northeastern Pacific. Zootaxa, **1383**: 23-43.

KOBJAKOVA, Z. I., 1936. Zoogeographical review of the Decapoda fauna from the Okhotsk Sea and the Sea of Japan. Trudy Lendingradskogo Obschestva Estesvoipytatelei, **65**: 185-228. [In Russian.]

KOMAI, T., 2001. *Lebbeus spongiaris*, a new species of deep-water shrimp (Crustacea: Decapoda: Caridea: Hippolytidae) from Izu Islands, Japan. Natural History Research, **6**(2): 57-65.

KOMAI, T., 2013. A new species of the hippolytid shrimp genus *Lebbeus* (Crustacea: Decapoda: Caridea) from lower bathyal zone in the Izu Islands, central Japan. Natural History Research, **12**(2): 81-89.

KOMAI, T., 2015. Reinstatement and redescription of *Lebbeus armatus* (Owen, 1839), long synonymized with *L. groenlandicus* (Fabricius, 1775), and description of one new species from the southwestern Sea of Okhotsk, Hokkaido, Japan (Crustacea: Decapoda: Caridea: Thoridae). Zootaxa, **3905**(4): 451-473.

KOMAI, T., K. I. HAYASHI & H. KOHTSUKA, 2004. Two new species of the shrimp genus *Lebbeus* White from the Sea of Japan, with redescription of *Lebbeus kuboi* Hayashi (Decapoda: Caridea: Hippolytidae). Crustacean Research, **33**: 103-125.

KOMAI, T. & H. KOMATSU, 2009. Deep-sea shrimps and lobsters (Crustacea: Decapoda: Penaeidae, Caridea, Polychelidae) from northern Japan, collected during the project "Research on deep-sea fauna and pollutants off Pacific coast of Northern Honshu, Japan 2005-2008". National Museum of Nature Science Monographs, **39**: 495-580.

KOMAI, T. & M. TAKEDA, 2004. A new hippolytid shrimp of the genus *Lebbeus* White, 1847 from the Sagami-Nada Sea, Central Japan, with further records of two little-known species (Crustacea: Decapoda: Caridea). Bulletin of the National Science Museum, Tokyo, (A), **30**(2): 77-86.

KOMAI, T., S. TSUCHIDA & M. SEGONZAC, 2012. Records of species of the hippolytid genus *Lebbeus* White, 1847 (Crustacea: Decapoda: Caridea) from hydrothermal vents in the Pacific Ocean, with descriptions of three new species. Zootaxa, **3241**: 35-63.

KRØYER, H., 1841. Udsigt over de nordiske arter af slægten *Hippolyte*. Naturhistorisk Tidsskrift, **3**: 570-579.

LEE, C., H. S. KO & K. H. LEE, 2013. New records of hippolytid shrimps, *Lebbeus speciosus* and *Lebbeus comanthi* (Crustacea: Decapoda: Caridea) from the East Sea, Korea. Anim. Syst. Evol. Divers., **29**(1): 84-91.

LIU, R. Y., 2008. Checklist of Marine Biota of China Sea: 1-717. (Science Press, Beijing).

MATSUZAKI, K., M. HIBINO & T. KOMAI, 2015. A new species of the caridean shrimp genus *Lebbeus* White, 1847 (Crustacea: Decapoda: Thoridae) from the southwestern Sea of Okhotsk, Hokkaido, Japan. Zootaxa, **4032**(3): 309-318.

NYE, V., J. COPLEY, S. PLOUVIEZ & C. L. VAN DOVER, 2012. A new species of *Lebbeus* (Crustacea: Decapoda: Caridea: Hippolytidae) from the Von Damm Vent Field, Caribbean Sea. Journal of the Marine Biological Association of the United Kingdom, **93**(3): 741-751.

NYE, V., 2013. New species of hippolytid shrimps (Crustacea: Decapoda: Caridea: Hippolytidae) from a southwest Indian Ocean seamount. Zootaxa, **3637**(2): 101-112.

OWEN, R., 1839. Crustacea. In: The zoology of Captain Beechey's Voyage to the Pacific Ocean and Bering's Strait, performed in H.M.S. Blossom in the years 1825-1828: 144 pp. (H. G. Bohn, London).

RATHBUN, M. J., 1902. Descriptions of new decapod crustaceans from the west coast of North America. Proceedings of the United States National Museum, **24**: 885-905.

RATHBUN, M. J., 1906. The Brachyura and Macrura of the Hawaiian islands. Bulletin of the United States Fish Commission, **23**: 827-930, plates 1-24.

SCHIAPARELLI, S., S. T. AHYONG & D. BOWDEN, 2015. Evidence of niche conservatism and host fidelity in the polar shrimp *Lebbeus kiae* n. sp. (Decapoda: Caridea: Thoridae) from the Ross Sea, Antarctica. Hydrobiologia, **761**(1): 45-69.

SCHMITT, W. L., 1921. The marine decapod Crustacea of California with special reference to the decapod Crustacea collected by the Unites States Bureau of Fisheries steamer "Albatross" in connection with the biological survey of San Francisco Bay during the years 1912-1913: 1-470, plates 1-50. (University of California press, Berkeley).

URITA, T., 1942. Decapod crustaceans from Saghalien, Japan. Bulletin of the Biogeographic Society of Japan, **12**: 1-78.

WICKSTEN, M. K., 1990. Key to the hippolytid shrimp of the eastern Pacific Ocean. Fishery Bulletin, **88**: 587-598.

WICKSTEN, M. K., 2012. Decapod Crustacea of the Californian and Oregonian zoogeographic provinces. Zootaxa, **3371**: 1-307.

XU, P. & X. Z. LI, 2015. Report on the Hippolytidae Bate (*sensu lato*) from China seas. Zoological Systematics, **40**(2): 107-165.

ZARENKOV, N. A., 1976. On the fauna of decapods adjacent to South America. Biologiya Morya, **5**: 8-18. [In Russian.]

First received 21 December 2018.
Final version accepted 26 August 2019.

[When citing this volume, please refer to *Crustaceana* 93 (2020), 1233-1546]

Prof. Ruiyu Liu Memorial Issue

SPECIES OF *CARIDINA NILOTICA* GROUP IN CHINA, WITH DESCRIPTION OF ONE NEW SPECIES (CRUSTACEA, DECAPODA, ATYIDAE)

BY

YIXIONG CAI[1])
National Biodiversity Centre, National Parks Board, 1 Cluny Road, Singapore 259569, Singapore

ABSTRACT

The atyid shrimps of *Caridina nilotica* species group from China are reviewed. Taxonomic diagnosis and discussions are given to three species so far known from China, viz. *Caridina gracilipes* De Man, 1892, *Caridina macrophora* Kemp, 1918 and *Caridina elongapoda* Liang & Yan, 1977. Detailed morphological description and illustrations are provided for a new species. *Caridina sheni* sp. nov. is morphologically similar to *C. gracilipes* but could be distinguished by its much longer dactylus of third and fifth pereiopods, and lower number of ventral rostral teeth.

Key words. — Freshwater shrimps, taxonomy, *Caridina sheni*, China

RÉSUMÉ

Les crevettes Atyidae du groupe d'espèces *Caridina nilotica* de Chine sont révisées. La diagnose taxonomique et des discussions sont données pour trois espèces connues jusqu'à présent de Chine, soit *Caridina gracilipes* De Man, 1892, *Caridina macrophora* Kemp, 1918 et *Caridina elongapoda* Liang & Yan, 1977. La description morphologique détaillée et des illustrations sont fournies pour une nouvelle espèce. *Caridina sheni* sp. nov. est morphologiquement similaire à *C. gracilipes* mais s'en distingue par son dactyle beaucoup plus long des troisièmes et cinquièmes péréiopodes et un plus petit nombre de dents rostrales ventrales.

Mots clés. — Crevettes d'eau douce, taxonomie, *Caridina sheni*, Chine

INTRODUCTION

Perhaps no species of *Caridina* sensu lato have caused as much taxonomic confusion as members of the *Caridina nilotica* species group. Yeo et al. (1999) listed the species of *Caridina* described that are relevant to *C. nilotica* (P. Roux,

[1]) e-mail: cai_yixiong@nparks.gov.sg

 | DOI:10.1163/9789004464353_012

1833), and briefly discussed problems associated with the taxonomy of this species complex. They refrained from the description of new species due to uncertain validity associated with morphologically alike forms. Since then, our understanding of this species complex has been progressing, although slowly, thanks to several efforts made to redescribe problematic species based on original type material or/and material from type localities, e.g. Richard & Clark (2005) on *C. nilotica*, Cai et al. (2006) on *C. grandirostris* Stimpson 1860 and *C. leucosticta* Stimpson, 1860, Cai & Shokita (2006) on *C. gracilipes* De Man, 1892, Cai et al. (2007) on *C. peninsularis* Kemp, 1918, Richard & Clark (2009) on *C. africana* Kingsley 1883, *C. nilotica natalensis* Bouvier, 1925, Richard & Clark (2010) on *C. brachydactyla* De Man, 1908 and Cai (2014) on *C. macrophora* Kemp, 1918.

To date, only three species of the *Caridina nilotica* species group have so far been reported from China, i.e., *Caridina gracilipes*, *C. macrophora*, and *C. elongapoda* Liang & Yan, 1977. Kemp (1918) reported *C. nilotica gracilipes* from Shanghai ditches. Yu (1936) reported *Caridina nilotica gracilipes* De Man, 1892 from Hai-kiu-sche (Hai Kou), Hainan Island, and commented that no diversity could be found in their characteristics compared to those inhabiting Shanghai although the habitats are quite different: the specimens recorded by Kemp were obtained in freshwater whereas those captured from Hainan in pure salt water. Yu (1938) subsequently briefly described *Caridina nilotica gracilipes* based on specimens from Hainan, and included specimens from "Chefoo" (Yantai, Shandong) and "Amoy" (Xiamen, Fujian) under this species. Liang & Yan (1977) described a new subspecies *Caridina nilotica elongapoda* from Fujian, which Liang (2004) synonymized under *Caridina longirostris* H. Milne Edwards, 1837 [in H. Milne Edwards, 1834-1840]. Liang & Yan (1983) added *Caridina macrophora* to the Chinese fauna, with material found in Hainan and Guangxi. Cai (2014) redescribed *C. macrophora* in detail and resurrected *C. elongapoda* as a valid species.

The aim of the present study is to review and update the status of the *Caridina nilotica* group species found in China based on all available specimens. Diagnosis and taxonomic discussion are provided for previously recorded species, detailed description and illustrations are provided for one new species. Specimens examined are deposited in the National Zoological Museum of China, Institute of Zoology, Chinese Academy of Sciences, Beijing (IZCAS); Beijing Natural History Museum, Beijing (BNHM); Shanghai Ocean University (ex Shanghai Fisheries University), Shanghai (SFU); National Taiwan Ocean University, Keelung, Taiwan (NTOU); the Zoological Reference Collection, Lee Kong Chian Natural History Museum (ex Raffles Museum of Biodiversity Research), National University of Singapore (ZRC); Muséum national d'Histoire naturelle, Paris (MNHN); Naturalis Biodiversity Centre, Leiden, (ex Rijksmuseum van Natuurlijke Historie, RMNH). The abbreviation cl is used for carapace length measured in mm from the post-orbital margin to the posterior margin of the carapace.

SYSTEMATIC ACCOUNT

Family ATYIDAE De Haan, 1849
Caridina H. Milne Edwards, 1837

Caridina gracilipes De Man, 1892
(figs. 1-3)

Caridina wyckii var. *gracilipes* De Man, 1892: 387, pl. 24: fig. 29-29e [type localities: Sulawesi (Celebes), and Selajar, Indonesia].
Caridina nilotica gracilipes — De Man, 1908: 270, fig. 7a, b; Yu, 1936: 88; 1938: 277, figs. 1, 2; Ueno, 1935: 272, fig. 2; Yu, 1974: 52, figs. 2, 3; Liang & Zheng, 1988: 15; Liang & Zhou, 1993: 231; Liang, 2004: 198 (part).
Non *Caridina nilotica gracilipes* — Kemp, 1918: 275; Liu, 1955: 28, figs. 11-17; Liang, 2004: 198 (part), fig. 95 (= *Caridina sheni* sp. nov.).
Caridina acuticaudata Dang, 1975: 70, fig. 4 [type locality: Hoa Binh, North Vietnam].
Caridina gracilipes — Cai & Shokita, 2006: 250; Cai, 2014: 207, figs. 1-3.
Caridina longirostris — Hung et al., 1993: 493, figs. 1f, 8a-g; Li & Liang, 2002: 708.

Material examined.— **Guangdong**: 1 female, cl 3.5 mm, 1 ovigerous female, cl 4.5 mm, BNHM, Zhapo village, Yangjiang County, Guangdong Province, southern China, coll. S. Yang & X. Sun, 4 Nov. 1987. **Fujian**: 1 male, cl 3.1 mm, 3 females, cl 2.5-5.8 mm, ZRC 2020.0446, river near Jimei Town, Xiamen, Fujian Province, southern China, coll. Y. Cai & N. K. Ng, 18 Nov. 1998; 5 females, cl 3.4-4.5 mm, IZCAS, Shanshilujiao Lake, Pingtan County, Pingtan Island, Fujian Province, China, 27 Nov. 1984; 1 female, cl 4.0 mm, IZCAS, Longhai County, Fujian Province, southern China, 27 Oct. 1975. **Hainan:** see Cai (2014: 208) for details. **Taiwan**: 1 male, cl 3.5 mm, 1 ovigerous female, cl 6.0 mm, NTOU 20443, Kankou Hsi, Hengchun, Pingtung County, Taiwan, coll. T. Y. Chan, 2 Jun. 1989; 4 males, cl 3.5-4.0 mm, 1 female, cl 4.7 mm, 3 ovigerous females, cl 4.7-5.6 mm, NTOU 20401, Kangkou Bridge, Manchou Village, Pingtung County, Taiwan, coll. Q. M. Chen & W H. Lui, no date; 6 males, cl 3.5-3.8 mm, 12 females, cl 2.5-5.8 mm, 35 ovigerous females, cl 4.7-6.2 mm, ZRC, Kangkou Hsi, Manchou Village, Pingtung County, Taiwan, coll. Y. Cai et al., 23 Nov. 1997; 1 male, cl 2.9 mm, 1 ovigerous female, cl 4.3 mm, Tungho River, Taitung County, Taiwan, coll. Y. Cai et al., 24 Nov. 1997; 7 males, cl 2.8-3.7 mm, 5 females, cl 2.6-4.8 mm, 1 ovigerous female, 4.3 mm, ZRC 2020.0447, Gaoping River, Kaosiung, Taiwan, coll. Y. Cai et al., 22 Nov. 1997; 6 males, cl 3.1-3.5 mm, 8 females, cl 2.5-4.5 mm, NTOU 20404, Lotung Town, Yilan County, Taiwan, coll. J. Y. Shy & M. S. Hung, 15 Mar. 1990.

Comparative material examined.— Lectotype: 1 ovigerous female, cl 5.3 mm, syntypes of *Caridina* var. *gracilipes* De Man, 1892, RMNH.CRUS.D.1317, Maros River, Sulawesi, Indonesia, coll. Max Weber, Sep.-Oct., 1888. Paralectotypes: 2 females, cl 4.2-4.4 mm, RMNH.CRUS.D.1317, data same as lectotype; 38 specimens, RMNH (ZMA De 102636), Balingnipa Brook, Sulawesi, Indonesia, coll. Max. Weber, 1888; 5 specimens, RMNH (ZMA De 102637), Palima Tjenrana River, Sulawesi, Indonesia, coll. M. Weber, 1888; 4 females (2 ovigerous), RMNH (ZMA De 102624), Bonea River, Salajar, Indonesia; 1 male, RMNH (ZMA De 102638), Pampanua, Tjenrana-River, Sulawesi, Indonesia, coll. M. Weber, 1888. Others: 2 males, cl 3.7-3.8 mm, 3 females, cl 2.8-3.7 mm, ZRC, Sungai Batang, on road from Palopo to Wotu, Sulawesi, Indonesia, coll. M. Kottelat, 22 Feb. 1999.

Diagnosis.— Rostrum long, upturned anteriorly, reaching beyond end of scaphocerite; rostral formula: 1-3 + 11-27 + 1-3/8-18; with toothed dorsal margin portion subequal to the length of untoothed portion. Sixth abdominal somite 1.9 times as long as fifth somite, as long as telson. Telson terminating in a projection. Pre-

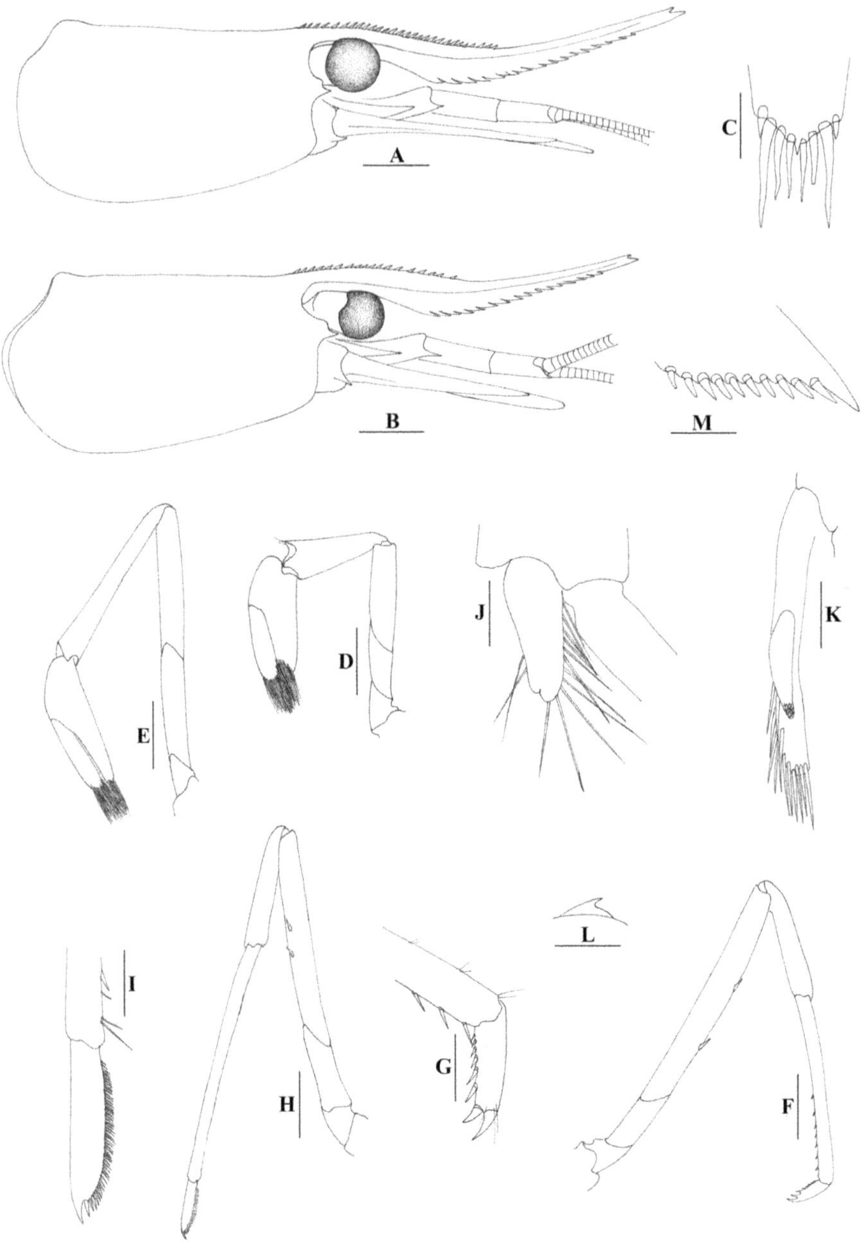

Fig. 1. *Caridina gracilipes* De Man, 1892. A, B, Cephalothorax and cephalic appendages; C, distal portion of telson; D, first pereiopod; E, second pereiopod; F, third pereiopod; G, dactylus of third pereiopod; H, fifth pereiopod; I, dactylus of fifth pereiopod; J, endopod of male first pleopod; K, appendix masculina and appendix interna of male second pleopod; L, preanal carina; M, uropodal diaeresis. Scales: A, B = 1 mm; C, G, I-K, M = 0.2 mm; D-F, H, L = 0.5 mm (A, J, K, male, cl 3.8 mm; B-I, L, M, male, 3.7 mm, ZRC, Sungai Batang, KMB on road from Palopo to Wotu, Sulawesi, coll. M. Kottelat, 22 Feb. 1999).

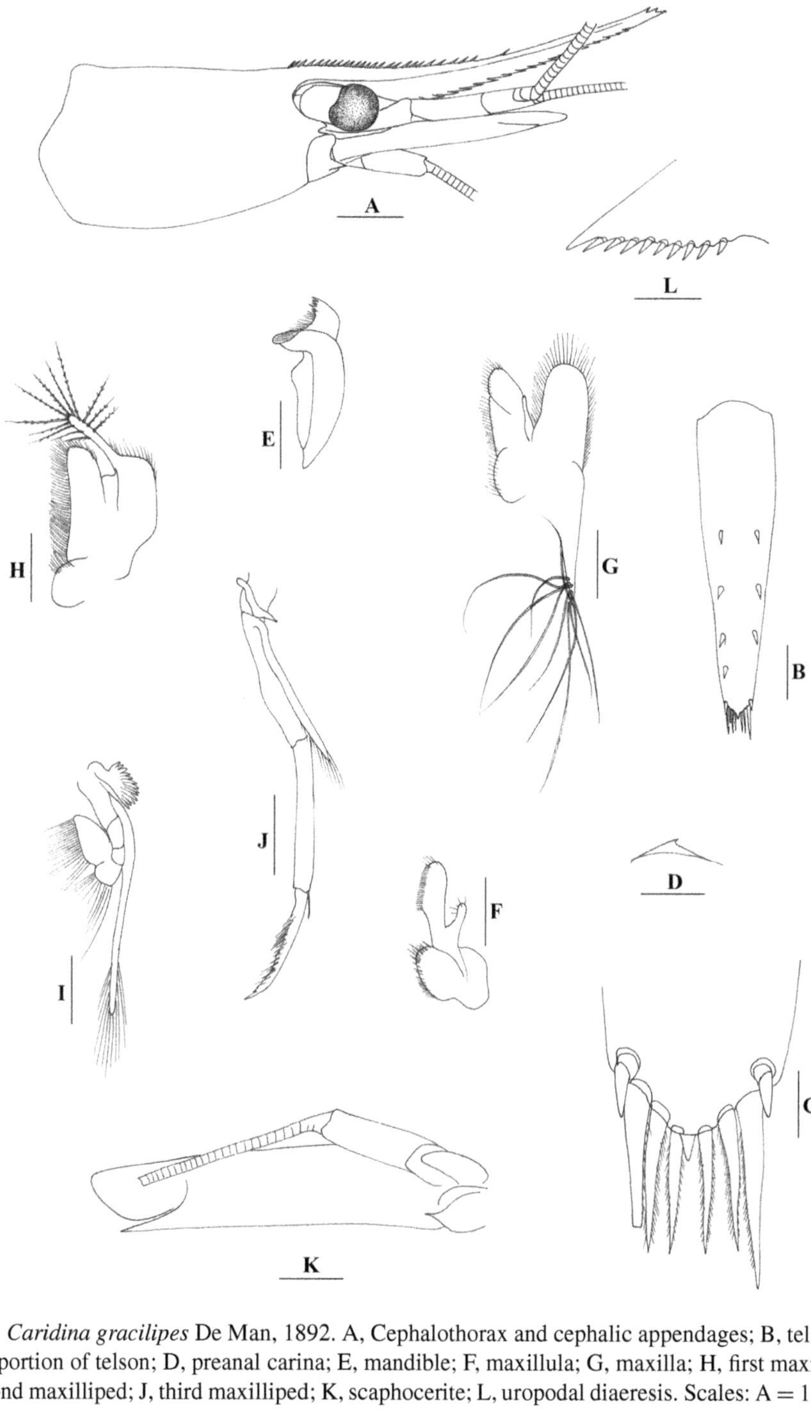

Fig. 2. *Caridina gracilipes* De Man, 1892. A, Cephalothorax and cephalic appendages; B, telson; C, distal portion of telson; D, preanal carina; E, mandible; F, maxillula; G, maxilla; H, first maxilliped; I, second maxilliped; J, third maxilliped; K, scaphocerite; L, uropodal diaeresis. Scales: A = 1.0 mm; B, D-K = 0.5 mm; C, L = 0.2 mm (male, cl 3.6 mm, ZRC, Kangkou Hsi, Manchou Village, Pingtung County, Taiwan, coll. Y. Cai et al., 23 Nov. 1997).

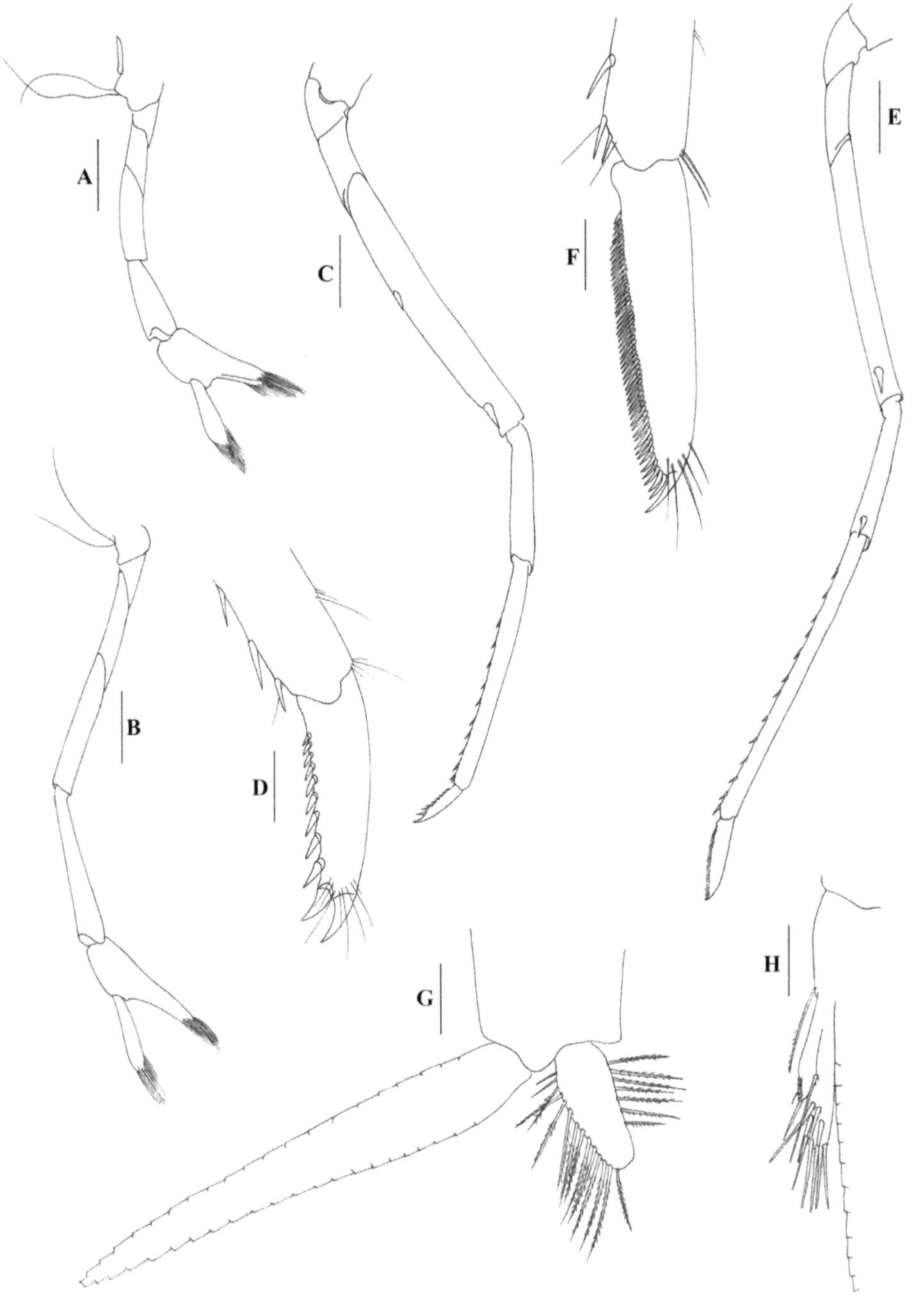

Fig. 3. *Caridina gracilipes* De Man, 1892. A, First pereiopod; B, second pereiopod; C, third pereiopod; D, dactylus of third pereiopod; E, fifth pereiopod; F, dactylus of fifth pereiopod; G, male first pleopod; H, appendix interna and appendix masculina of male second pleopod. Scales: A-C, E = 0.5 mm; D, E, G, H = 0.2 mm (male, cl 3.6 mm, ZRC, Kangkou Hsi, Manchou Village, Pingtung County, Taiwan, coll. Y. Cai et al., 23 Nov. 1997).

anal carina with a spine. Antennular peduncle 0.8-1.05 times as long as carapace. Scaphocerite 4.4 times as long as wide. First pereiopod with carpus shorter than chela, 2.0-2.5 times as long as high; chela 2.0-2.7 times as long as broad; fingers distinctly longer than palm. Second pereiopod with carpus 1.4 times as long as chela, 5.4-6.3 times as long as high; chela 2.5-3.2 times as long as broad; fingers 1.6 times as long as palm. Third pereiopod with propodus 3.8-4.4 times as long as dactylus; dactylus 3.3 times as long as wide (spines included), with 8-10 accessory spines on flexor margin. Fifth pereiopod with propodus 3.1-3.8 times as long as dactylus, dactylus 4.1 times as long as wide (spinules included), with 36-57 spinules on flexor margin. Endopod of male first pleopod subtriangular, 0.25 times length of exopod, no appendix interna, or with a vestige of it. Appendix masculina of male second pleopod half-length of endopod. Uropodal diaeresis with 8-12 movable spinules. Eggs 0.42-0.53 × 0.23-0.33 mm in diameter.

Habitat.— Lowland freshwater rivers with seawater influence.

Remarks.— Cai & Shokita (2006) recorded this species from the Philippines, given an informative diagnosis and compared it with *C. elongapoda* Liang & Yang, 1977 and *C. longirostris* H. Milne Edwards, 1837. These authors confirmed *C. gracilipes* as a valid species. Cai (2014) provided detailed figures based on specimens from Hainan. Dang (1975) described *Caridina acuticaudata* from Hoa Binh, North Vietnam, which was subsequently synonymized with *Caridina longirostris* by Li & Liang (2002). Although no Vietnamese material of this species available, however, based on the detailed figures illustrated of rostrum, dactylus of third pereiopod, dactylus of fifth pereiopod, male appendix interna, and preanal carina (Dang, 1975: Hinh h), the species should be referred to *C. gracilipes*. Kemp (1918) reported *C. gracilipes* from Shanghai, and Liu (1955) reported *C. gracilipes* from Hebei. Characteristics of both records are very distinct from *C. gracilipes* and should be referred to *Caridina sheni* sp. nov. (see remarks under *Caridina sheni* sp. nov.). In China, the species only occurs in Guangdong, Fujian, Hainan and Taiwan, representing the northern limit of its distribution range. For comparative purpose, two male specimens from Sulawesi and one male from Taiwan were illustrated here in details.

Distribution.— Sulawesi, Philippines, Borneo, Malay Peninsula, India (Richard & Chandran, 1994) and southern China (Fujian, Guangdong, Guangxi, Hainan, Taiwan).

Caridina macrophora Kemp, 1918

(fig. 4)

Caridina nilotica macrophora Kemp, 1918: 277, fig. 9a-f [type locality: Tale Sap Lake, Peninsular Siam (Thailand)].

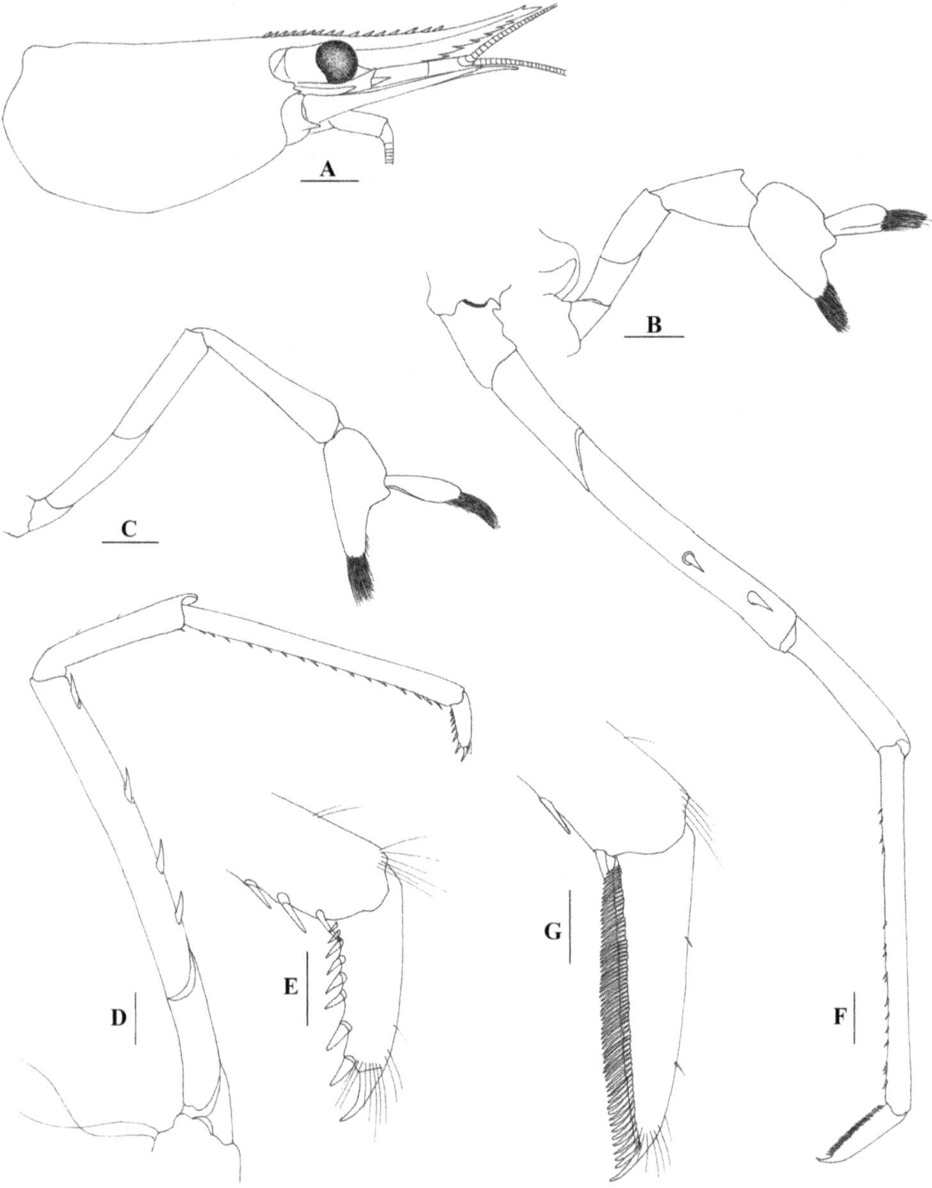

Fig. 4. *Caridina macrophora* Kemp, 1918. A, Cephalothorax and cephalic appendages; B, first pereiopod; C, second pereiopod; D, third pereiopod; E, dactylus of third pereiopod; F, fifth pereiopod; G, dactylus of fifth pereiopod. Scales: A = 1 mm; B-D, F = 0.5 mm; E, G = 0.2 mm (female, cl 4.7 mm, ZRC.2012.1155, stream across Dongtai Bridge, Wanquan River, Huishan Town, Hainan, 19°04.23′N 110°15.41′E, pH 6.5, coll. Y. Cai & N. K. Ng, 2 Dec. 1998).

Caridina subnilotica Dang, 1975: 69, fig. 3 [type locality: Hanoi, North Vietnam].
Caridina nilotica macrophora — Liang & Yan, 1983: 211; Liang & Zhou, 1993: 231; Liang, 2004: 200, figs. 96.
Caridina macrophora — Cai, 2014: 209, figs. 4, 5.

Material examined.— **Guangxi**: 1 female, cl 3.7 mm, 1 ovigerous female, cl 4.9 mm, eggs 0.8 × 0.5 mm, IZCAS, stream in a small hill near main town of Fang Cheng County, Guangxi, southern China, coll. Y. Cai & S. Yang. 2 Apr. 1992. **Hainan**: see Cai (2014: 209) for details.

Comparative material examined.— **Thailand**: 1 male, cl 2.9 mm, 1 female, cl 3.8 mm, 1 ovigerous female, cl 4.3 mm, eggs 0.95 × 0.55 mm, MNHN Na 769, syntype of *Caridina nilotica* var. *macrophora* Kemp, 1918, exchanged from Indian Museum in 1921, Tale Sap, Siam (Thailand). **Vietnam**: 14 males, cl 2.5-3.2 mm, 3 females, cl 3.0-3.8 mm, 7 ovigerous females, cl 3.7-4.2 mm., eggs 0.9 × 0.52 mm, ZRC, stream in Harianoi Province, Vietnam, coll. H. H. Ng, 22 Apr. 1998. **Cambodia**: 6 males, cl 2.0-2.3 mm, 10 females, 2.2-3.3 mm, 8 ovigerous females, cl 3.5-4.1 mm, ZRC, Phnom Penh, Central Market (confluence of Tonle Sap river & Mekong River), 11°33.56′N, 104°56.28′E, Cambodia, coll. H. H. Tan, 22 May 1999.

Diagnosis.— Rostrum long, upturned anteriorly, reaching beyond end of scaphocerite; rostral formula: 2-3 + 11-19 + 1-2/6-20. Sixth abdominal somite 2.2 times as long as fifth somite, as long as telson. Telson not terminating in a projection. Preanal carina small, rounded, without spine. Antennular peduncle 0.86 times as long as carapace. Scaphocerite 4.5 times as long as wide. First pereiopod with carpus shorter than chela, 1.6 times as long as high; chela 1.8 times as long as broad; fingers as long as or slightly longer than palm. Second pereiopod with carpus 1.2 times as long as chela, 4.3 times as long as high; chela 2.2 times as long as broad; fingers slightly longer than palm. Third pereiopod with propodus 3.8 times as long as dactylus; dactylus 3.5 times as long as wide (spines included), with 6-10 accessory spines on flexor margin. Fifth pereiopod with propodus 3.1 times as long as dactylus, dactylus 3.7 times as long as wide (spinules included), with 34-46 spinules on flexor margin. Endopod of male first pleopod subtriangular, 0.25 times length of exopod, no appendix interna. Appendix masculina of male second pleopod half-length of endopod. Uropodal diaeresis with 7-9 movable spinules. Eggs 0.80-0.95 × 0.50-0.60 mm in diameter.

Habitat.— Lowland freshwater habitats: ponds, larger rivers, streams.

Remarks.— Cai (2014) described and illustrated the species in detail based on specimens from Hainan, discussed the differences from its close congener *Caridina aruensis* J. Roux, 1911, and synonymized *Caridina subnilotica* Dang, 1975 from Hanoi, North Vietnam with *C. macrophora*. Although both species do not have an appendix interna at the male first pleopods, *C. macrophora* could be easily separated from *C. gracilipes* by the much larger egg size, as well as the absence of a spine at the preanal carina. In China, the species only occurs in Guangxi and Hainan, representing the northern limit of its distribution range.

Distribution.— Thailand, Cambodia, Vietnam and southern China.

Caridina elongapoda Liang & Yan, 1977
(fig. 5)

Caridina nilotica elongapoda Liang & Yan, 1977: 220, figs. 5-8. [type locality: Xinzai, Gulei village, Zhangpu County, Fujian, southern China] — Liang & Yan, 1988: 15.
Caridina longirostris — Liang, 2004: 195, fig. 94.
Caridina elongapoda — Cai & Shokita, 2006: 249; Cai, 2014: 214.

Material examined.— **Fujian**: 2 males, cl 3.8-4.1 mm, IZCAS, Mingjiang River, Fuzhou, Fujian Province, southern China, no date. **Guangdong**: 6 males, cl 3.3-3.9 mm, 8 females, cl 3.6-5.5 mm, ZRC, Nan'ao County, Guangdong Province, southern China, coll. Y. Cai, Nov. 1998. **Hong Kong**: 1 female, cl 4.4 mm, Tin Liu, Sai Kung, New Territories, Hong Kong, ZRC, coll. R. Yam, 16 Feb. 2001; 1 female, Kap Man Hang, New Territories, coll. R. Yam, 11 Nov. 2000; 1 female, cl 3.7 mm, ZRC, Tai Ho, New Territories, coll. R. Yam, Apr. 2001. **Guangxi**: 2 males, cl 3.7-4.2 mm, 2 females, cl 3.8-4.0 mm ZRC.2020.0448, mountain stream in Shengle Hill, near Dongxin Town, Fangcheng County, Guangxi, southern China, coll. Y. Cai & S. Yang, Apr. 1992. **Hainan**: see Cai (2014) for details.

Diagnosis.— Rostrum horizontal or slightly upturned at distal half, reaching to end of antennular peduncle, or slightly beyond distal end of scaphocerite. Rostral formula: 2-3 + 16-22 + 1-3/7-13. Sixth abdominal somite 1.9 times as long as fifth somite, slightly shorter than telson. Telson terminating in a posteromedian projection. Preanal carina triangular, without spine. Antennular peduncle 0.9 times as long as carapace. Scaphocerite 4.0 times as long as wide. First pereiopod with carpus 1.7-2.2 times as long as high; chela 1.9-2.1 times as long as broad, with fingers 1.0-1.8 times as long as palm. Second pereiopod with carpus 5.5 times as long as high. Third pereiopod with propodus 4.0 times as long dactylus; dactylus 3.1 times as long wide (spines included), with 4-8 accessory spines along flexor margin. Fifth pereiopod with propodus 4 times as long as dactylus; dactylus 3.0 times as long as wide (spinules included), with 36-47 denticulate spinules along flexor margin. Endopod of male first pleopod triangular, with an elongated appendix at its distal end. Appendix masculina of male second pleopod reaching to half-length of endopod. Uropodal diaeresis with 12-14 movable denticles. Eggs 0.38-0.42 × 0.20-0.25 mm in diameter.

Habitat.— Lowland streams and rivers with seawater influence.

Remarks.— Taxonomic updates of the species have been discussed by Cai (2014). In China, *Caridina elongapoda* can be easily distinguished from other members of the *Caridina nilotica* species group by the presence of an appendix interna at the male first pleopod.

Liang (2004) examined two female specimens from Taiwan for his *C. longirostris* (= *Caridina elongapoda*). This record however needs to be confirmed with male specimens. Females of *Caridina elongapoda* are morphologically similar to *C. gracilipes*, which is a common species in Taiwan.

Distribution.— Southern China, Peninsular Malaysia, Borneo and Philippines.

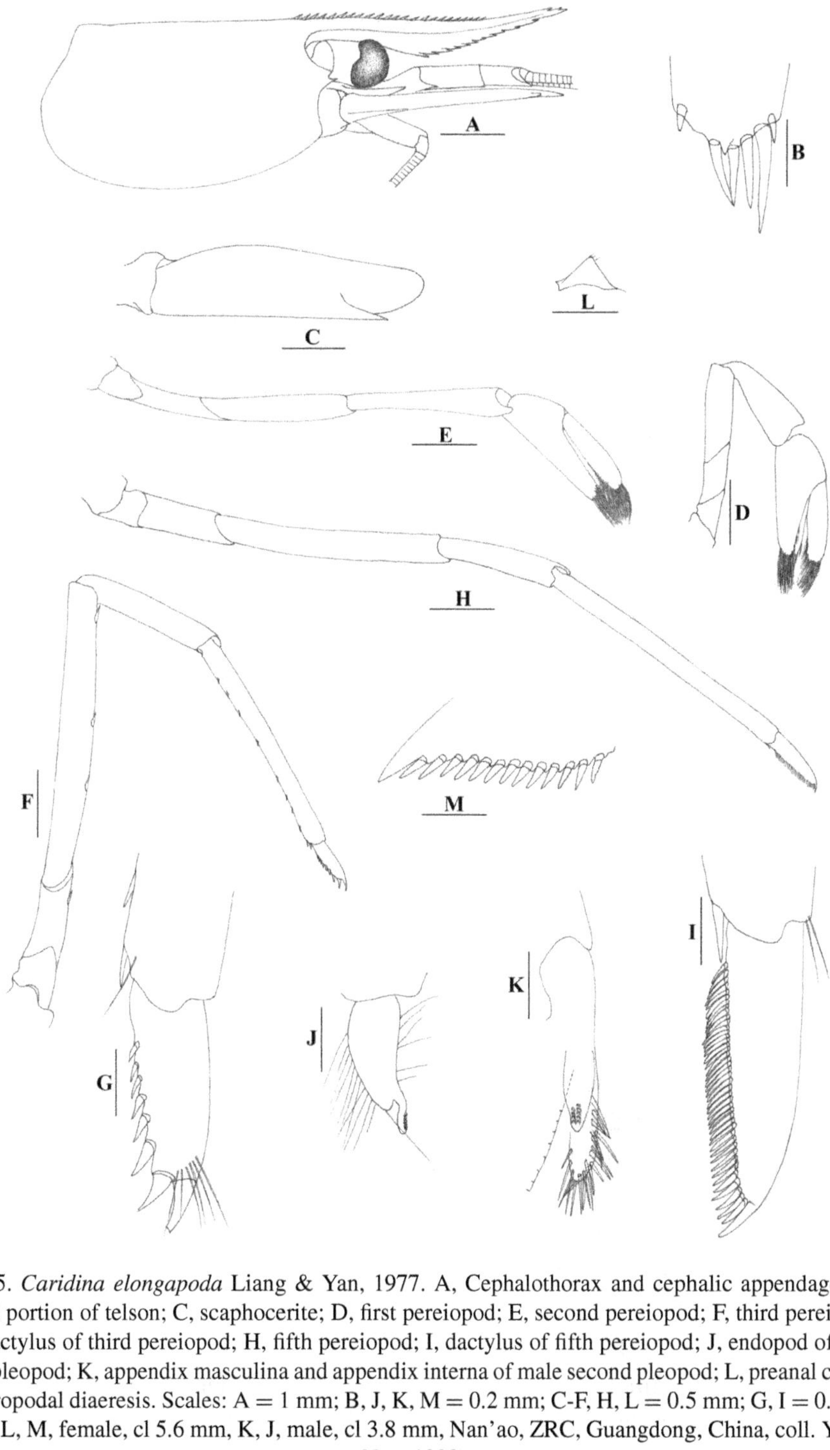

Fig. 5. *Caridina elongapoda* Liang & Yan, 1977. A, Cephalothorax and cephalic appendages; B, distal portion of telson; C, scaphocerite; D, first pereiopod; E, second pereiopod; F, third pereiopod; G, dactylus of third pereiopod; H, fifth pereiopod; I, dactylus of fifth pereiopod; J, endopod of male first pleopod; K, appendix masculina and appendix interna of male second pleopod; L, preanal carina; M, uropodal diaeresis. Scales: A = 1 mm; B, J, K, M = 0.2 mm; C-F, H, L = 0.5 mm; G, I = 0.1 mm (A-I, L, M, female, cl 5.6 mm, K, J, male, cl 3.8 mm, Nan'ao, ZRC, Guangdong, China, coll. Y. Cai, Nov. 1998).

Caridina sheni sp. nov.
(figs. 6-7)

Caridina nilotica gracilipes — Kemp, 1918: 275; Liu, 1955: 27, figs. 11-17; Liang, 2004: 198 (part), fig. 95.

Material examined.— Holotype: ovigerous female, cl 5.1 mm, eggs 0.5 × 0.3 mm, IZCAS DE 5051, Lake Baiyangdian at south of Dashu Liuzhuan village, Renqiu County, **Hebei** Province, northern China, 25 May 1957.— Paratypes: **Yunnan**: 2 ovigerous females, cl 5.5-6.0 mm, eggs 0.50 × 0.30 mm, ZRC.2020.0449, reservoir near Gongjiagang village, Jianshui County, Yunnan Province, southeastern China, coll. L. Chen, Aug. 1993; 3 males, cl 2.9-3.2 mm, 11 females, 2.5-3.6 mm, ZRC.2020.0450, spring water at Jinxiandeng near Langshan village, Kunming, Yunnan Province, Southeast China, coll. B. Sket, 21 Nov. 1995; 1 male, cl 2.9 mm, 1 female, cl 3.0 mm, ZRC.2020.0454, Shi Cheng, Da Wan Lakeside of Dianchi lake, Yunnan Province, southeastern China, 24°47.118′N 102°45.381′E; 1858 m asl, pH 7.67, water temperature 17.8°C, [O_2] 2.59 mg/l, conductivity 470.00 μs/c, coll. Y. Cai et al., 10 Apr. 2005; 12 females, cl 3.5-5.3 mm, ZRC.2020.0455 Bai Yu Kou, You Yi Chun, Dianchi Lake, Yunnan Province, southeastern China, 24°48.779′N 102°39.325′E, 1750 m asl, pH 8.70, water temperature 13.3°C, [O_2] 2.86 mg/l, conductivity 424.00 μs/cm, coll. Y. Cai et al., 13 Apr 2005; 2 females, cl 3.4-4.9 mm, ZRC.2020.0456, pond at Haigen Minzhucun, Kunming, 29 Jan. 1994, coll. N. Zhang. **Hubei**: 9 females, cl 3.5-4.5 mm, 14 females, cl 3.7-4.5 mm, egg 0.45-50 × 2.7-3.0 mm, ZRC.2020.0451, lake Donghu, Wuchang, Hubei Province, central China, 8 Aug. 1979; 1 male, cl 2.6 mm, 1 female, cl 4.4 mm, ZRC.2020.0452, lake Tanxun, Wuhan, Hubei Province, central China, 17 Sep. 1988; 1 male, cl 2.9 mm, IZACAS DE 5052, 1 female, cl 3.1 mm, IZCAS DE 5053, 1 female, cl 3.7 mm, IZCAS DE 5054, 1 female, cl 4.3 mm, IZCAS DE 5055, 1 ovigerous female, cl 3.9 mm, IZCAS DE 5056, lake Donghu, Wuhan, Hubei Province, central China, 18 Aug 1979. **Zhejing**: 6 females, cl 3.4-3.7 mm, BNHM, Zhujiajian village, Zhoushan Island, Zhejiang Province, eastern China, coll. S. Yang, 1987. **Jiangxi**: 1 female, cl 3.7 mm, ZRC.2020.0453, Qingnan Lake, Nanchang, Jiangxi Province, eastern China, Aug. 1993.

Description.— Rostrum long, upturned anteriorly, slightly shorter or longer than carapace length, reaching beyond end of scaphocerite; rostral formula: 1-3 + 9-16 + 1-3/6-10. Antennal spine lower than inferior orbital angle. Pterygostomian angle broadly rounded.

Sixth abdominal somite 0.67 times carapace length, 1.9 times as long as fifth somite, as long as telson. Telson 3.3 times as long as wide, terminating in a posteromedian projection, with 3-4 pairs of dorsal spinules and 1 pair of dorsolateral spinules; distal margin with 3 pairs of spines, lateral pair of distal spines distinctly longer than intermediate pair of spines. Preanal carina small, with a small spine.

Eyes well developed, anterior end reaching to 0.8 times length of basal segment of antennular peduncle. Antennular peduncle 0.7-0.8 times as long as carapace; basal segment of antennular peduncle longer than sum of second and third segment length, anterolateral angle pointed, reaching 0.4 times length of the second segment, second segment distinctly longer than third segment. Stylocerite reaching to 0.8 times length of basal segment of antennular peduncle. Scaphocerite 3.6 times as long as wide.

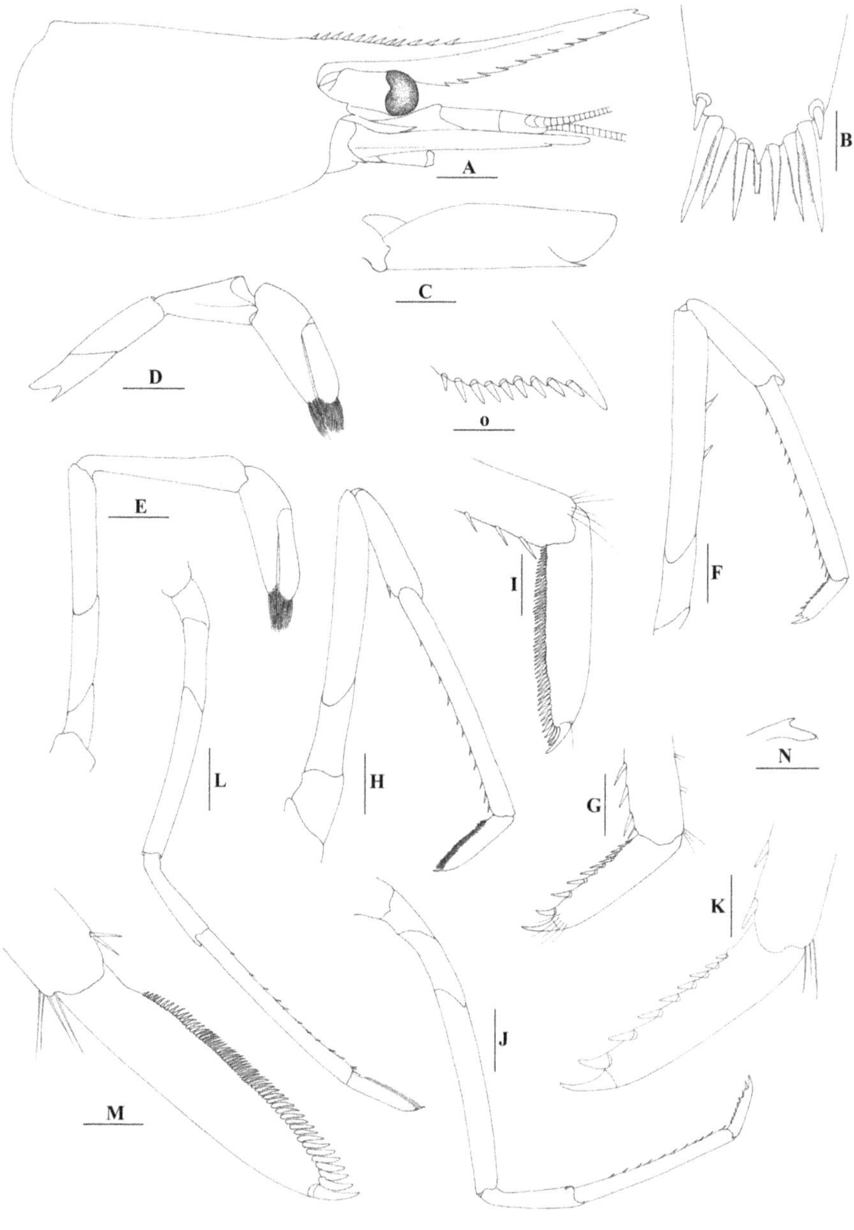

Fig. 6. *Caridina sheni* sp. nov. A, Cephalothorax and cephalic appendages; B, distal portion of telson; C, scaphocerite; D, first pereiopod; E, second pereiopod; F, third pereiopod; G, dactylus of third pereiopod; H, fifth pereiopod; I, dactylus of fifth pereiopod; J, third pereiopod; K, dactylus of third pereiopod; L, fifth pereiopod; M, dactylus of fifth pereiopod; N, preanal carina; O, uropodal diaeresis. Scales: A, C = 1 mm; B, G, I, K, M, O = 0.2 mm; D-F, H, J, L, N = 0.5 mm (A-I, N, O, ovigerous female, cl 5.1 mm, holotype, IZCAS DE 5051, Lake Baiyangdian at the south of Dashu Liuzhuan village, Renqiu County, Hebei, 25 May 1957; J-M, female, cl 3.7 mm, BNHM, Zhujiajian village, Zhoushan Island, Zhejiang, coll. S. Yang, 1987).

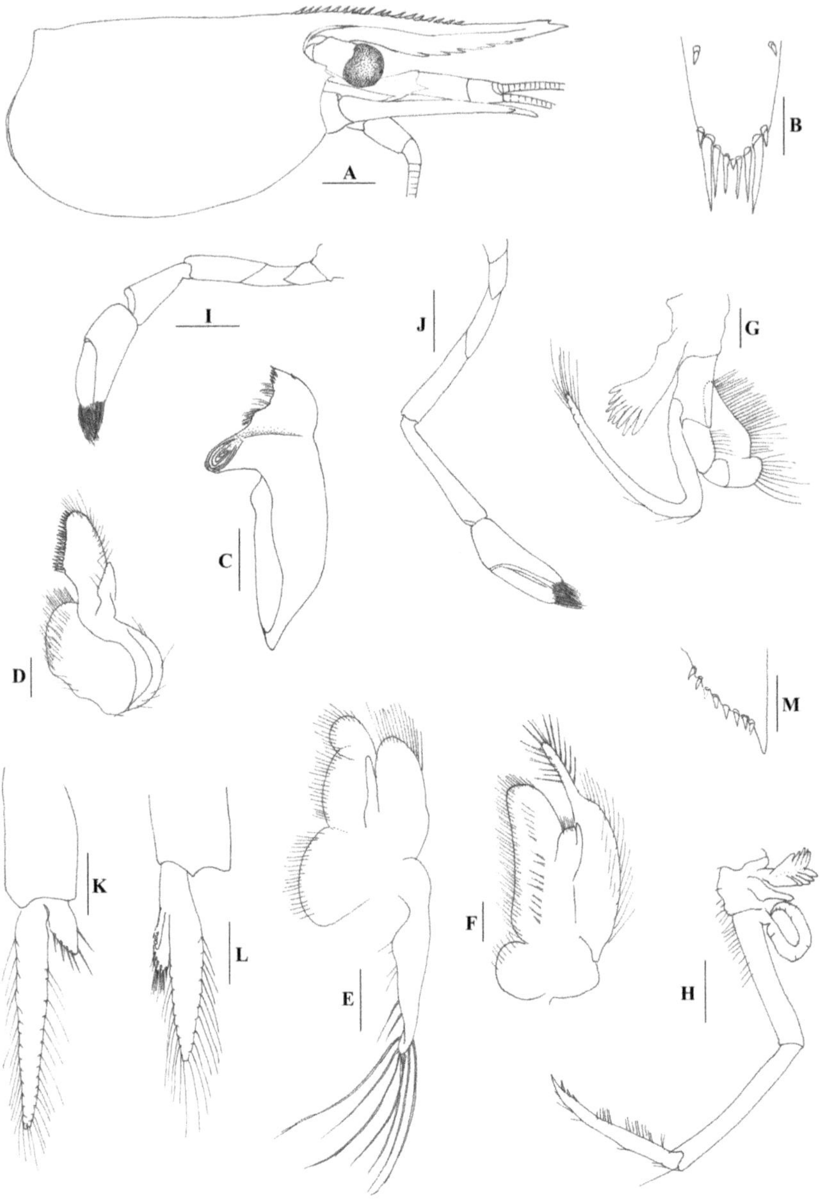

Fig. 7. *Caridina sheni* sp. nov. A, Cephalothorax and cephalic appendages; B, distal portion of telson; C, mandible; D, maxillula; E, maxilla; F, first maxilliped; G, second maxilliped; H, third maxilliped; I, first pereiopod; J, second pereiopod; K, endopod of male first pleopod; L, appendix masculina and appendix interna of male second pleopod; M, uropodal diaeresis. Scales: A = 1 mm; C-J = 0.5 mm; K, L = 0.25 mm; B = 0.2 mm (A, B, I, J, M, female, cl 3.7 mm, BNHM, Zhujiajian village, Zhoushan Island, Zhejiang, coll. S. Yang, 1987; C-H, ovigerous female, cl 5.1 mm, holotype, IZCAS DE 5051, Lake Baiyangdian at the south of Dashu Liuzhuan village, Renqiu County, Hebei, 25 May 1957; K, L, male, cl 2.9 mm, ZRC.2020.0454, Shi Cheng, Da Wan Lakeside of Dianchi lake, Yunnan, coll. Y. Cai et al., 10 Apr. 2005).

Incisor process of mandible ending in 4-5 irregular teeth, molar process truncated. Lower lacinia of maxillula broadly rounded, upper lacinia elongated, with a number of distinct teeth on inner margin, palp slender. Upper endites of maxilla subdivided, palp short, scaphognathite tapering posteriorly with some long, curved setae at posterior end. Palp of first maxilliped terminating in a finger-like projection. Second maxilliped typical of genus. Third maxilliped reaching to end of second segment of antennular peduncle, with ultimate segment distinctly shorter than penultimate segment.

Epipods on first 4 pereiopods. First pereiopod reaching to end of first segment of antennular peduncle; merus 2.5-2.6 times as long as broad, slightly shorter than carpus; carpus excavated anteriorly, shorter than chela, 2.0-2.2 times as long as high; chela 2.5-2.6 times as long as broad; fingers 1.4 times as long as palm. Second pereiopod reaching middle of second segment of antennular peduncle; merus shorter than carpus, 4.4-5.2 times as long as broad; carpus slightly longer than chela, 4.9-5.3 times as long as high; chela 3.0-3.1 times as long as broad; fingers 1.4-1.6 times as long as palm. Third pereiopod reaching to end of antennular peduncle, propodus 10 times as long as broad, 2.6-3.1 times as long as dactylus; dactylus slender, 3.9-4.0 times as long as wide (spines included), with 8-12 accessory spines on flexor margin. Fifth pereiopod reaching slightly beyond end of second segment of antennular peduncle, propodus 11 times as long as broad, 2.5-2.9 times as long as dactylus, dactylus 3.9-4.4 times as long as wide (spinules included), with 44-51 spinules on flexor margin.

Endopod of male first pleopod subtriangular, 0.25 times length of exopod, no appendix interna. Appendix masculina of male second pleopod half-length of endopod.

Uropodal diaeresis with 8-9 movable spinules.

Eggs 0.45-0.50 × 0.26-0.30 mm in diameter.

Habitat.— *Caridina sheni* sp. nov. was mostly collected from inland open freshwater bodies such as lakes or reservoirs, with no seawater influence.

Etymology.— Since there has been a species of *Caridina* named after Prof Liu Ruiyu, namely *Caridina liui* Liang & Yan, 1986 from Guizhou, southern China, the author would like to name the new species after the late Prof. Dr. C. J. Shen, mentor of Prof Liu Ruiyu (see Dai, 1997 for biography of C. J. Shen) to commemorate both pioneer Chinese carcinologists.

Remarks.— Examination of specimens from various localities in the mainland China reveals an undescribed species, hereby named as *C. sheni*. This species has been known as *C. gracilipes* since Kemp (1918) reported the species based on specimens from Shanghai. With regard to the rostrum, the small eggs, and the slender dactylus of the last three pereiopods, *C. sheni* is similar to *C. gracilipes*. It, however, can be easily be separated from *C. gracilipes* by the following characters: the longer dactylus in third pereiopod (propodus 2.6-3.1 times as long as dactylus vs. 3.8-4.4 times); having more spines on dactylus of the third pereiopod (8-15 vs. 8-9); and the longer dactylus of the fifth pereiopod (propodus 2.5-2.9 times as long as dactylus vs. 3.1-3.8 times); generally the lower number of ventral rostral teeth (6-10 vs. 8-18); shorter antennular peduncle (the proportion of antennular peduncle length and carapace length 0.7-0.8 times vs. 0.8-1.05 times); the wider scaphocerite (3.6 times as long as wide vs. 4.4 times). On the basis of Kemp's (1918: 275) description, his specimens of *C. n. gracilipes* should be referred to *C. sheni* as well. Liu (1955: 27) described *C. nilotica gracilipes* from northern China (Wenanwa and Baiyangdian Lake of Hebei Province). According to his description and the illustration, his material should also be referred to *C. sheni*.

Yu (1938) reported *Caridina gracilipes* from China. He stated that the specimens he examined included some from "Chefoo" (Yantai, Shandong Province, northern China). This record cannot be confirmed as Yu's specimens are no longer available and our available material shows that currently the northern limit of the *C. gracilipes* is at Fujian, southern China. Yu's specimens from "Chefoo" could be *C. sheni* instead.

Distribution.— China (Hebei, Yunnan, Hubei, Zhejiang and Jiangxi).

ACKNOWLEDGEMENTS

I am grateful to Peter K. L. Ng for his support during the course of this study, to Charles H. J. M. Fransen and Jasmine Richard for critically reading an earlier version of the manuscript.

REFERENCES

Cai, Y., 2014. Atyid shrimps of Hainan Island, southern China, with description of a new species of *Caridina* (Crustacea, Decapoda, Atyidae). In: D. C. J. Yeo, N. Cumberlidge & S. Klaus (eds.), Advances in freshwater decapod systematics and biology. Crustaceana Monographs, **19**: 207-231, figs. 1-11.

Cai, Y., P. K. L. Ng & S. Choy, 2007. Freshwater shrimps of the family Atyidae (Crustacea: Decapoda: Caridea) from Peninsular Malaysia and Singapore. The Raffles Bulletin of Zoology, **55**(2): 277-309, figs. 1-17.

Cai, Y., P. K. L. Ng, S. Shokita & K. Satake, 2006. On the species of Japanese atyid shrimps (Decapoda: Caridea) described by William Stimpson (1860). Journal of Crustacean Biology, **26**(3): 392-419, figs. 1-15.

Cai, Y. & S. Shokita, 2006. Report on a collection of freshwater shrimps (Crustacea: Decapoda: Caridea) from the Philippines, with descriptions of four new species. The Raffles Bulletin of Zoology, **54**(2): 245-270.

Dai, A.-Y., 1997. Shen Jia-Rui (C. J. Shen) (1902-1975). Crustaceana, **70**(4): 497-503.

DANG, N. T., 1975. Phan loai tom cua nuoc ngot mien bac vietnam. *Tap san Sinh Vat-Dia Hoc* (The identities of North Vietnamese freshwater shrimps and crabs. Journal of Biology-Geology), **13**(3): 65-78, figs. 1-8. [In Vietnamese with French summary.]

DE MAN, J. G., 1892. Decapoden des Indischen Archipels. In: M. WEBER (ed.), Zoologische Ergebnisse einer Reise in Niederländisch Ost-Indien, **2**: 265-527, pls. 15-29.

DE MAN, J. G., 1908. On the *Caridina nilotica* (Roux) and its varieties. Records of the Indian Museum, **2**: 255-283, pls. 1-20.

HUNG, M. S., T. Y. CHAN & H. P. YU, 1993. Atyid shrimps (Decapoda: Caridea) of Taiwan, with descriptions of three new species. Journal of Crustacean Biology, **13**(3): 481-503, figs.1-14.

KEMP, S., 1918. Zoological results of a tour in the Far East. Crustacea Decapoda and Stomatopoda. In: N. ANNANDALE (ed.), Zoological results of a tour in the Far East. Memoirs of the Asiatic Society of Bengal, **6**: 219-297.

LI, S.-Q. & X.-Q. LIANG, 2002. Caridean prawns of northern Vietnam (Decapoda: Atyidae, Palaemonidae). Acta Zootaxonomica Sinica, **27**(4): 707-716.

LIANG, X. & J. ZHOU, 1993. Study on new atyid shrimps (Decapoda, Caridea) from Guangxi, China. Acta Hydrobiologica Sinica, **17**(3): 231-239, figs. 1-4. [In Chinese with English abstract.]

LIANG, X.-Q., 2004. Fauna sinica: invertebrata (vol. 36). Crustacea: Decapoda: Atyidae: 1-375. (Science Press, Beijing). [In Chinese with English abstract.]

LIANG, X.-Q. & S.-L. YAN, 1977. New species and subspecies of *Caridina* (Decapoda, Caridea) from Fukien, China. Acta Hydrobiologica Sinica, **6**(2): 219-225, figs. 1-13. [In Chinese with English abstract.]

LIANG, X.-Q. & S.-L. YAN, 1983. New species and new records of freshwater shrimps (Crustacea Decapoda) from Hainan Island, China. Oceanologia et Limnologia Sinica, **14**(3): 211-216, figs.1-3. [In Chinese with English abstract.]

LIANG, X.-Q. & S.-L. YAN, 1986. Study on *Caridina* (Decapoda, Caridea) from Guizhou Province, China. Oceanologia et Limnologia Sinica, Suppl. 1986: 196-206, figs. 1-5. [In Chinese with English abstract.]

LIANG, X.-Q. & M.-Q. ZHENG, 1988. Notes on *Caridina* from Fujian, China. Acta Zootaxonomica Sinica, **13**(1): 15-19, figs. 1-9. [In Chinese with English abstract.]

LIU, J. Y., 1955. Economic shrimps of northern China: i-iii, 1-74, pls. 1-24. (Science Press, Beijing). [In Chinese.]

MILNE EDWARDS, H., 1837. Histoire naturelle des crustacés, comprenant l'anatomie, la physiologie et la classification de ces animaux, **2**: 1-532. (Libraire Encyclopedique de Roret, Paris).

RICHARD, J. & M. R. CHANDRAN, 1994. A systematic report on the fresh water prawns of the atyid genus *Caridina* H. Milne Edwards 1837, from Madras (Tamilnadu, India). Journal of the Bombay Natural History Society, **91**: 241-259.

RICHARD, J. & P. F. CLARK, 2005. *Caridina nilotica* (P. Roux, 1833) (Crustacea: Decapoda: Caridea: Atyidae) from East Africa, with descriptions of four new species. Proceedings of the Biological Society of Washington, **118**(4): 706-730.

RICHARD, J. & P. F. CLARK, 2009. African *Caridina* (Crustacea: Decapoda: Caridea: Atyidae): redescriptions of *C. africana* Kingsley, 1882, *C. togoensis* Hilgendorf, 1893, *C. natalensis* (Bouvier, 1925) and *C. roubaudi* Bouvier, 1925 with descriptions of 14 new species. Zootaxa, **1995**: 1-75.

RICHARD, J. & P. F. CLARK, 2010. *Caridina* H. Milne Edwards, 1837 (Crustacea: Decapoda: Caridea: Atyoidea: Atyidae) — freshwater shrimps from eastern and southern Africa. In: S. DE GRAVE & C. H. J. M. FRANSEN (eds.), Contributions to shrimp taxonomy. Zootaxa, **2372**: 305-337.

ROUX, J., 1911. Nouvelles espèces de décapodes d'eau douce provenant de Papouasie. Notes from the Leyden Museum, **33**: 81-106, figs. 1-5.

UENO, M., 1935. Inland water fauna of Formosa. I. Crustacea Decapoda. Transactions of the Natural History Society of Formosa, **25**: 270-276.

YEO, D. C. J., Y. CAI & P. K. L. NG, 1999. The freshwater and terrestrial decapod Crustacea of Palau Tioman, peninsular Malaysia. The Raffles Bulletin of Zoology, **6**: 197-244, figs. 1-20.

YU, H. P., 1974. On the Atyidae shrimps (Crustacea, Decapoda, Atyidae) from Taiwan. Aquaculture, **2**: 49-58.

YU, S. C., 1936. Report on the Macrurous Crustacea collected during the "Hainan biological expedition" in 1934. The Chinese Journal of Zoology, **2**: 85-99, figs. 1-7.

YU, S. C., 1938. Studies on Chinese *Caridina* with descriptions of five species. Bulletin of the Fan Memorial Institute of Biology, Zoology Series, **8**(3): 271-310, figs. 1-16.

First received 10 November 2019.
Final version accepted 17 September 2020.

[When citing this volume, please refer to *Crustaceana* 93 (2020), 1233-1546]

Prof. Ruiyu Liu Memorial Issue

NEW RECORDS OF THE SPONGE-SYMBIOTIC SHRIMP GENUS *ANCHISTIOIDES* FROM THE INDO-WEST PACIFIC (DECAPODA, CARIDEA)

BY

CHARLES H. J. M. FRANSEN[1])

Taxonomy & Systematics Group, Naturalis Biodiversity Center, P.O. Box 9517, NL-2300 RA Leiden, The Netherlands

ABSTRACT

The family Anchistioididae comprises a single genus *Anchistioides* with three species: *A. compressus* and *A. willeyi* from the tropical Indo-West Pacific, and *A. antiguensis* from the tropical West Atlantic. The two Indo-West Pacific species are known to be symbionts of sponges. Material collected in recent years throughout the Indo-West Pacific yielded range extensions as well as new host records. This material is herein described and their infraspecific morphological variation discussed.

Key words. — Crustacea, Decapoda, Caridea, *Anchistioides*, sponge symbiont, new records

RÉSUMÉ

La famille Anchistioididae comprend un seul genre *Anchistioides* et trois espèces : *A. compressus* et *A. willeyi* de l'Indo-Ouest Pacifique tropical et *A. antiguensis* de l'Atlantique Ouest tropical. Les deux espèces indo-occidentales du Pacifique sont connues pour être des symbiotes d'éponges. Le matériel collecté dans les dernières années dans l'Indo-Ouest Pacifique a permis d'élargir leur répartition ainsi que l'observation de nouveaux enregistrements hôtes. Ce matériel est décrit ici et la variation morphologique infraspécifique est discutée.

Mots clés. — Crustacea, Decapoda, Caridea, *Anchistioides*, symbiotes d'éponges, nouvelles observations

INTRODUCTION

The family Anchistioididae Borradaile, 1915 comprises a single genus *Anchistioides* Paulson, 1875 with three species: *Anchistioides compressus* Paulson, 1875

[1]) e-mail: charles.fransen@naturalis.nl

 | DOI:10.1163/9789004464353_013

and *A. willeyi* (Borradaile, 1900) from the tropical Indo-West Pacific, and *A. antiguensis* (Schmitt, 1924) from the tropical West Atlantic. Both *A. compressus* and *A. willeyi* have been recorded as symbionts of sponges. The host of *A. antiguensis* has long been unknown. It was first recorded from sponges by Williams (1984) and recently confirmed by De Grave & Anker (2017) and Soledade et al. (2017).

Several specimens of the genus were found by J. E. Garcia Hernandez when studying sponges in the collections of the Florida Museum of Natural History from the Hawaiian Islands and sent over for identification. These specimens were compared with material of the genus in the collections of Naturalis Biodiversity Center, Leiden, The Netherlands. Most of this material has been collected during expeditions to the Seychelles (Netherlands Indian Ocean Programme, 1992-1993), the Maldives (The Marine Research and Higher Education Centre University of Milano-Bicocca, 2014), Indonesia (SW Sulawesi, 2000; Bali, 2001; Marine Biodiversity Workshop, Lembeh Strait, 2012), the Philippines (Cebu, 1999) and Vanuatu (SANTO 2006 Global Biodiversity Survey). This material yielded several range extensions as well as new host records.

As descriptions of the Indo-West Pacific species are generally brief and based on few specimens, more comprehensive descriptions are presented here. The relatively large number of specimens available allowed for describing the morphological variation as well.

Abbreviations used: pocl., postorbital carapace length; UF, Florida Museum of Natural History; RMNH, Rijksmuseum van Natuurlijke Historie (at present Naturalis Biodiversity Center).

SYSTEMATIC ACCOUNT

Superfamily PALAEMONOIDEA Rafinesque, 1815
Family ANCHISTIOIDIDAE Borradaile, 1915
Genus *Anchistioides* Paulson, 1875
Anchistioides compressus Paulson, 1875 (figs. 1-3, 10)

Anch.[*istioides*] *compressus* Paulson, 1875: 115, pl. 19 figs. 5-5n [type locality: Red Sea].

Amphipalaemon Seurati Nobili, 1906b: 259 [type locality: Tearia, Tuamotu Islands according to Bruce, 1967]; Nobili, 1907: 364, pl. I fig. 3.

Anchistioides compressus — Kemp, 1925: 339; Gordon, 1935: 340, fig. 22; Holthuis, 1952: 18; Bruce, 1967: 570; Miyake & Fujino, 1967: 279, figs. 2, 3, pl. 3; Bruce, 1976: 463; Bruce, 1979a: 241; Bruce, 1981: 2; Li, 1996: 231; Hayashi, 1999: 390, fig. 378a-c; Hayashi, 2006: 290, fig. 520; Li et al., 2007: 45, fig. 8; Bruce, 2010: 50.

Material examined.— **Hawaiian Islands, U.S.A.**: 1 female pocl. 1.95 mm (R = 10/8), 1 male pocl. 2.30 mm (R = 9/6) (UF 047910), stn FFS-0022, French Frigate Shoals, Hawaiian Islands, U.S.A., GPS 23.862°-166.185°, depth 40 feet, 12.x.2006, fore reef, under overhang in calcareous sponge (UF 1459), collected by G. Paulay, S. McKeon, B. Zgliczynski, R. Brainard, J. Maragos, J.

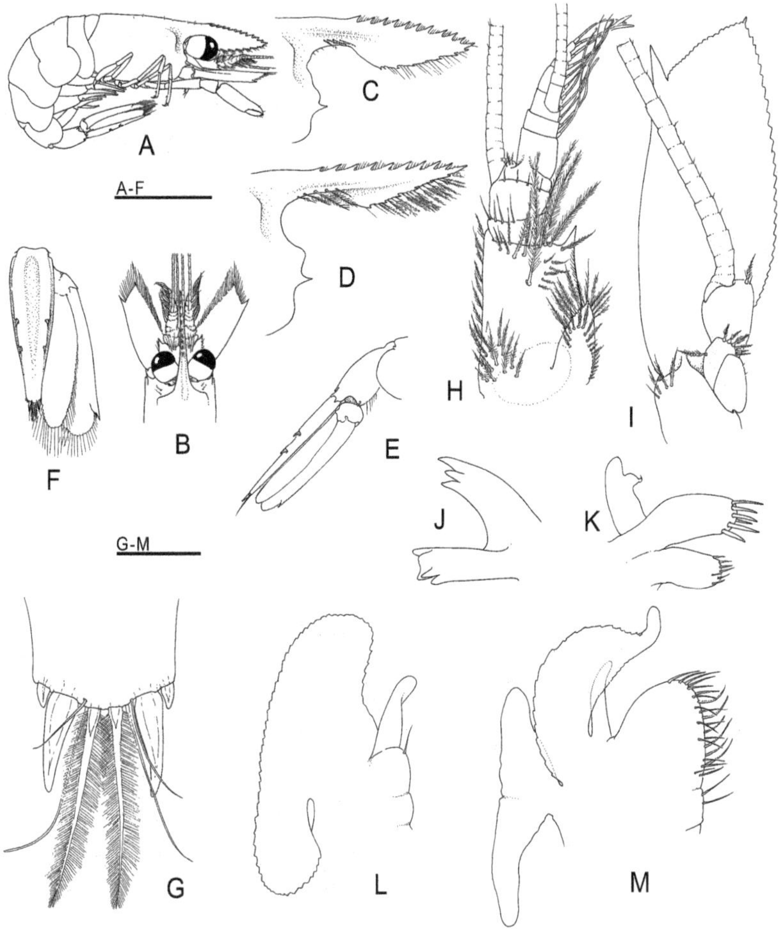

Fig. 1. *Anchistioides compressus* Paulson, 1875, Hawaiian Islands, U.S.A., UF47910. A, C, E, G-M, non-ovigerous female, pocl. 1.95 mm; B, D, F, male, pocl. 2.30 mm. A, Habitus; B, anterior appendages, dorsal view; C, D, rostrum, lateral view; E, sixth abdominal somite and caudal fan, lateral view; F, telson and uropods, dorsal view; G, telson, distal part, dorsal view; H, antennular peduncle, dorsal view; I, antennular peduncle, ventral view; J, mandible; K, maxillula; L, maxilla; M, first maxilliped. Scale: A, B = 2 mm; C-F = 1 mm; G = 0.125 mm; H, I = 0.5 mm; J-M = 0.25 mm.

Stermer, A. Hall & S. Godwin. Two males pocl. 2.25 mm (R = 9/7) and 2.27 mm (R = 10/7) (UF 47907), stn FFS-0188, French Frigate Shoals, Hawaiian Islands, U.S.A., GPS 23.817°-166.289°, depth 12-18 m, 25.x.2006, lagoon patch reef underside overhangs, in sponge (UF 1528), collected by S. Godwin, R. Most, G. Paulay, J. Stermer & S. McKeon. — **Indonesia**: One male pocl. 3.1 mm (R = 10/7), RMNH.CRUS.D.50394: SW Sulawesi, Spermonde Archipelago, Kudingareng Keke; 03.viii.2000; 20 m depth; in *Coelocarteria singaporense* (Carter, 1883) (det. N. de Voogd); collected by A. Janssen Cs1/030800/KK/AJ/4. One female pocl. 2.2 mm (R = 10/7), RMNH.CRUS.D.50393: SW Sulawesi, Spermonde Archipelago, Bone Lola; 28.ix.2000; 17 m depth; scuba; in *Acanthella cavernosa* Dendy, 1922 (det. N. de Voogd); collected by A. Jansen nr. Ac23/280900/BL/AJ/4. One ovigerous female pocl. 4.0 mm (R = 10/7), RMNH.CRUS.D.50392: SW Sulawesi, Spermonde

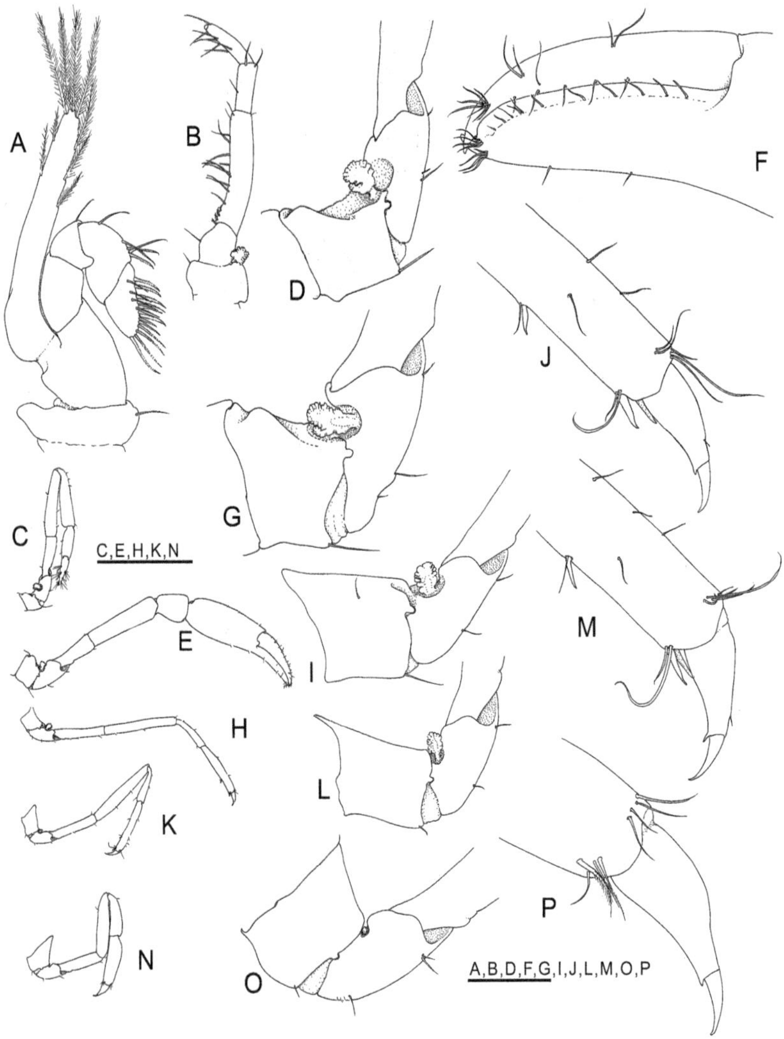

Fig. 2. *Anchistioides compressus* Paulson, 1875, Hawaiian Islands, U.S.A., UF47910. non-ovigerous female, pocl. 1.95 mm. A, Second maxilliped; B, third maxilliped; C, first pereiopod; D, idem, detail coxa and basis; E, second pereiopod; F, idem, fingers of chela; G, idem, detail coxa and basis; H, third pereiopod; I, idem, detail coxa and basis; J, idem, dactylus and distal part propodus; K, fourth pereiopod; L, idem, detail coxa and basis; M, idem, dactylus and distal part propodus; N, fifth pereiopod; O, idem, detail coxa and basis; P, idem, dactylus and distal part propodus. Scale: A, B, D-G, I, L, O = 0.25 mm; C, E, H, K, N = 1 mm; J, M, P = 0.125 mm.

Archipelago, Samalona; 05.ix.2000; 7 m depth; scuba; in *Melophlus sarasinorum* Thiele, 1899 (det. N. de Voogd); collected by A. Jansen nr. Ms13/050900/SA/AJ/4. One male, pocl. 2.5 mm, RMNH.CRUS.D.57858: stn LEM.12, Kelapadua, 1°26′8.3754″N 125°12′34.0914″E, 4.ii.2012, depth 15 m, in orange sponge, collected by C. H. J. M. Fransen. Two males pocl. 3.0 mm (R = 12/8) and 3.3 mm (R = 10/7), 1 damaged specimen pocl. 3.5 mm (R = 11/8), RMNH.CRUS.D.50391: BAL.02; Bali, Sanur, off Kesumasari Beach, Palung Semawang, N of tidal channel; 08°42′35″S

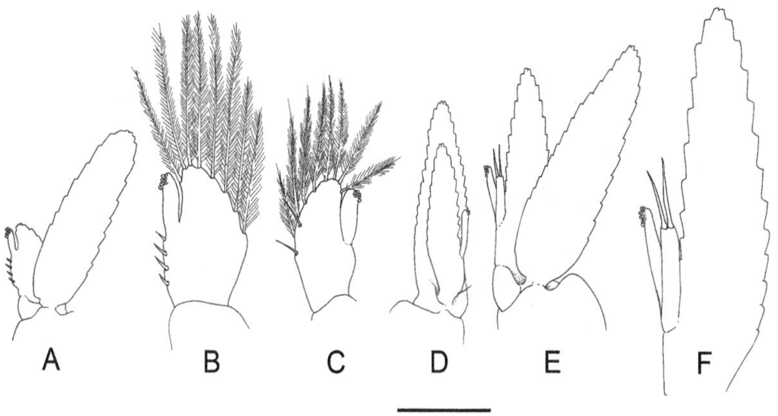

Fig. 3. *Anchistioides compressus* Paulson, 1875, Hawaiian Islands, U.S.A., UF47910. A, B, E, F, male, pocl. 2.30 mm; C, D, non-ovigerous female, pocl. 1.95 mm. A, First pleopod; B, C, idem, endopod; D, E, second pleopod; F, idem, endopod. Scale: A, D, E = 0.5 mm; B, C, F = 0.25 mm.

115°16′13″E; reef slope, sandy base; scuba-diving; 14 m depth; in *Stylissa massa* (Carter, 1887) (det. N. de Voogd); 31.iii.2001; collected by N. de Voogd. — **Philippines**: One male pocl. 1.8 mm (R = 10/7), RMNH.CRUS.D.48540: stn CEB.08: Cebu Strait, W of Bohol, S side of Cabilao Island, Cabacungan Point; 9°51.55′N 123°45.95′E; overhanging wall with caves; scuba;11.xi.1999; depth 20 m; in greyish sponges on dead *Antipathes* spec.; leg. C. H. J. M. Fransen. — **Vanuatu**: One male pocl. 2.37 mm (R = 12/6), RMNH.CRUS.D.57859: stn EP34; E Aoré Island, off Aimbué Bay, 15°33.2/33.3′S 167°12.8/12.9′E; 14.x.2006; tangle net; 40-60 m; Santo Marine Biodiversity Survey. — **Maldives**: One non-ovigerous female, pocl. 2.5 mm, RMNH.CRUS.D.57860: stn MAL.10; Magoodhoo Atoll, dive site route 66, 3°04′29.2″N 72°57′58.6″E, 10.v.2014, depth 22 m, in *Leucetta chagosensis* Dendy, 1913 (photo and collected), collected by C. H. J. M. Fransen, photo MAL10.003-004, 025-032. One non-ovigerous female, pocl. 2.9 mm, RMNH.CRUS.D.57861: stn MAL.21; Magoodhoo Atoll, dive site Free-Climbing, 3°3′53.9″N 72°55′16.53″E; 16.v.2014, depth 10 m, in *Leucetta chagosensis*, collected by C. H. J. M. Fransen, photo MAL21.039-044.

Description.— Carapace smooth (fig. 1A), with robust marginal antennal tooth, hepatic tooth absent, with blunt postorbital tubercle and indistinct longitudinal ridge running from antennal tooth obliquely dorsally to posterior margin of carapace. Rostrum (fig. 1A-C) deep at medial part, with 9-12 large dorsal teeth of about equal size and at equal distances, all anterior of orbit; ventral lamina with 6-8 shallow teeth in distal part; lateral carina indistinct, proximally bordered with row of long plumose setae ventrally. Inferior orbital angle slightly produced and bluntly rounded.

Abdominal pleura (fig. 1A) of first to fifth somite rounded. Sixth abdominal somite (fig. 1E) with posterodorsal angles rectangular, bearing two sharp dorsal projections, posteroventral angle sharply produced. Telson (fig. 1A, F) almost twice as long as sixth abdominal somite, 3.5 times as long as proximal width, tapering distally, distal margin broadly rounded, about 0.3 times as wide as proximal width, with 2 pairs of dorsal spines, pairs set close together at around half length of telson; tip of telson (fig. 1G) with short, stout distolateral spines, stout intermediate spines, about 4 times as long as distolateral spines and about as long as dorsal spines, median pair of long plumose setae, twice as long as intermediate spines.

Eye (fig. 1A, B) with hemispherical cornea; eyestalk short, about as long as cornea, tapering proximally.

Antennula (fig. 1H) with basal segment about 1.4 times as long as proximal width, with distinct distolateral tooth almost reaching distal margin of intermediate segment, anterior margin slightly produced; medioventral tooth distinct, acute, situated at about 0.4 of length of basal segment; stylocerite twice as long as wide, distally acute, about half length of basal segment, with many plumose setae on lateral and dorsodistal surface. Intermediate segment short, about half as long as wide. Distal segment slightly longer than intermediate segment. Upper flagellum biramous, with proximal 3 segments fused, swollen, short free ramus with 3 segments, fused part and short ramus with many long aesthetascs, longer free ramus about 1.5 times length carapace. Lower flagellum as long as longer free ramus of upper flagellum.

Antenna (fig. 1I) with basicerite unarmed. Carpocerite short and stout, only slightly longer than wide. Scaphocerite about 3 times as long as maximal width; lateral border slightly convex, ending in strong distolateral tooth. Anterior margin of lamella oblique; with distomedial acute tooth, overreaching distolateral tooth. Antennal peduncle reaching to one-fourth of scaphocerite length.

Thoracic sternites unarmed.

Mandible (fig. 1J) with incisor process slender, with three distal teeth; molar process also rather slender, with blunt distal teeth, without setal bristles.

Maxillula (fig. 1K) with both upper and lower lacinia slender; upper lacinia with about 7 distal spines; lower lacinia with distal simple and serrulate setae; palp bilobed, lower lobe with small ventral tubercle with single short recurved simple seta.

Maxilla (fig. 1L) with basal endite obsolete, indistinctly bilobed; distal lobe with single simple seta; lower lobe unarmed; palp distinct, not reaching distal margin of scaphognathite, without setae; scaphognathite with distal and proximal lobes well developed.

First maxilliped (fig. 1M) with coxal and basal endites partly fused, fringed with many simple and serrulate setae along entire median margin; exopod with short flagellum without distal setae; caridean lobe normal; palp slender, simple, not reaching beyond distal margin of caridean lobe; epipod large, triangular, bilobed.

Second maxilliped (fig. 2A) with endopod normally developed; dactylar segment triangular, fringed with serrulate setae medially; distomedian margin of propod not developed, with few long serrulate setae; carpal segment triangular, unarmed; meral segment oblong, unarmed; ischial and basal segments fused; exopod broad, with distal and subdistal long plumose setae; coxal segment medially rounded, with single seta, lateral margin with small epipod.

Third maxilliped (fig. 2B) short, reaching distal margin of carpocerite; coxa slightly produced medially, with lateral plate. Basis medially slightly produced, with rudimentary exopod; ischiomeral segment oblong, about 5 times as long as wide, medially concave with short plumose setae proximomedially and longer serrulate setae along rest of medial margin; penultimate segment 0.4 times length of ischiomeral segment, unarmed; ultimate segment as long as penultimate segment, with few simple and serrulate setae.

First pereiopod (fig. 2C) slender, chela almost as long as carpus, with fingers slightly shorter than palm, with several groups of serrulate setae, cutting edges entire; carpo-propodal joint with cleaning apparatus medially; carpus unarmed, slightly shorter than merus; merus almost twice as long as ischium, about 5.5 times as long as wide; basis (fig. 2D) shorter than unarmed ischium, with rudimentary exopod; coxa unarmed.

Second pereiopod (fig. 2E) with chela short and stout, fingers (fig. 2F) curved inward, about 0.75 of palm length, cutting edges entire, tips hooked, fixed finger proximally broader than more slender dactylus; carpus short, slightly longer than distal width; merus straight, oblong, about 4 times as long as wide; ischium half as long as merus; basis (fig. 2G) short with rudimentary exopod.

Third (fig. 2H-J) and fourth (fig. 2K-M) pereiopods rather short and slender; dactylus about 0.25 length of propodus, flexor margin of corpus sinuous, tapering distally, ending in small, acute accessory tooth; unguis simple, acute, slightly curved, about third of corpus length; propodus 6 times as long as wide, with ventrodistal pair of simple short spines, with one spine along ventral margin at about 0.7 of propodus length. Carpus unarmed, 0.75 of propodus length; merus 1.4 times propodus length, unarmed, ischium as long as propodus, unarmed; basis with rudimentary exopod.

Fifth pereiopod (fig. 2N-P) shorter and more robust than third and fourth pereiopods; dactylus distinctly larger than those of third and fourth pereiopods; propodus without spines; basis with very small rudimentary exopod.

Male first pleopod (fig. 3A) with endopod much shorter than exopod; endopod (fig. 3B) with distinct appendix interna with terminal hooked setae; median margin with row of few short simple setae; distolateral margin with long plumose setae.

Female first pleopod (fig. 3C) with endopod short and broad, with appendix interna almost reaching distal margin of endopod; second pleopod (fig. 3D) with endopod slightly shorter than exopod, with appendix interna medially.

Male second pleopod (fig. 3E) slightly shorter than exopod, with appendix interna and appendix masculina medially; appendix masculina (fig. 3F) almost as long as appendix interna, with 2 long simple setae distally.

Uropods (fig. 1E, F) feebly convex laterally, ending in large acute distolateral tooth with strong mobile spine just medially; both endopod and exopod overreaching telson.

Size.— Small species with maximum postorbital carapace length of 3.5 mm.

Colouration.— Kemp (1925) noted one of the specimens he described to be 'absolutely transparent' when alive. Bruce (1979a) noted the colouration to be transparent, with "conspicuous red spots along second pereiopod". Miyake & Fujino (1967) also recorded the ovigerous specimen to be 'transparent with some red pigments' and that the 'eggs in life are green'.

Hosts.— Miyake & Fujino (1967) were the first to note that the species was living inside a sponge of the genus *Caminus* and on the bottom of the osculum in *Haliclona oculata* (Linnaeus, 1759) (Miyake & Fujino, 1967, pl. 3). This latter one must have been misidentified, as *Haliclona* (*Haliclona*) *oculata* (Linnaeus, 1759) is a species only occurring in the Atlantic. Bruce (1976, 1979a, 1981) recorded the species from unidentified sponges. The species is now recorded as a symbiont of *Coelocarteria singaporense* (Carter, 1883), *Acanthella cavernosa* Dendy, 1922, *Melophlus sarasinorum* Thiele, 1899, *Leucetta chagosensis* Dendy, 1913 and *Stylissa massa* (Carter, 1887) for the first time.

Distribution.— The species is known from the Red Sea (Paulson, 1775), Kenya and Zanzibar (Bruce, 1976); Andaman Islands (Kemp, 1925), Tuamotu Islands (Nobili, 1906 according to Bruce, 1967), Great Barrier Reef, Australia (Bruce, 1981), South China Sea (Bruce, 1979a; Li, 1996); Japan (Miyake & Fujino, 1967; Hayashi, 1999, 2006). The species is now recorded for the first time from the Philippines, Indonesia, Maldives, Vanuatu and the Hawaiian Islands.

Remarks.— The present specimens fit the original description of the species by Paulson (1875) and subsequent descriptions. Nobili (1907) first illustrated the blunt postorbital tubercle on the carapace. Kemp (1925) mentions a lateral plate of the sixth abdominal somite with 1 or 2 teeth. In the present material there are always 2 teeth present. Features not previous recorded are (1) the row of plumose setae below the lateral carina at the base of the rostrum, (2) the fifth pereiopods being usually more robust than the third and fourth pereiopod, especially with regard to the propodus, and (3) the presence of rudimentary exopods on the third maxilliped and pereiopods. In most specimens these rudimentary exopods were present. In the specimens from Bali (RMNH.CRUS.D.50391) often only the pits where the rudimentary exopods were attached were visible. I do not recall having seen these structures elsewhere within the Caridea.

Anchistioides willeyi (Borradaile, 1900) (figs. 4-9, 10C)

Palaemonopsis willeyi Borradaile, 1900: 410, pl. 36 figs. 7-7b, pl. 37 figs. 7c-e [type locality: Ralun, New Britain].

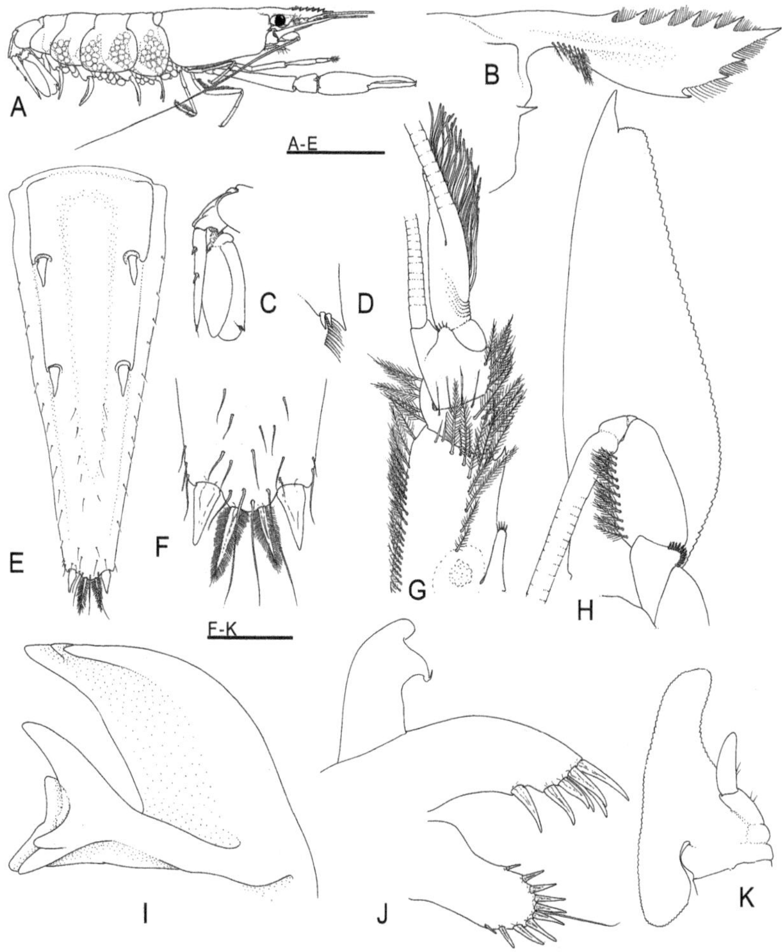

Fig. 4. *Anchistioides willeyi* (Borradaile, 1900), Vanuatu, ovigerous female, pocl. 9.7 mm, type with robust second chela with fingers about as long as palm, RMNH.CRUS.D.57864. A, Habitus, lateral; B, rostrum, lateral; C, sixth abdominal somite and tail fan, lateral; D, distolateral part of exopod of uropod; E, telson, dorsal view; F, idem, tip; G, antennular peduncle, dorsal view; H, antennal peduncle, ventral view; I, mandible; J, maxillula; K, maxilla. Scale: A = 8 mm; B = 2 mm; C = 4 mm; D = 1 mm; E, G, H, K = 0.5 mm; F, I, J = 0.125 mm.

Amphipalaemon willeyi — Nobili, 1901: 5; Borradaile, 1915: 2009; Borradaile, 1917: 407, pl. 59 fig. 13.

Amphipalaemon cooperi Borradaile, 1915: 209 [type locality: S. Nilandu Atoll, Maldive Islands].

Amphipalaemon gardineri Borradaile, 1915: 209 [type locality: N. Malé Atoll, Maldive Islands]; Borradaile, 1917: 407, pl. 59 fig. 14.

Amphipalaemon australiensis Balss, 1921: 11, figs. 3-6 [type locality: Cape Jaubert, "45 Meilen WSW, 66 Fuss Tiefe"].

Anchistioides australiensis — Gordon, 1935: 341, figs. 23d, 24d; Bruce, 1971: 24, fig. 6; Chace & Bruce, 1993: 132.

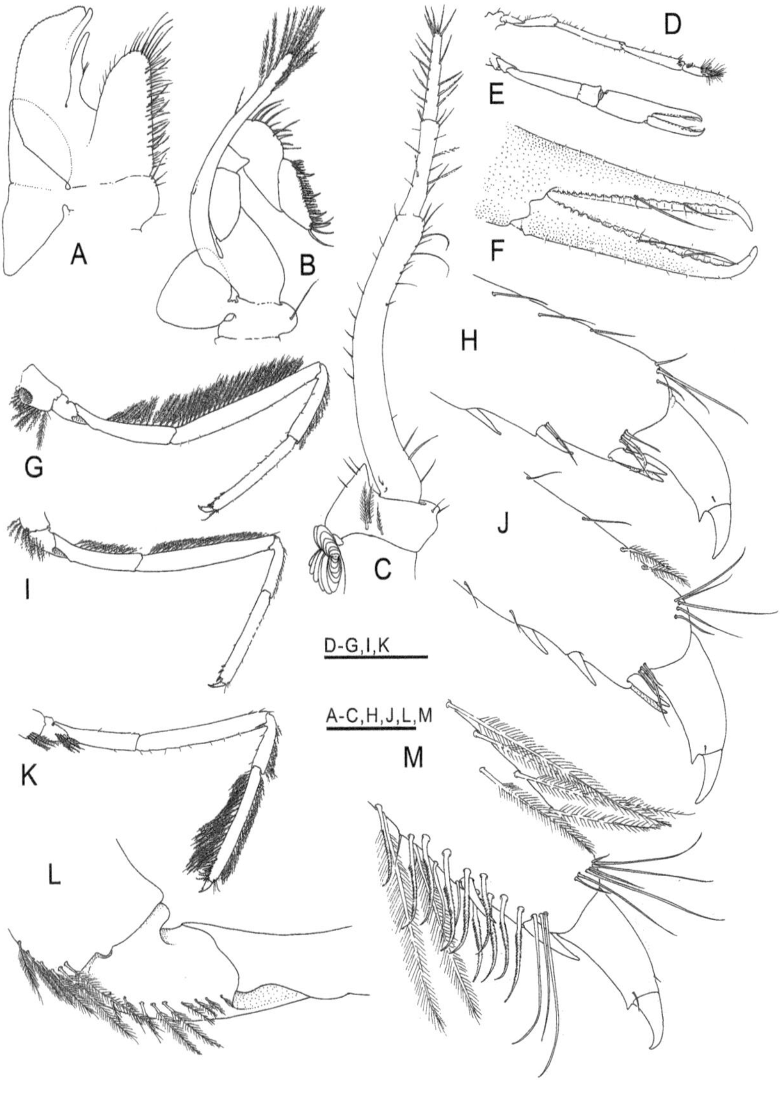

Fig. 5. *Anchistioides willeyi* (Borradaile, 1900), Vanuatu, ovigerous female, pocl. 9.7 mm, type with robust second chela with fingers about as long as palm, RMNH.CRUS.D.57864. A, First maxilliped; B, second maxilliped; C, third maxilliped; D, first pereiopod; E, second pereiopod; F, idem, fingers of chela; G, third pereiopod; H, idem, distal part propodus and dactylus; I, fourth pereiopod; J, idem, distal part propodus and dactylus; K, idem, basis; L, fifth pereiopod; idem, distal part propodus and dactylus. Scale: A-C = 1.25 mm; D = 4 mm; E = 8 mm; F, G, I, K = 2 mm; H, J, L, M = 0.25 mm.

Anchistioides willeyi — Gordon, 1935: 344, figs. 23a, 24a; Holthuis, 1952: 214, 215, figs. 106, 107; Bruce, 1971: 22, fig. 8; Bruce, 1975: 25, colour fig. 7; Bruce, 1976: 464; Bruce, 1978: 285, fig. 44; Bruce, 1979a: 241; Bruce, 1979b: 339, 342; Bruce, 1981: 2; Bruce, 1986: 469; Bruce, 1990: 211; Bruce, 1991a: 400, fig. 70; Bruce, 1991b: 269, figs. 3 g, 29, 30; Fransen, 1994: 86, pl. 1A;

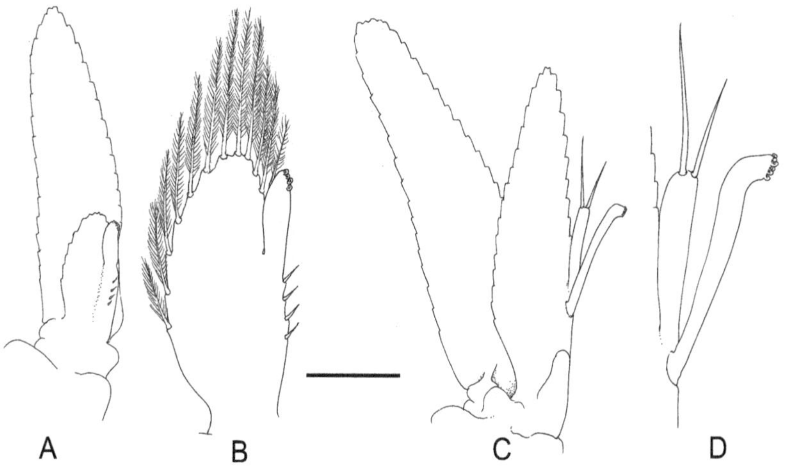

Fig. 6. *Anchistioides willeyi* (Borradaile, 1900), Vanuatu, male, pocl. 5.7 mm, type with robust second chela with fingers about as long as palm, RMNH.CRUS.D.57862. A, First pleopod; B, idem, endopod; C, second pleopod; D, idem, detail appendix masculina and appendix interna. Scale: A, C = 0.5 mm; B, D = 0.25 mm.

Bruce & Coombes, 1995: 137, fig. 12 g, h; Jones & Morgan, 2002: 69, colour fig.; Li et al., 2007: 45, fig. 9; Bruce, 2010: 50.
Anchistioides gardineri — Gordon, 1935: 345, figs. 23b, 24b.
Anchistioides cooperi — Gordon, 1935: 345, figs. 23c, 24c.
Anchistioides aff. *willeyi* — Bruce, 1976: 464.

Material examined.— **Indonesia**: One ovigerous female pocl. 8.3 mm (R = 8/4), 3 females pocl. 8.4 mm (R = 8/4), 8.6 mm (R = 9/4), 3.0 mm (R = 8/3), RMNH.CRUS.D.49822: Sulawesi, Spermonde Archipelago; 2001; in sponge *Niphates olemda* (de Laubenfels, 1954); collected and donated by N. de Voogd. One female pocl. 6.7 mm (R = 9/4), RMNH.CRUS.D.46487: SW Sulawesi, Spermonde Archipelago, Barang Lompe, NW side; depth 5 m; scuba diving; 23.ix.1994; in brown sponge with yellow-green tissue; leg. C. H. J. M. Fransen. One specimen, pocl. 6.4 mm, ZMA.CRUS.D.202808: Anchorage off Pulu Jedan, E coast of Aru-Islands (Pearl-banks), 23-26.xii.1899, collected by O. K. v. Arvell. One specimen, pocl. 7.9 mm, ZMA.CRUS.D.202809: Anchorage off Pulu Jedan, E coast of Aru-Islands (Pearl-banks), 23-26.xii.1899, collected by O. K. v. Arvell. One ovigerous female, pocl. 10.5 mm, RMNH.CRUS.D.4759: 4°25′S 114°31′E, 4.x.1908, in sponge, Gier Expedition No. 3, No. 12. — **Philippines**: Two females pocl. 6.5 mm (R = 9/4) and 5.4 mm (R = 8/3), RMNH.CRUS.D.48324: Sta. CEB.08: Cebu Strait, W of Bohol, S side of Cabilao Island, Cabacungan Point; 9°51.55′N 123°45.95′E; overhanging wall with caves; scuba-diving; 11.xi.1999; depth 20 m; in sponge on dead antipatharian; collected by C. H. J. M. Fransen. — **Seychelles**: One male, carapace damaged (R = 8/4), RMNH.CRUS.D.45465: NIOP-E stn. 720, E of Bird Island; 3°45′S 55°14′E; sediment mainly *Halimeda*, depth: 45 m; 1.2 m Agassiz trawl; in tubes of brown sponge *Spheciospongia inconstans* (Dendy, 1887) (det. R. van Soest); 20.xii.1992. One male pocl. 8.13 mm (R = 8/4), 1 female pocl. 8.13 mm (R = 5/3 regenerated), RMNH.CRUS.D.42823: NIOP-E stn 717: Bird Island, off E coast; 3°43′S 55°13′E; edge of bank, depth ca. 10 m; scuba-diving; in brown sponge *Spheciospongia inconstans* (det. R. van Soest); 20/21.xii.1992; collected by C. H. J. M. Fransen Photo 6/24-29. One female pocl. 3.44 mm (R = 7/3), RMNH.CRUS.D.42800: NIOP-E stn 738: Seychelles, SE of Mahé; 4°45′S 55°33′E; depth: 35-45 m; rectangular dredge; in sponge *Spheciospongia* (det. R. van Soest); 24.xii.1992, collected

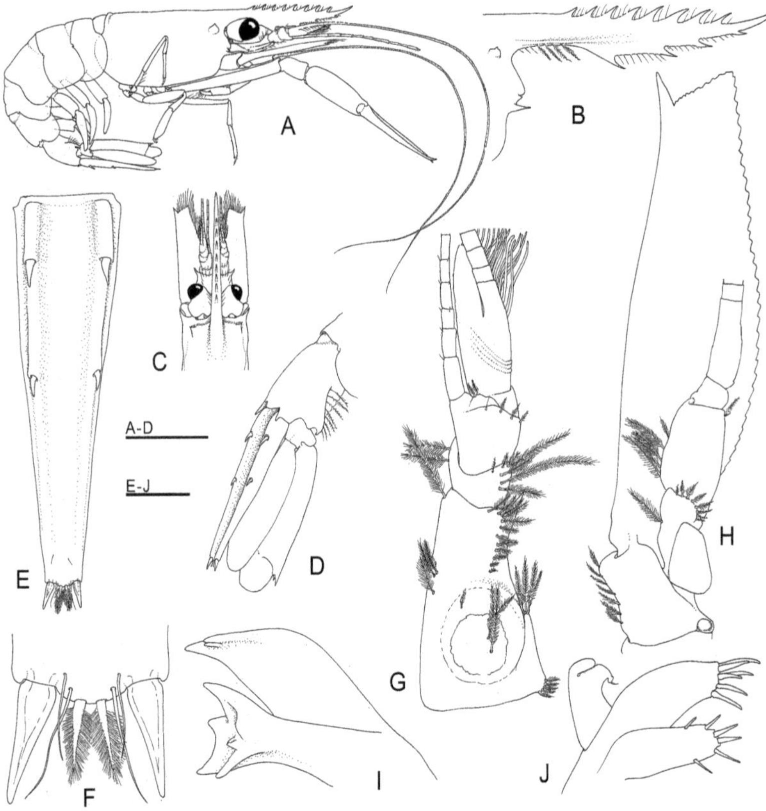

Fig. 7. *Anchistioides willeyi* (Borradaile, 1900), Vanuatu, non-ovigerous female, pocl. 3.0 mm, type with slender second chela with fingers distinctly longer than palm, RMNH.CRUS.D.57865. A, Habitus, lateral; B, rostrum, lateral; C, anterior appendages, dorsal; D, sixth abdominal somite and tail fan, lateral; E, telson, dorsal; F, idem, tip; G, antennular peduncle, dorsal; H, antennal peduncle, ventral; I, mandible; J, maxillula. Scale: A, C = 2 mm; B, D = 1 mm; E, G, H = 0.5 mm; F = 0.125 mm; I, J = 0.25 mm.

by C. H. J. M. Fransen Photo 25/17-19. — **Vanuatu**: One ovigerous female pocl. 8.1 mm (R = 6/3), 1 male pocl. 5.7 mm (R = 6/3), RMNH.CRUS.D.57862: stn AT14; W Mavéa Island, Palikulo Bay, 15°23.7/24′S 167°12.9/13.5′E; 19.ix.2006; Agassiz trawl; depth 102-120 m; Santo Marine Biodiversity Survey. One male pocl. 6.1 mm (R = 5/3), RMNH.CRUS.D.57862: stn FP46; Aoré Island, Aimbuei Bay, 15°32.4′S 167°12.7′E; 2.x.2006; tangle net; depth 45-50 m; Santo Marine Biodiversity Survey. One ovigerous female pocl. 9.7 mm (R = 6/3), 1 male pocl. 4.7 mm (R = 6/3), RMNH.CRUS.D.57864: stn AT80; Segond Channel, 15°31.7′S 167°10.8′E; 12.x.2006; Agassiz trawl; depth 36-43 m; Santo Marine Biodiversity Survey. One non-ovigerous female pocl. 3.0 mm (R = 8/4), RMNH.CRUS.D.57865: stn FR28; Santo, W Palikulo Bay, 15°28.7′S 167°14.1′E; slowly declining coral slope; sandy reefflat; depth 3-34 m depth; in yellow sponge; 22.ix.2006; Santo Marine Biodiversity Survey CF99.— **Marshall Islands**: One male, pocl. 7.9 mm, RMNH.CRUS.D.10239: Jaluit, 8.ii.1948, collected by H. Edmondson.

Comparative material of *Anchistioides antiguensis* (Schmitt, 1924). — **Caribbean Sea**: Two non-ovigerous females, pocl. 5.7 and 4.5 mm, RMNH.CRUS.D.9251: Bermuda, St. George Island, The Reach, ii-v.1936, collected by J. F. G. Wheeler. One male, pocl. 3.4 mm, RMNH.CRUS.D.9252:

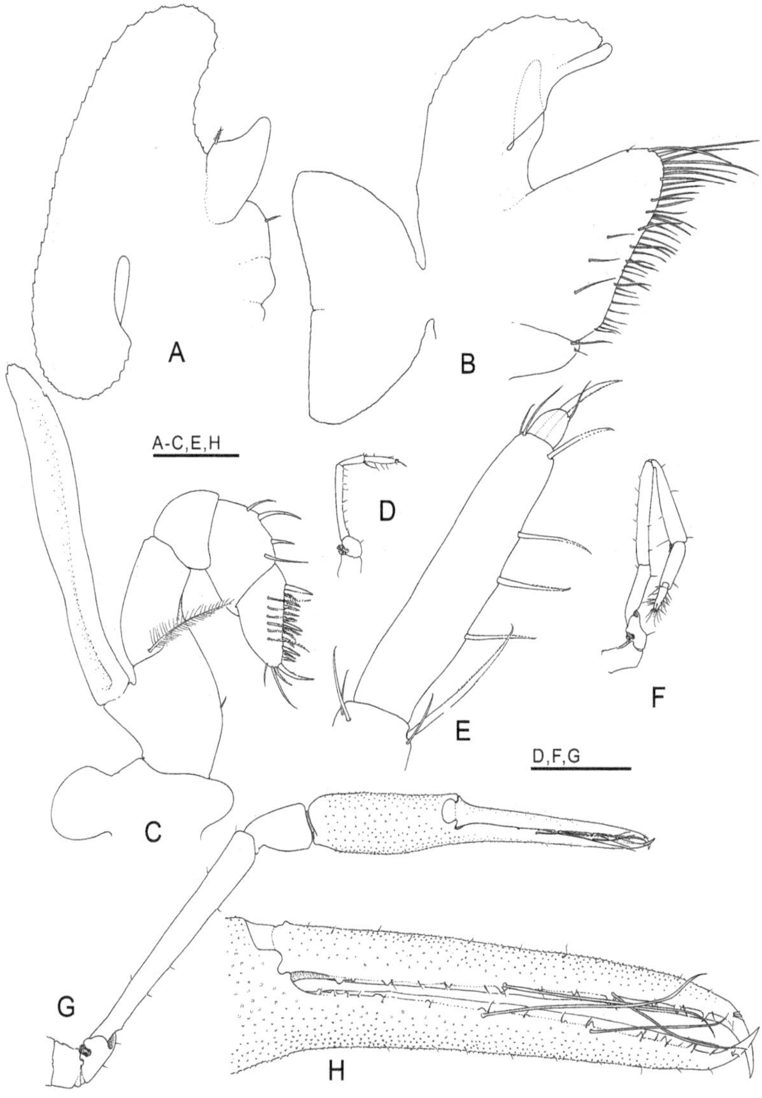

Fig. 8. *Anchistioides willeyi* (Borradaile, 1900), Vanuatu, non-ovigerous female, pocl. 3.0 mm, type with slender second chela with fingers distinctly longer than palm, RMNH.CRUS.D.57865. A, Maxilla; B, first maxilliped; C, second maxilliped; D, third maxilliped; E, idem, distal segment; F, first pereiopod; G, second pereiopod; H, idem, fingers of chela. Scale: A-C = 0.25 mm; D, F, G = 1 mm; E = 0.125 mm; H = 0.5 mm.

U.S.A., Florida, Tortugas, 21.vi.1931, collected by W. L. Schmitt. One male, pocl. 5.4 mm, RMNH.CRUS.D.27698: The Bahamas, Bimini, 1 km off the coast, 17.viii.1949, collected by P. Wagenaar Hummelinck. One non-ovigerous female, pocl. 5.5 mm, RMNH.CRUS.D.30353: R/V

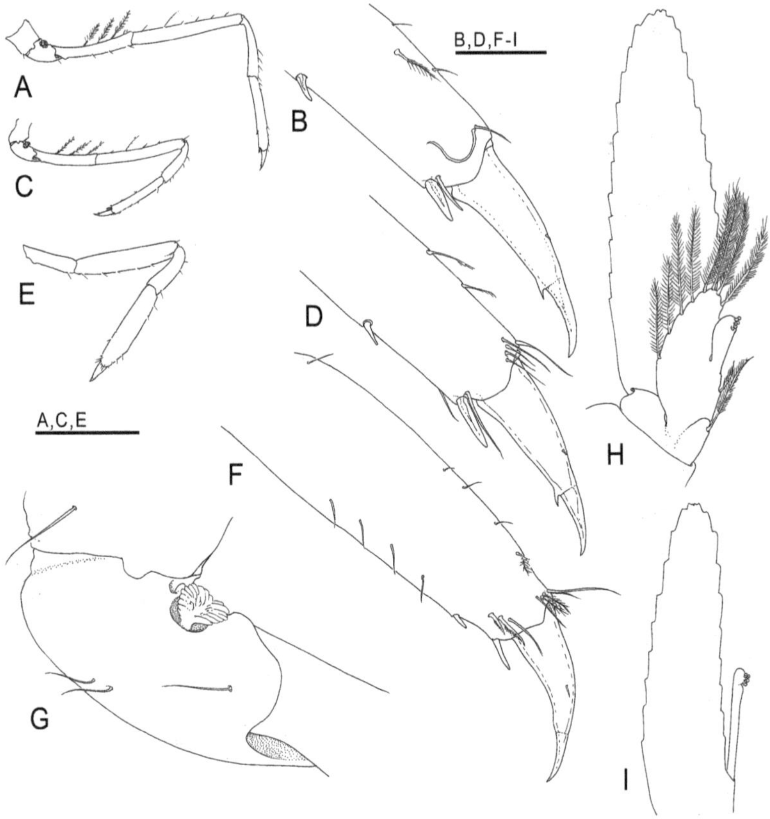

Fig. 9. *Anchistioides willeyi* (Borradaile, 1900), Vanuatu, non-ovigerous female, pocl. 3.0 mm, type with slender second chela with fingers distinctly longer than palm, RMNH.CRUS.D.57865. A, Third pereiopod; B, idem, distal part propodus and dactylus; C, fourth pereiopod; D, idem, distal part propodus and dactylus; E, fifth pereiopod; F, idem, distal part propodus and dactylus; G, idem, basis; H, first pleopod; I, second pleopod, endopod. Scale: A, C, E = 1 mm; B, D, F, I = 0.25 mm; G = 0.125 mm; H = 0.5 mm.

"Pillsbury" stn 1386, off the S coast of the Dominican Republic, 18°21.4′N 69°06.0′W — 18°22.8′N 69°06.6′W, 9.vii.1971, depth 149 m.

Description.— Carapace (figs. 4A, 7A) smooth, with robust marginal antennal tooth, hepatic tooth absent, with blunt but distinct postorbital tubercle. Rostrum (figs. 4B, 7B) deep at medial part, with 5-9 large dorsal teeth of about equal size and at equal distances, all anterior of orbit, distal part of dorsal lamina without teeth; ventral lamina with 3-4 robust teeth in distal part; lateral carina indistinct, proximally bordered with row of long plumose setae ventrally. Inferior orbital angle slightly produced and bluntly rounded.

Abdominal pleura (fig. 4A, 7A) of first to fifth somite rounded. Sixth abdominal somite (figs. 4C, 7D) with dorsal submedial carinae; posterodorsal angles rectan-

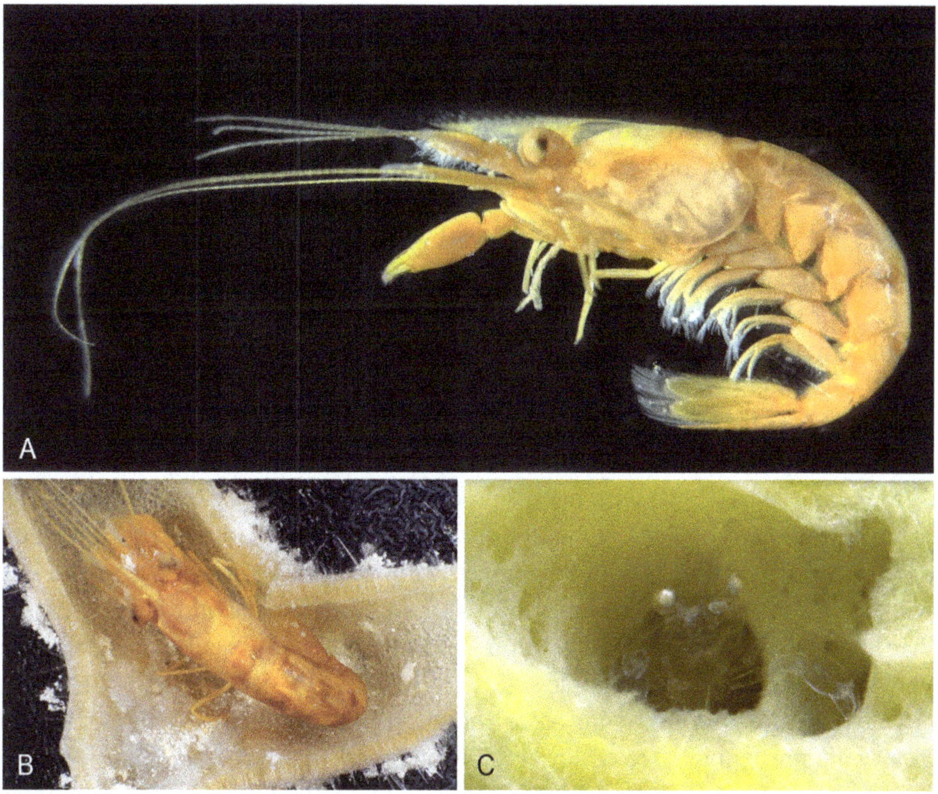

Fig. 10. *Anchistioides compressus* Paulson, 1875. A, Hawaiian Islands, U.S.A., UF47910, preserved specimen in lateral view; B, Hawaiian Islands, U.S.A., UF1528, preserved non-ovigerous female, pocl. 1.95 mm in host sponge; C, Maldives, RMNH.CRUS.D.57861, alive in host sponge *Leucetta chagosensis* Dendy, 1913. (Photographs A, B, by Jaaziel E. Garcia Hernandez; C, by Charles H. J. M. Fransen).

gular with or without sharp dorsal projections; posteroventral angle sharply produced. Telson (figs. 4C-D, 7D-E) almost twice as long as sixth abdominal somite, 2.75 times as long as proximal width, tapering distally; with submarginal longitudinal elevations in proximal half, with 2 pairs of dorsal spines at about 0.2 and 0.5 of telson length; distal margin (figs. 4F, 7F) rounded, about 0.3 times as wide as proximal width, with short, stout distolateral spines, about as long as dorsal spines, median pair of long plumose setae slightly longer than distolateral pair of spines.

Eye (figs. 4A, 7A) with hemispherical cornea; eyestalk short, about as long as cornea, tapering proximally.

Antennula (figs. 4G, 7G) with basal segment about 1.6 times as long as proximal width, with distinct though small distolateral tooth, anterior margin slightly produced; medioventral tooth distinct, acute, situated at about 0.6 of length of basal segment; stylocerite twice as long as wide, distally rounded, about half length of basal segment, with few setae distally. Intermediate segment short, about half as long as wide. Distal segment slightly longer than intermediate segment. Upper flagellum biramous, with proximal segments fused, short free ramus with about 6 segments, fused part and short ramus with many long aesthetascs, longer free ramus twice as long as carapace. Lower flagellum slender, about as long as carapace.

Antenna (fig. 4H, 7H) with basicerite unarmed. Carpocerite short and stout, about twice as long as wide. Scaphocerite 3.2 times as long as maximal width; lateral border slightly convex (fig. 4H) to slightly concave (fig. 7H), ending in strong distolateral tooth. Anterior margin of lamella small (fig. 4H) to moderately broad (fig. 7H), distomedially rounded (fig. 4H) or angular (fig. 7H), not overreaching distolateral tooth. Antennal peduncle reaching to one-third of scaphocerite length.

Thoracic sternites unarmed.

Mandible (figs. 4I, 7I) with incisor process broad, with three distal teeth; molar process proximally slender, with large protruding distal teeth, without setal bristles.

Maxillula (figs. 4J, 7J) with both upper and lower lacinia slender; upper lacinia with about 7 distal spines; lower lacinia with row of marginal teeth, slightly smaller than those of distal lacinia; palp bilobed, lower lobe with small ventral tubercle with single short, recurved simple seta.

Maxilla (figs. 4K, 8A) with basal endite obsolete, indistinctly bilobed; distal lobe with few simple setae; lower lobe unarmed; palp distinct, one- or two-segmented, not reaching distal margin of scaphognathite, with few short simple setae; scaphognathite with distal and proximal lobes well developed.

First maxilliped (figs. 5A, 8B) with coxal and basal endites partly fused, fringed with many simple and serrulate setae along entire median margin; exopod with short flagellum without distal setae; caridean lobe normal; palp slender, simple, not reaching beyond distal margin of caridean lobe, with few simple setae on medial margin; epipod large, triangular, bilobed.

Second maxilliped (figs. 5B, 8C) with endopod normally developed; dactylar segment triangular, fringed with many serrulate setae and few strong spines medially; distomedial margin of propod slightly developed, with few long serrulate setae; carpal segment triangular, unarmed; meral segment oblong, unarmed; ischial and basal segments fused; exopod broad, with distal and subdistal long plumose setae; coxal segment medially rounded, lateral margin with large rounded epipod.

Third maxilliped (figs. 5C, 8D-E) short, not reaching distal margin of carpocerite, with arthrobranch; coxa slightly produced medially, with lateral plate; basis medially slightly produced, without rudimentary exopod, completely fused with ischiomeral segment, combined segment oblong, curved, about 7 times as long as wide, with few simple setae proximomedially and distomedially; penultimate segment 0.3 times length of basal and ischiomeral segments combined, with few serrulate and plumose setae; ultimate segment as long as penultimate segment, with few serrulate setae.

Pereiopods usually without rudimentary exopod on basis, sometimes present in small specimens.

First pereiopod (figs. 5D, 8F) slender, chela about half as long as carpus, with fingers about as long as palm, with several groups of serrulate setae, cutting edges entire; cleaning apparatus present on carpo-propodal joint; carpus otherwise unarmed, slightly shorter than merus; merus twice as long as ischium, about 9 times as long as wide; basis shorter than unarmed ischium, without exopod; coxa unarmed.

Second pereiopod (figs. 5E, 8G) with chela long and stout, surface with many small tubercles, with fingers (figs. 5F, 8H) straight, about as long as palm (fig. 5E) to distinctly longer than palm (fig. 8G), cutting edges entire, with row of tubercles and pair of very long forward directed setae just medially of cutting edges; tips strongly hooked, fixed finger proximally slightly broader than dactylus; carpus short, about as long as distal width; merus straight, oblong, slightly increasing in width distally, about 4.5 times as long as wide; ischium short, about 0.3 times merus length; basis short.

Third (figs. 5G-H, 9A-B) and fourth (figs. 5I-J, 9C-D) pereiopods long and slender; dactylus about 0.14-0.18 length of propodus, flexor margin of corpus sinuous, tapering distally, ending in small, acute accessory tooth; unguis simple, acute, curved, about half of corpus length; propodus 9 times as long as width, with ventrodistal pair of simple short spines, and one or two spines along ventral margin in distal part; carpus, 0.65-0.75 of propodus length; merus 1.4-1.5 times propodus length; ischium as long as propodus; ischium, merus and carpus with row of many (fig. 5G, I) or few (fig. 9A, C) long plumose setae medially; basis without armature; coxa with or without lateral row of long plumose setae.

Fifth pereiopod (figs. 5K-M, 9E-F) as third and fourth, with ischium and merus without plumose setae, with propodus slightly longer, with (fig. 5L) or without (fig. 9E) rows of long plumose setae along both ventral and dorsal margins, and with (fig. 5M) or without (fig. 9F) oblique rows of strong serrulate setae in distoventral part, and with single ventral spine in ventrodistal part. Sometimes propodus more robust than in third and fourth pereiopods (fig. 9E)

Male first pleopod (fig. 6A) with endopod much shorter than exopod; endopod (fig. 6B) with distinct appendix interna with terminal hooked setae; median margin with row of few short simple setae; distolateral margin with long plumose setae.

Female first pleopod (fig. 9H) with endopod short and broad, with appendix interna almost reaching distal margin of endopod.

Male second pleopod (fig. 6C) slightly shorter than exopod, with appendix interna and appendix masculina medially; appendix masculina (fig. 6D) almost as long as appendix interna, with 2 long simple setae distally.

Female second pleopod (fig. 9I) with endopod with appendix interna.

Uropods (figs. 4C, 7D) with exopod feebly convex laterally, ending in large, acute distolateral tooth with one (fig. 7D), seldom two (fig. 4D) mobile spines just medially. Both endopod and exopod about as long as telson.

Size.— Large species with postorbital carapace length up to 13 mm.

Colouration.— Bruce (1971) writes that the colouration was noted as 'completely transparent in life' and by Bruce (1979a) as 'transparent, yellowish over pleura'.

Hosts.— *Anchistioides willeyi* was recorded living inside sponges for the first time by Holthuis (1952). Various authors recorded the species from unidentified sponges. Bruce (1976) recorded is from *Callyspongia fibrosa* (Ridley & Dendy, 1886), *Leuconia* spec., *Lissodendoryx isodyctialis* (Carter, 1882), and *Phloedictyon* spec. (= *Oceanapia* Norman, 1869). The species is now recorded as a symbiont of *Niphates olemda* (de Laubenfels, 1954) and *Spheciospongia inconstans* (Dendy, 1887) for the first time.

Distribution.— The species is known from the Red Sea (Marin, 2009), Maldives (Borradaile, 1915: type locality of *A. cooperi* and *A. gardineri*), Seychelles (Fransen, 1994), Kenya and Zanzibar (Bruce, 1976), Madagascar (Bruce, 1978), Indonesia (Holthuis, 1952; present material), Philippines (Holthuis, 1952; present material) South China Sea (Bruce, 1979a; Li, 1996), Northern Territories, Australia (Bruce & Coombes, 1995), the Great Barrier Reef, Australia (Bruce, 1971, 1981, 2010), Cape Jaubert, Western Australia (Balss, 1921, as *Amphipalaemon australiensis*); Arafura Sea, West New Guinea (Bruce, 1971, as *Anchistioides australiensis*); New Caledonia (Bruce, 1991b); Chesterfield Islands (Bruce, 1991a) and New Britain (Papua New Guinea) (Borradaile, 1900: type locality of *A. willeyi*). The species has been recorded from shallow waters to a depth of 127 m (Bruce, 1990). The species is now recorded from Vanuatu and the Marshall Islands for the first time.

Remarks.— The specimens fit the rather brief description by Borradaile (1900). Borradaile mentions 4 pairs of dorsal spines on the telson and also indicates these in his fig. 7c of pl. 37. Of the most distal two pairs, only one spine is drawn, whereas the pits of the other 3 are indicated. Borradaile (1917, pl. 59 fig. 13) shows at least the distalmost pair of dorsal spines again. This has not been observed in the present material nor has this been noted by others recording the species since. Holthuis (1952) provided a more comprehensive description, also indicating the variation in the dentition and relative length of the rostrum, and in the development of the postorbital tubercle as well as the relative length of the fingers of the second chelae in relation to the palm (1.08-1.44). He observed that the relative lengths of rostrum and fingers of the second chela are shorter in larger specimens than in smaller specimens. Bruce (1978, 1991a, b) recognized two types, which he thought might represent two separate species. In his 1978 paper, Bruce described one type with a relatively short rostrum not overreaching the scaphocerite and provided with 6-7 teeth dorsally and 3-4 ventrally, combined with a robust chela on the second pereiopods with the fingers about as long as the palm. The other type has a longer rostrum, overreaching the scaphocerite and provided with 8-13 dorsal teeth and 6 ventral teeth, combined with a slender chela on the second pereiopods with the fingers much longer than the palm. In the material of Bruce (1978) these differences cannot be related to sex nor size. In his material from New Caledonia, Bruce (1991b) recognized the two types with regard to the form of the second chela, however, not in combination with the dentition of the rostrum. In the New Caledonian material, specimens had a rostral formula of 6-10/3-4. The same holds for the specimens described herein, with a rostral formula of 5-9/3-4. Both types of second chela can be recognized in the present material as well. The type with the robust chela on the second pereiopod, with the fingers about as long as the palm, is depicted in figs. 4-6, the type with the slender chela of the second pereiopod with fingers longer than the palm is depicted in figs. 7-9.

Bruce (1971) was the first to describe the remarkable fringe of plumose setae along the dorsal borders of ischium and merus of the ambulatory pereiopods. Balss (1921, fig. 3), however, without describing this feature, already showed it in a drawing (as *Anchistioides australiensis* (Balss, 1921)). Balss (1921) described *A. australiensis* based on two ovigerous female specimens from Cap Jaubert, Australia, with the rostrum dorsally armed with 6 and ventrally with 3 teeth and overreaching the scaphocerite, whereas in *A. willeyi* (Borradaile) 6 dorsal and 4 ventral rostral teeth have been described and the rostrum not overreaching the scaphocerite. He also mentions the relatively short dactyli of the ambulatory pereiopods in *A. australiensis* compared to the long and slender ones of *A. willeyi* (Borradaile). Gordon (1935) reviewed the genus *Anchistioides*. She considered *A. australiensis* (Balss) to be a synonymn of *A. willeyi* (Borradaile). This was followed by Holthuis (1952), who reported that the description of *A. australiensis* was based upon not yet fully grown specimens (16 and 24 mm total length) of *A. willeyi*. Bruce (1971) described a very large specimen of *Anchistioides* (total length 50 mm), which he identified as *A. australiensis*, as it closely corresponded to Balss's description and illustrations. This specimen was found near the type locality of *A. australiensis* (Balss). He thus regarded *A. australiensis* a valid species, distinct from *A. willeyi*. This single large specimen misses the dorsal telson spines and has a much larger posterolateral tooth on the sixth abdominal somite than in other *Anchistioides* specimens known. Bruce regarded these characters an anomaly of this individual specimen. Moreover, the distal lamina of the scaphocerite in Bruce's specimen is reduced, and differs from that recorded for

A. willeyi. However, it has been shown that there is a great deal of variation in the width and form of the distal margin of the lamina in *A. willeyi* (cf. Holthuis, 1952; Bruce, 1971). I am, therefore, inclined to follow Gordon (1935) and Holthuis (1952) in regarding *A. australiensis* a synonym of *A. willeyi*, and accordingly I treat the specimen described as *A. australiensis* by Bruce (1971) as an aberrant specimen of *A. willeyi*.

In the present material, there is great deal of variation with respect to the remarkable fringe of plumose setae along the borders of the ischium, merus, carpus and propodus of the ambulatory pereiopods. In most specimens this fringe is present (fig. 5G, I, L) and the pereiopods are all slender. In some specimens, this fringe is not well developed (fig. 9A, C, E) and the fifth pereiopod is more robust than the third and fourth. These fringes of plumose setae along the ambulatory pereiopods might be functional during swimming. Pelagic swarming was suggested for this species by Bruce & Coombes (1995) as they collected two small specimens as the surface during the night by deploying lights. This pelagic swarming at night has been described extensively for *Anchistioides antiguensis*, the Atlantic sister species of *A. willeyi*, by Wheeler & Brown (1936) and Wheeler (1944). In the material of *A. antiguensis* studied for comparison here, no signs of these fringes of plumose setae were observed. In a male specimen (RMNH.CRUS.D.9251 and RMNH.CRUS.D.27698) however, the propodus of the fifth pereiopod is broadened like in some specimens of *A. willeyi* (fig. 9E, F). Bruce & Coombes (1995) noted the striking autapomorphic characters of *A. antiguensis* and *A. willeyi*, which they thought might be related to the pelagic feeding habits of these species. One being the fingers of the second chela with a pair of very long rigid simple setae (figs. 5f, 8H; Schmitt, 1924, pl. 3 fig. 60) that become erect on the opened fingers, but depressed when the fingers are closed. The other being the strange morphology of the molar process of the mandibles, with large protruding distal teeth but without setal bristles. Although Wheeler & Brown (1936) found a certain periodicity in the pelagic swarming of *A. antiguensis*, they could not relate this to tides, temperature, reproductive period, nor directly to lunar control. Wheeler (1944) expected that the periodicity of the swarming might be related to the periodicity of the species' food, the polychaete worm *Perinereis anderssoni* Kinberg, 1865 (as *Perinereis melanocephala* McIntosh, 1885), of which he found setae in the stomach content of the shrimp. Sponges as a host for *A. antiguensis* were first recorded by Williams (1984) and recently confirmed by De Grave & Anker (2017) and Soledade et al. (2017).

Another varying feature is the development of the supraorbital tubercle. This can be obtuse (fig. 4B) to acute (fig. 7B).

The rudimentary exopods on third maxillipeds and pereiopods are usually absent. In smaller specimens these can be present, however.

KEY TO THE INDO-WEST PACIFIC SPECIES OF *Anchistioides* PAULSON, 1875

1. Rostrum with 6-7 shallow teeth ventrally; supraorbital tubercle obtuse; stylocerite well developed, distally acute; scaphocerite with lamina distally broad, distomedially pointed, overreaching distolateral tooth; second pereiopod chelae glabrous, fingers distinctly shorter than palm with entire cutting edges; ambulatory pereiopods devoid of plumose setae; sixth abdominal somite with two equally developed posterolateral teeth; telson with anterior pair of dorsal spines more close to the distal pair than to anterior margin of telson, with pair of distolateral spines . . . . *A. compressus*
– Rostrum with 3-4 protruding teeth ventrally; supraorbital tubercle usually acute; stylocerite reduced to a mere tubercle; scaphocerite with distal lamina small, distomedially rounded or angular, not overreaching distolateral tooth; second pereiopod chelae minutely tuberculate, fingers slightly to much longer than palm, with cutting edges with series of rounded teeth over entire length; ambulatory pereiopods with rows of plumose setae; sixth abdominal somite with one distinct posterolateral tooth with a small tubercle medially; telson with anterior pair of dorsal spines more close to anterior margin of telson than to distal pair of dorsal spines, without pair of distolateral spines . . . . *A. willeyi*

CONCLUSIONS AND DISCUSSION

The species *A. compressus* has now been recorded for the first time from the Philippines, Indonesia, the Maldives, Vanuatu and the Hawaiian Islands. The sponge species *Coelocarteria singaporense* (Carter, 1883), *Acanthella cavernosa* Dendy, 1922, *Melophlus sarasinorum* Thiele, 1899, *Leucetta chagosensis* Dendy, 1913 and *Stylissa massa* (Carter, 1887) constitute new host records. *Anchistioides willeyi* is herein recorded from Vanuatu and the Marshall Islands and as a symbiont of *Niphates olemda* (de Laubenfels, 1954) and *Spheciospongia inconstans* (Dendy, 1887) for the first time.

Within the species of *Anchistioides*, there seems to be a great deal of morphological variation. This variation is especially prominent in the rostrum, second pereiopods and ambulatory pereiopods. It has been suggested that in *A. willeyi* the morphological types recognized might refer to separate species (Bruce, 1978, 1991a, b). Based on a large number of specimens in the present study it seems, however, that this is mere infraspecific variation. A molecular analysis will be needed to draw more definite conclusions, though. It is here suggested that part of the morphological variation observed might be related to pelagic swarming. Pelagic swarming has been described for *A. antiguensis* by Wheeler & Brown (1936) and Wheeler (1944) and was suggested for *A. willeyi* by Bruce & Coombes (1995). The fringes of plumose setae on the ambulatory pereiopods in part of the specimens of *A. willeyi* and the broadening of the propodus in the fifth pereiopod in part of the material of both *A. willeyi* and *A. antiguensis*, might be related to sessile and pelagic phases in their biological cycle.

ACKNOWLEDGEMENTS

It is with great pleasure that I dedicate this contribution to Prof. Dr. Ruiyu Liu. His long years of intense and unconditional dedication to the study of Indo-West Pacific shrimp taxonomy and systematics yielded important contributions to the knowledge of this group. The author had the opportunity to meet Prof. Liu when he was visiting Prof. dr. L. B. Holthuis at the Rijksmuseum van Natuurlijke Historie (now Naturalis Biodiversity Center) in Leiden in 1992. He is remembered as a dignified and friendly colleague.

Jaaziel E. Garcia Hernandez is acknowledged for making the specimens from the Hawaiian Islands available for study.

The Seychelles Expedition (1992-1993) was carried out as part of the Netherlands Indian Ocean Programme, coordinated by Dr. J. van der Land (Naturalis Biodiversity Center).

The Naturalis–USC Cebu Strait Expedition (1999) was coordinated by Dr. B. W. Hoeksema (Naturalis Biodiversity Center) and Prof. Dr. Thomas Heeger and Prof. Dr. Filipina B. Sotto (University of San Carlos).

Fieldwork on Bali (2000) was carried out within the Fauna Malesiana Marina research programme of the National Museum of Natural History — Naturalis at Leiden and coordinated by Dr. B. W. Hoeksema (Naturalis Biodiversity Center) in cooperation with Dr. Anna E. W. Manuputty and Ir. M. I. Yosephine Tuti of the Research and Development Centre for Oceanology, Jakarta (PPPO-LIPI).

The SANTO 2006 Global Biodiversity Survey was initiated by the Muséum national d'Histoire naturelle, Paris (MNHN), Institut de Recherche pour le Développement (IRD) and ProNatura International. Dr. Philippe Bouchet acted as expedition leader for the marine theme (Bouchet et al., 2011).

The Lembeh Strait Marine Biodiversity Workshop (2012) was organized by Universitas Sam Ratulangi (UNSRAT in Manado) and the Bitung field station of the Research Centre of Oceanography (PPO-LIPI) in cooperation with Naturalis Biodiversity Center. Prof. dr. Suharsono, former director of the Research Centre for Oceanography (PPO-LIPI), is acknowledged for his support. Prof. Dr. Markus T. Lasut (Universitas Sam Ratulangi), Dr. Bert W. Hoeksema (Naturalis Biodiversity Center) and Ir. Yosephine Tuti (PPO-LIPI) acted as workshop coordinators.

Dr. Francesca Benzoni organized the Marine Biodiversity Workshop at the Marine Research and Higher Education Centre University of Milano-Bicocca, Magoodhoo Island, Faafu Atoll, Republic of Maldives, 2014.

REFERENCES

BALSS, H., 1921. Results of Dr. E. Mjöbergs Swedish Scientific Expeditions to Australia 1910-13. XXIX. Stomatopoda, Macrura, Paguridea und Galatheidae. Bihang till Kungliga Svenska Vetenskapsakademiens Handlingar, **61**: 1-24.

BORRADAILE, L. A., 1900. On the Stomatopoda and Macrura brought by Dr. Willey from the South Seas. In: A. WILLEY, Zoological results based on material from New Britain, New Guinea, Loyalty Islands and elsewhere, collected during the years 1895, 1896, and 1897, by Arthur Willey, D. Sc. Lond., Hon. M. A. Cantab.: 395-428, pls. 36-39. (University Press, Cambridge).

BORRADAILE, L. A., 1915. Notes on carides. The Annals and Magazine of Natural History, (8) **15**: 205-213.

BORRADAILE, L. A., 1917. The Percy Sladen Trust Expedition to the Indian Ocean in 1905, under the leadership of Mr. J. Stanley Gardiner, M.A. No. VIII. — On the Pontoniinæ. Transactions of the Linnean Society of London, (2) (Zoology) **17**: 323-396.

BOUCHET, P., H. LE GUYADER & O. PASCAL (eds.), 2011. The natural history of Santo. Patrimoines Naturels, **70**: 1-572.

BRUCE, A. J., 1967. The results of the re-examination of the type specimens of some pontoniid shrimps in the collection of the Muséum national d'Histoire naturelle, Paris. Bulletin du Muséum national d'Histoire naturelle, (2) **39**: 564-572.

BRUCE, A. J., 1971. Records of some rare pontoniid shrimps from Australian waters, with remarks upon the mouthparts of some species of the genus *Periclimenes* Costa, 1844. Zoologische Verhandelingen, Leiden, **114**: 1-32.

BRUCE, A. J., 1975. Coral reef shrimps and their color patterns. Endeavour, **34**: 23-27.

BRUCE, A. J., 1976. A report on a small collection of pontoniine shrimps from the northern Indian Ocean. Journal of the Marine Biological Association of India, **16** [for 1974]: 437-454.

BRUCE, A. J., 1978. A report on a collection of pontoniine shrimps from Madagascar and adjacent seas. Zoological Journal of the Linnean Society, London, **62**: 205-290.

BRUCE, A. J., 1979a. Records of some pontoniine shrimps from the South China Sea. Cahiers de l'Indo-Pacifique, **1**: 215-248.

BRUCE, A. J., 1979b. The evolution and zoogeography of shallow-water tropical shrimps. New Zealand Department of Scientific and Industrial Research, New Zealand, Information Series, **137**: 337-355.

BRUCE, A. J., 1981. Pontoniine shrimps of Heron Island. Atoll Research Bulletin, **245**: 1-33.

BRUCE, A. J., 1986. Observations on the family Gnathophyllidae Dana, 1852 (Crustacea: Decapoda). Journal of Crustacean Biology, **6**(3): 463-470.

BRUCE, A. J., 1990. Crustacea Decapoda: deep-sea palaemonid shrimps from New Caledonian waters. In: A. CROSNIER (ed.), Résultats des Campagnes MUSORSTOM, **6**. Mémoires du Muséum national d'Histoire naturelle, (A) (Zoologie) **145**: 149-215.

BRUCE, A. J., 1991a. Crustacea Decapoda: further deep-sea palaemonoid shrimps from New Caledonian waters. In: A. CROSNIER (ed.), Résultats des Campagnes MUSORSTOM, **9**. Mémoires du Muséum national d'Histoire naturelle, (A) (Zoologie) **152**: 299-411.

BRUCE, A. J., 1991b. Shallow-water palaemonoid shrimps from New Caledonia (Crustacea: Decapoda). In: B. RICHER DE FORGES (ed.), Le benthos des fonds meubles des lagons de Nouvelle-Calédonie, **1**: 221-279. (ORSTOM, Paris).

BRUCE, A. J., 2010. Pontoniine shrimps (Crustacea: Decapoda: Palaemonidae) from the CReefs 2009 Heron Island Expedition, with a review of the Heron Island pontoniine fauna. Zootaxa, **2541**: 50-68.

BRUCE, A. J., 2012. Notes on some Indo-Pacific Pontoniinae, LI. *Periclimenaeus quadridentatus* (Rathbun, 1906) and *P. crassipes* (Calman, 1939) (Decapoda, Pontoniinae), with the designation of *P. calmani* and *P. serenei* spp. nov. Crustaceana, **85**(4-5): 513-531.

BRUCE, A. J. & K. E. COOMBES, 1995. The palaemonoid shrimp fauna (Crustacea: Decapoda: Caridea) of the Cobourg Peninsula, Northern Territory. The Beagle, Records of the Museums and Art Galleries of the Northern Territory, **12**: 101-144.

CHACE JR., F. A. & A. J. BRUCE, 1993. The caridean shrimps (Crustacea: Decapoda) of the Albatross Philippine Expedition, 1907-1910, part 6: Superfamily Palaemonoidea. Smithsonian Contributions to Zoology, **543**: 1-152.

De Grave, S. & A. Anker, 2017. An annotated checklist of marine caridean and stenopodidean shrimps (Malacostraca: Decapoda) of the Caribbean coast of Panama. Nauplius, **25**: e2017015. DOI:10.1590/2358-2936e2017015.

Felder, D. L., F. Alvarez, J. W. Goy & R. Lemaitre, 2009. Decapoda (Crustacea) of the Gulf of Mexico, with comments on the Amphionidacea. In: D. L. Felder & D. K. Camp (eds.), Gulf of Mexico. Origin, waters and biota: 1019-1104. (Texas A&M University Press, College Station, TX).

Fransen, C. H. J. M., 1994. Marine palaemonoid shrimps of the Netherlands Seychelles Expedition 1992-1993. Zoologische Verhandelingen, Leiden, **297**: 85-152.

Gordon, I., 1935. On new or imperfectly known species of Crustacea Macrura. The Journal of the Linnean Society, London (Zoology) **39**(266): 307-351.

Hayashi, K.-I., 1999. Family Palaemonidae, subfamily Palaemoninae — Genera *Brachycarpus* and *Exopalaemon*. Prawns, shrimps and lobsters of Japan (108). Aquabiology, Tokyo, **21**(5): 389-393. [In Japanese.]

Hayashi, K.-I., 2006. Family Palaemonidae, subfamily Pontoniinae — Commensalism of Japanese species 3. Family Anchistioididae — Genus *Anchistioides*. Prawns, Shrimps and Lobsters of Japan (147). Aquabiology, Tokyo, **28**(3): 287-291.

Holthuis, L. B., 1952. The Decapoda of the Siboga Expedition. Part XI. The Palaemonidae collected by the Siboga and Snellius Expeditions with remarks on other species II. Subfamily Pontoniinae. Siboga Expeditie Monograph, **39a**: 1-253.

Holthuis, L. B., 1955. The recent genera of the caridean and stenopodidean shrimps (class Crustacea, order Decapoda, supersection Natantia) with keys for their determination. Zoologische Verhandelingen, Leiden, **26**: 1-157.

Jones, D. & G. Morgan, 2002. A field guide to crustaceans of Australian waters (2nd ed.). (Western Australian Museum, New Holland Publishers, Sydney).

Kemp, S., 1925. Notes on Crustacea Decapoda in the Indian Museum. XVII on various Caridea. Records of the Indian Museum, **27**: 249-342.

Li, X., 1996. The pontoniine shrimps from Nansha Islands, China. 1. Marine fauna and flora and biogeography of the Nansha Islands and neighbouring waters, **2**: 222-233. [In Chinese.]

Li, X., R. Liu, X. Liang & G. Chen, 2007. Fauna Sinica Invertebrata, **44**. Crustacea Decapoda Palaemonoidea: i-ii, 1-381. (Science Press, Beijing).

Marin, I., 2009. On the presence of *Anchistioides willeyi* (Borradaile, 1899) (Decapoda: Caridea: Anchistioididae) in the Red Sea. Zootaxa, **2243**: 53-56.

Miyake, S. & T. Fujino, 1967. On four species of Pontoniinae (Crustacea, Decapoda, Palaemonidae), found in Porifera inhabiting the coastal regions of Kyushu, Japan. Journal of the Faculty of Agriculture, Kyushu University, **14**(2): 275-291.

Nobili, G., 1901. Decapodi raccolti dal Dr. Filippo Silvestri nell'America meridionale. Bolletino dei Musei di Zoologia ed Anatomia comparata della R. Università di Torino, **16**(402): 1-16.

Nobili, G., 1906. Diagnoses préliminaires de Crustacés, Décapodes et Isopodes nouveaux recueillis par M. le Dr G. Seurat aux îles Touamotou. Bulletin du Muséum national d'Histoire naturelle, (1) **12**: 256-270.

Nobili, G., 1907. Recerche sui crostacei della Polinesia: decapodi, stomatopodi, anisopodi e isopodi. Memorie della Reale Accademia delle Scienze di Torino, (2) **57**: 351-430.

Paulson, O., 1875. Studies on Crustacea of the Red Sea with notes regarding other seas. Part 1 Podophthalmata and Edriophthalmata (Cumacea): i-xiv, 1-144, pls. 1-22. (Kiev). [In Russian.]

Rafinesque, C. S., 1814. Précis des Découvertes et travaux somiologiques de Mr. C. S. Rafinesque-Schmaltz entre 1800 et 1814 ou choix raisonné de ses principales découvertes en zoologie et en botanique, pour servir d'introduction á ses ouvrages futurs: 1-56. (Royale Typographie Militaíre, Palerme).

SOLEDADE, G. O., G. G. SANTOS, U. PINHEIRO & A. O. ALMEIDA, 2017. New records of association between caridean shrimps (Decapoda) and sponges (Porifera) in Abrolhos Archipelago, northeastern Brazil. Nauplius, **25**: e2017027. DOI:10.1590/2358-2936e2017027.

WHEELER, J. F. G., 1944. Lunar periodicity in animals. Revue Agricole de l'Ile Maurice, **23**(4): 151-156.

WHEELER, J. F. G. & F. A. BROWN, 1936. The periodic swarming of *Anchistioides antiguensis* (Schmitt) (Crustacea Decapoda) at Bermuda. Journal of the Linnean Society of London, (Zoology) **39**: 413-428.

WILLIAMS, A. B., 1984. Shrimps, lobsters, and crabs of the Atlantic coast of the eastern United States, Maine to Florida. (Smithsonian Institution Press, Washington, D.C.).

First received 16 October 2019.
Final version accepted 11 January 2020.

DECAPODA, AXIIDEA

[When citing this volume, please refer to *Crustaceana* 93 (2020), 1233-1546]

Prof. Ruiyu Liu Memorial Issue

A NEW SPECIES OF THE GENUS *EUCALLIAXIOPSIS* SAKAI, 2011 FROM THE SOUTH CHINA SEA (DECAPODA, AXIIDEA, EUCALLIACIDAE)

BY

WENLIANG LIU[1,2,4]) and GARY C. B. POORE[3,4])

[1]) Yangtze River Delta Estuary Wetland Station, Shanghai Key Lab for Urban Ecological Processes and Eco-Restoration, School of Ecological and Environmental Science, East China Normal University, Shanghai 200241, P.R. China

[2]) Institute of Eco-Chongming, Shanghai 200062, P.R. China

[3]) Museums Victoria, GPO Box 666, Melbourne, Vic. 3001, Australia

ABSTRACT

Eucalliaxiopsis liui sp. nov. is described from a specimen taken from the South China Sea. The new species differs from other species of *Eucalliaxiopsis* in the combination of a telson with a transverse posterior margin, without a transverse row of spiniform setae, and a hook on the upper margin of the carpus of the cheliped. *Eucalliaxiopsis liui* is unique in having a biramous female pleopod 2, its endopod slender, with finger-like appendix interna, and a slender segmented exopod.

RÉSUMÉ

Eucalliaxiopsis liui nov. sp. est décrite à partir d'un spécimen provenant du sud de la Mer de Chine. La nouvelle espèce diffère des autres espèces de *Eucalliaxiopsis* par l'association d'un telson avec le bord postérieur transversal sans rangée transversale de soies spiniformes, et un crochet sur le bord supérieur du carpe du chélipède. *Eucalliaxiopsis liui* est unique par son pléopode femelle 2 biramé, son endopode plus étroit avec un appendice interne en forme de doigt et un exopode segmenté plus mince.

INTRODUCTION

Study of the axiidean fauna (Crustacea, Decapoda) of the South China Sea has revealed a eucalliacid ghost shrimp belonging to a new species of *Eucalliaxiopsis* Sakai, 2011, herein described. The specimen was collected by the late Professor Ruiyu Liu (J. Y. Liu), to whom this paper is dedicated, during a visit to Hainan Province in 1955. The family Eucalliacidae Manning & Felder, 1991 and the genus

[4]) Corresponding authors; e-mails: wlliu@sklec.ecnu.edu.cn; gpoore@museum.vic.gov.au

 | DOI:10.1163/9789004464353_014

Eucalliaxiopsis were rediagnosed as part of a major revision of all callianassoid families by Poore et al. (2019) and revisited with descriptions of new species by Poore (2020).

MATERIAL AND METHODS

The holotype is deposited in the Institute of Oceanology, Chinese Academy of Sciences, Qingdao, China (IOCAS). The drawings were made with the aid of drawing tube mounted on a Zeiss Stemi Sv11 compound microscope.

SYSTEMATICS

Infraorder AXIIDEA de Saint Laurent, 1979
Family EUCALLIACIDAE Manning & Felder, 1991
Genus *Eucalliaxiopsis* Sakai, 2011
Eucalliaxiopsis liui sp. nov.
(figs. 1-3)

Material examined.— Holotype, ♀ (carapace length, 9.0 mm), Yinggehai, Hainan Province, South China Sea, in tow net, 18°30′36.85″N, 108°41′12.40″E, coll. Ruiyu Liu, 12 December 1955, MBM136788/55K-479 (IOCAS).

Description.— Carapace 0.8 times as long as pleomeres 1 and 2 combined (figs. 1, 2A). Frontal margin of carapace with broadly triangular acute rostrum, extending to third visible length of eyestalks, flanked by weak anterolateral angles. Carapace lacking dorsal, oval, cardiac prominence. Linea thalassinica strong, parallel to midline of carapace. Cervical groove distinct, disjunct near linea thalassinica. Hepatic boss strong, in anterior third of carapace ventral to linea thalassinica. Distinct suture (linea anomurica) ventral to hepatic boss extending posteroventrally to ventral margin of carapace, there continuing in complex network of sutures. Cardiac sulcus well defined, incomplete in midline, extending anteroventrally to linea anomurica.

Abdomen (figs. 1, 2B), smooth dorsally; dorsal length ratio (along midline) of pleomeres 1-6, 1.0 : 1.9 : 1.2 : 0.9 : 1.1 : 1.3. Pleomere 1 dorsally and posterolaterally sclerified, with anterodorsal section separated by shallow groove; pleomere 2 glabrous dorsally; pleonal terga 3-5 (fig. 2B) each with pair of longitudinal, anteriorly convergent grooves, extending to quarter, third and third of dorsal length from posterior margin of tergites 3-5 respectively; each with pair of dorsolateral transverse or oblique rows of plumose setae; pleomere 6 0.75 times as long as wide, bearing pair of short oblique grooves near posterior margin, lateral row of plumose setae in posterior third.

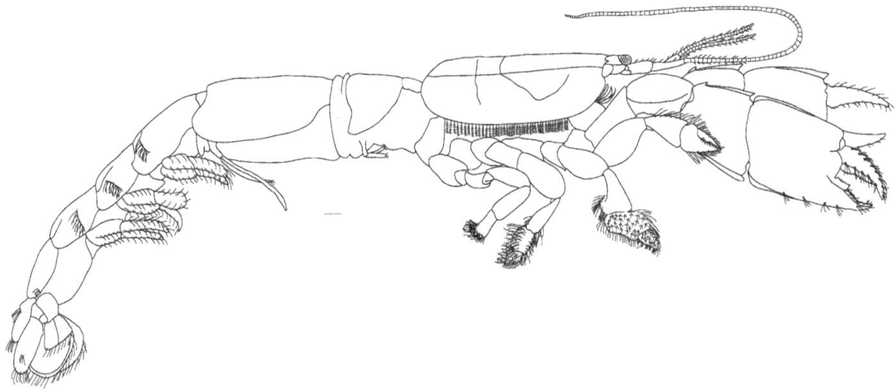

Fig. 1. *Eucalliaxiopsis liui* sp. nov., holotype female. Entire animal, lateral view. Scale = 1 mm.

Eyestalk (fig. 2A) twice as long as wide, with parallel slightly convex lateral margin, oblique-transverse distal margin, with acute medioapical lobe; overreaching distal margin of antennular peduncle article 1; cornea subterminal, disk-shaped, almost as wide as peduncle, swollen in lateral view, with scattered purple ommatidea (alcohol specimen).

Antennular peduncle as long as but slightly heavier than antennal peduncle (fig. 2A); article 2 1.3 times as long as basal article; article 3 about 0.9 times as long as article 2; articles 2 and 3 with ventrolateral row of long, ventrally directed setae,

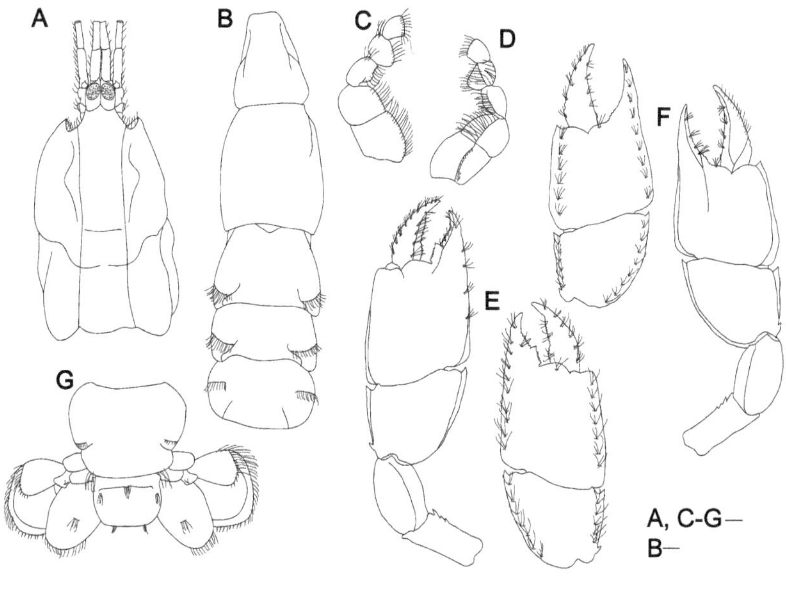

Fig. 2. *Eucalliaxiopsis liui* sp. nov., holotype female. A, Carapace, dorsal view; B, pleomere, dorsal view; C, D, left maxilliped 3, outer and inner view; E, right cheliped, lateral and mesial view; F, left cheliped, lateral and mesial view; G, telson and uropods, dorsal view. Scale = 1 mm.

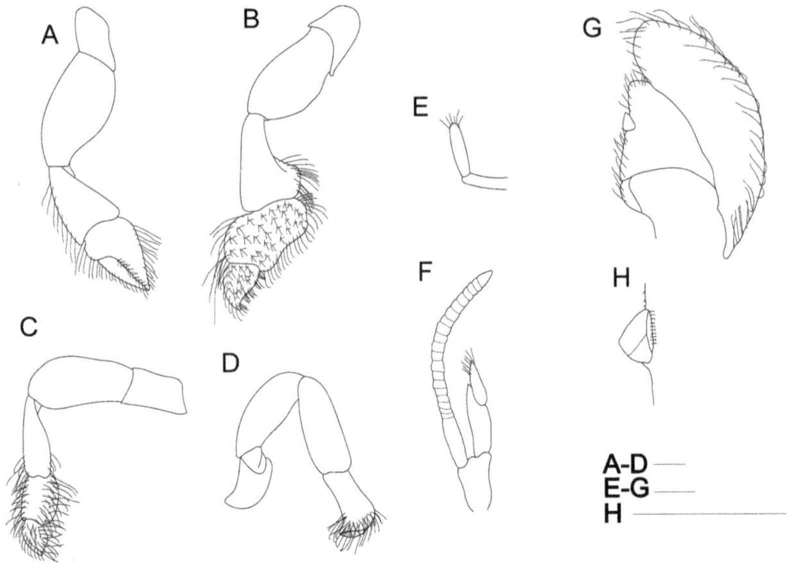

Fig. 3. *Eucalliaxiopsis liui* sp. nov., holotype female. A-D, Pereopods 2-5, lateral view; E, pleopod 1, posterior view; F, pleopod 2, posterior view; G, pleopod 3, posterior view; H, appendix interna of pleopod 3, posterior view. Scale = 1 mm.

continued onto ventral ramus of flagellum; rami of flagellum of about equal length, about 6 times as long as peduncle of article 3; dorsal ramus with sparse short setae, subterminal articles of dorsal ramus heavier than those of ventral ramus and with thick line of ventral aesthetascs.

Antennal peduncle as long as antennular peduncle (fig. 2A); article 1 with dorsolateral carina bearing regular line of fine setae above laterally produced excretory pore; article 2 0.6 times as long as basal article, broad, scaphocerite acute; article 4, longer than combined length of first three; article 5 narrower, slightly shorter than article 4; flagellum sparsely setose, extending posteriorly to middle of pleomere 1.

Maxilliped 3 (fig. 2C, D) without exopod; with long dense setation on mesial margin; ischium almost as long as wide, proximomesial lip rounded, inner face with crista dentata of 13 sharp teeth, basal three longer than others; merus subquadrate, almost as long as wide, lower margin convex, obtusely angled; carpus strongly flexed in proximal third with setose lobe on flexor margin; propodus subquadrate, with lower border expanded proximally, 1.2 times as wide as long; dactylus expanded terminally, as long as wide, fringed with dense field of close-set, stiff serrated setae on terminal margin.

Pereopods 1 almost equal in size, dissimilar in dentition of fixed fingers. Major cheliped (fig. 2E) ischium about 2.7 times as long as wide, upper margin almost straight, lower margin with three distal denticles; merus as long as ischium,

1.6 times as long as wide; upper and lower margins convex, unarmed; carpus increasing in width distally, slightly shorter than merus, lower margin arcuate, carinate, ending in sharp tooth, upper margin carinate, terminating in sharp tooth, with subproximal triangular spine; propodus upper margin 0.9 times width, length (including fixed finger) about 1.5 times width, inner surface of palm smooth, with tufts of setae below upper margin and above lower margin; lateral face with unarmed ridge extending posteriorly from just below gape on to fixed finger; upper and lower margins keeled, keel of lower becoming ill-defined beyond mid-length and absent on fixed finger; fixed finger triangular, cutting margin with well-defined denticular triangular tooth reaching mid-length; dactylus curved, longer than fixed finger, cutting margin with a row of small denticles on basal third.

Minor cheliped (fig. 2F) slightly smaller, similar in shape, ischium more slender, merus narrower, carpus and propodus 0.9 times as wide as major cheliped; carpus upper margin with subproximal triangular spine; fixed finger cutting edge with 3 denticles; dactylus more slender than in major cheliped.

Pereopod 2 (fig. 3A) chelate; merus about twice as long as wide, upper margin slightly convex, lower margin convex proximally; carpus subtriangular, shorter than merus; chela slightly shorter than carpus, with dense setae on lower and upper margins; palm shorter than fingers, upper margin slightly convex; dactylus 1.9 times as long as upper margin of palm; carpus and chela fringed with short to long setae along margins; each prehensile margin with small sharp corneous spiniform setae, tips of dactylus and fixed finger corneous.

Pereopod 3 (fig. 3B) merus about 1.8 times as long as wide, upper margin weakly sinuous, lower margin convex; carpus subtriangular, shorter and wider than merus, broadly flared distally to produce strong lower lobe, width there about 0.6 length, lower lobe with terminal field of long arched setae, diminishing in length toward articulation with propodus; propodus subrectangular, upper margin slightly convex, 0.4 length of carpus, lower margin with proximally directed short rounded lobe, marginally setose, lateral face with tufts of short setae; dactylus teardrop-shaped, about 1.4 times as long as wide, terminating in narrow corneous tip hooked toward external side, lower margin sinuous, external face crossed by fields of short, slightly hooked setae in which longest setae are near upper margin, with separate, dense field of slightly heavier short weakly hooked setae along lower extreme of external face and inferior margin.

Pereopod 4 (fig. 3C) ischium rectangular; merus 2.0 times as long as ischium; carpus 0.6 length of merus; propodus lower margin with rounded distal angle, with 1 long seta near end of upper margin, reaching well beyond tip of dactylus; dactylus teardrop shaped; dense setation on outer surface of propodus and dactylus divided into upper and lower fields, setae slightly stronger in lower fields of both, densest on dactylus, especially on and near lower margin.

Pereopod 5 (fig. 3D) slender, all articles unarmed; ischium rectangular; merus about 1.9 times as long as ischium; carpus almost as long as merus; propodus about 0.4 length of carpus, lower distal corner projecting to form a beak-like chela with dactylus, lateral surface beset distally with dense setae; dactylus hooked toward external side of fixed finger, tips of dactylus and fixed finger obtuse.

Telson (fig. 2G) about 1.4 times as wide as long, widest at midlength, postero-lateral margin rounded; posterior margin almost straight, with row of setae near each lateral margin; dorsal surface with strong transverse ridge, row of setae in notch at midpoint, pair of rows of setae near lateral margins. Uropodal protopod bearing dorsal spine; endopod oval, 1.5 times as long as wide, overreaching telson, distal margin with dense fringe of setae; exopod with dorsal plate falling short of distal endopod margin, distal edge of plate lined with short spiniform setae grading anteriorly to thinner longer setae of exopod margin, exopod posterodistal margin broadly rounded, with long spiniform setae at posterodistal curvature, distal margin with dense fringe of setae.

Female pleopod 1 (fig. 3E), uniramous, 2-articled; article 1 shorter than article 2; article 2 bearing short setae distally. Female pleopod 2 (fig. 3F) biramous; endopod slender, tapering terminally, with a finger-like appendix interna on distomesial margin; appendix interna slightly shorter than endopod, bearing short setae distally; exopod unarmed, slender, segmented, 15 times as long as wide, 4.1 times as long as endopod. Female pleopods 3-5 (fig. 3G) biramous, foliaceous, rami wide; appendix interna (fig. 3H) stubby, weakly projecting beyond margin of endopod, bearing numerous small hooks along mesial margin.

Etymology.— The species is named in honour of the late Professor Ruiyu Liu (J. Y. Liu), the Institute of Oceanology, Chinese Academy of Sciences, for his great contribution to Chinese carcinology.

Distribution and habitat.— Known only from the type locality. Intertidal zone on sand.

Remarks.— *Eucalliaxiopsis* is a genus of ten species, seven from the tropical Western Atlantic and seven from the Indo-West Pacific (Poore, 2020). *Eucalliaxiopsis liui* sp. nov. is similar to several species of *Eucalliaxiopsis* but differs from all in the well-developed sculpture and setation on pleomeres 3-5 and the unusual female pleopod 2. Long transverse rows of pleomeral setae have been illustrated in several species of *Eucalliaxiopsis* [e.g., *E. caeraensis* (Rodriques & Manning, 1992), *E. inaequimana* (Dworschak, 2014), *E. macilhennyi* (Felder & Manning, 1994), *E. panglaoensis* (Dworschak, 2006)] and are probably a feature of all species, as they are to varying degrees in all callianassoids. In most species the dorsal tergite is separated from the posterolateral margins of pleomeres 3-5 by short shallow oblique grooves visible in dorsal view. These grooves are longer and more obvious in the new species, extending anterior of the lateral setal rows. *Eucalliaxiopsis liui* shares with *E. dworschaki* Poore, 2020 a subproximal hook on the carpus of the chelipeds but the new species differs from *E. dworschaki* in the telson having a transverse posterior margin (concave in *E. dworschaki*), lacking a row of spiniform setae on the telson (present), and in an ovate uropodal endopod (terminally excavate).

The female pleopod 2 of most species of *Eucalliaxiopsis* so far described has a broad triangular endopod with a linear appendix interna at the midpoint, and a laminar exopod of similar length. *Eucalliaxiopsis panglaoensis* (Dworschak, 2006) and *E. dworschaki* differ in that the endopod appears to lack an appendix interna, having a setose lobe instead (Dworschak, 2006: fig. 4i; Poore, 2020: fig. 17j), a situation seen too in another eucalliacid, *Pseudocalliax tooradin* (Poore & Griffin, 1979) (Poore, 2020: fig. 26i). In the single female of *E. liui*, pleopod 2 the appendix interna dominates the apex of the endopod and the exopod is segmented. This may be a teratological condition, a possibility only confirmed by the collection of more specimens.

Male chelipeds, somewhat sexually dimorphic in *Eucalliaxiopsis*, and male pleopods are unknown in the new species.

ACKNOWLEDGEMENT

The first author thanks his teacher, a renowned carcinologist, the late Professor Ruiyu Liu (J. Y. Liu) (Institute of Oceanology, Chinese Academy of Sciences) for his considerable guidance in carcinology research.

REFERENCES

DE SAINT LAURENT, M., 1979. Vers une nouvelle classification des Crustacés Décapodes Reptantia. Bulletin de l'Office Nationale de Pêche de Tunisie, **3**: 15-31.

DWORSCHAK, P. C., 2006. A new species of *Eucalliax* Manning & Felder, 1991 (Decapoda: Callianassidae) from the Philippines. Raffles Bulletin of Zoology, **54**: 349-359.

DWORSCHAK, P. C., 2014. The Axiidea (Crustacea, Decapoda) of Cocos (Keeling) and Christmas Islands, with description of a new species of *Eucalliax* Manning & Felder, 1991. Raffles Bulletin of Zoology Supplement, **30**: 230-245.

FELDER, D. L. & R. B. MANNING, 1994. Description of the ghost shrimp *Eucalliax mcilhennyi*, new species, from south Florida, with reexamination of its known congeners (Crustacea: Decapoda: Callianassidae). Proceedings of the Biological Society of Washington, **107**: 340-353.

MANNING, R. B. & D. L. FELDER, 1991. Revision of the American Callianassidae (Crustacea: Decapoda: Thalassinidea). Proceedings of the Biological Society of Washington, **104**: 764-792.

POORE, G. C. B., 2020. Indo-West Pacific and Australian species of Eucalliacidae with descriptions of four new species (Crustacea, Axiidea). Memoirs of Museum Victoria, **79**: 51-91.

POORE, G. C. B., P. C. DWORSCHAK, R. ROBLES, F. L. MANTELATTO & D. L. FELDER, 2019. A new classification of Callianassidae and related families (Crustacea: Decapoda: Axiidea) derived from a molecular phylogeny with morphological support. Memoirs of Museum Victoria, **78**: 73-146.

POORE, G. C. B. & D. J. G. GRIFFIN, 1979. The Thalassinidea (Crustacea: Decapoda) of Australia. Records of the Australian Museum, **32**: 217-321.

RODRIGUES, S. D. A. & R. B. MANNING, 1992. Two new callianassid shrimps from Brazil (Crustacea: Decapoda: Thalassinidea). Proceedings of the Biological Society of Washington, **105**: 324-330.

SAKAI, K., 2011. Axioidea of the world and a reconsideration of the Callianassoidea (Decapoda, Thalassinidea, Callianassida). Crustaceana Monographs, **13**: 1-520.

First received 8 March 2019.
Final version accepted 16 April 2020.

DECAPODA, SCYLLARIDAE

[When citing this volume, please refer to *Crustaceana* 93 (2020), 1233-1546]

Prof. Ruiyu Liu Memorial Issue

A NEW SLIPPER LOBSTER OF THE GENUS *GALEARCTUS* HOLTHUIS, 2002 (DECAPODA, SCYLLARIDAE) FROM MADAGASCAR

BY

CHIEN-HUI YANG and TIN-YAM CHAN[1])
Institute of Marine Biology and Center of Excellence for the Oceans,
National Taiwan Ocean University, Keelung 20224, Taiwan, R.O.C.

ABSTRACT

Comparison on the abundant material of *Galearctus timidus* (Holthuis, 1960) recently collected from the Philippines and Madagascar showed that there are differences in the development of carapace teeth, abdominal sculpture, pereiopods ornamentation and coloration between the specimens from these two localities. Molecular analysis also reveals high genetic divergence between the two forms. Since the type locality of *G. timidus* is from the Philippines, the Madagascan form is described as a new species.

Key words. — New species, lobster, *Galearctus*, Madagascar

RÉSUMÉ

Des comparaisons sur l'abondant matériel de *Galearctus timidus* (Holthuis, 1960), récemment collecté aux Philippines et à Madagascar, ont montré des différences dans le développement des dents de la carapace, de la sculpture abdominale, de l'ornementation des pléopodes et de la coloration entre les spécimens des deux localisations. Les analyses moléculaires ont aussi montré une forte divergence génétique entre les deux formes. Comme le type de *G. timidus* est originaire des Philippines, la forme de Madagascar est décrite comme une nouvelle espèce.

Mots clés. — Nouvelle espèce, Scyllaridé, *Galearctus*, Madagascar

INTRODUCTION

The slipper lobster genus *Galearctus* Holthuis, 2002, which was split from *Scyllarus* Fabricius, 1775, is characterized by having an arborescent sculpture on the abdomen, fourth antennal segment with only one carina, anterior protrusion of the thoracic sternum bifurcate with a minute median incision, and the third

[1]) Corresponding author; e-mail: tychan@mail.ntou.edu.tw

 | DOI:10.1163/9789004464353_015

pereiopod with a board propodus which is lacking a distinct anteroventral tooth. Seven species are now known in *Galearctus*, namely: *G. timidus* (Holthuis, 1960), *G. kitanoviriosus* (Harada, 1962), *G. umbilicatus* (Holthuis, 1977), *G. aurora* (Holthuis, 1982), *G. rapanus* (Holthuis, 1993), *G. lipkei* Yang & Chan, 2010 and *G. avulsus* Yang, Chen & Chan, 2011.

Galearctus timidus is the type species of the genus and is believed to be widely distributed in the Indo-West Pacific (Holthuis, 2002). It is distinct in having a high gastric tooth, two parallel grooves present on the propodus of third pereiopod, and having transverse grooves on the articulate surfaces of second to fifth abdominal tergites (Holthuis, 2002; Yang et al., 2011). However, the abundant material assignable to *G. timidus* recently collected from the Philippines and Madagascar showed that there are probably two forms present in this species; differing in the development of carapace teeth, abdominal sculpture, pereiopods ornamentation and coloration. Molecular analysis also reveals high genetic divergence between the two forms. Since the type locality of *G. timidus* is from the Philippines and re-examination of the holotype showed the Philippines material belonging to the same species, the Madagascan form is determined to be an undescribed species.

MATERIAL AND METHODS

The types of the new species are deposited in the Muséum national d'Histoire naturelle, Paris (MNHN). Other specimens examined are deposited at the National Taiwan Ocean University (NTOU), National Museum of Marine Biology and Aquarium, Taiwan (NMMBA), Naturalis Biodiversity Center, Leiden (RMNH), and Zoological Reference Collection, Lee Kong Chian Natural History Museum, University of Singapore (ZRC). The measurement given is carapace length (cl), measured along the dorsal midline from tip of the rostrum to posterior margin of carapace. Morphological terminology used mainly follows Holthuis (2002) and Yang et al. (2011).

Comparative material of *Galearctus timidus*.— **Philippines.** Snellius Expedition, stn 60, holotype male 18 mm cl (RMNH.CRUS.D.14507). Bohol, Balicasag Island, tangle nets, 200-300 m, Dec. 2000, 2 males 21.7-22.9 mm cl, 2 ovigerous females 25.1-27.7 mm cl (ZRC 2001 0325); 2 males 20.6-25.3 mm cl, 1 ovigerous female 26.8 mm cl (ZRC 2001 0331); Jul. 2003, 1 male 18.5 mm cl, 3 females 21.6-26.3 mm cl (ZRC 2008 0541); Mar. 2004, 1 ovigerous female 27.7 mm cl (NTOU M02237); stn L11, 9°30.9′N 123°40.8′E, 87-111 m, 26 Oct. 2003, 1 female 24.8 mm cl (NTOU M02238). Northwest coast of Panglao, 120-160 m, Jan.-Dec. 2006, 1 male 20.4 mm cl, 1 ovigerous female 27.2 mm cl, 1 female 30.1 mm cl (ZRC 2020 0291); Jan.-Dec. 2007, 1 female 25.2 mm cl (NTOU M02243). Maribohol Bay, 100-300 m, Nov. 2003-Apr. 2004, 1 male 18.6 mm cl (ZRC 2020 0292); 3 males 18.4-24.5 mm cl (NTOU M02244); 11 Mar. 2004, 1 male 20.0 mm cl (NTOU M00982). Macton Island, Punta Engaño, coral point trawler, 180-250 m, 2011, 1 male 14.7 mm cl (NTOU M02246). Looc, 7-8 m, 18 Mar. 2012, 1 male 24.2 mm cl (NTOU M02247). PANGLAO 2004, Apr. 2004, 2 females 24.3-26.7 mm cl (ZRC 2020 0293); stn L42, 9°31.2′N 123°40.7′E,

80-90 m, 2 Jul. 2004, 1 male 22.1 mm cl (ZRC 2020 0294); stn L44, 9°30.8′N 123°41.0′E, 85-100 m, 3 Jul. 2004, 1 female 21.6 mm cl (NTOU M02239); no special date, 1 male 27.0 mm cl (NTOU M02240), 1 male 25.2 mm cl (NTOU M02241), 1 male 21.3 mm cl (NTOU M02242).— **Taiwan.** Donggang fishing port, Pingtung County, commercial trawler, 28 Feb. 2012, 1 male 7.8 mm cl (NMMBA CD5592).

SYSTEMATICS

Genus *Galearctus* Holthuis, 2002

Galearctus yuanshi sp. nov.

(figs. 1, 3A)

Galearctus timidus — Holthuis, 2002: 559-564 (in part) [not Holthuis, 1960].

Material examined.— Holotype, Madagascar, ATIMO VATAE, stn CP3583, South Pointe Barrow, 25°31′S 44°16′E, 296-302 m, 10 May 2010, 1 ovigerous female 24.4 mm cl (MNHN-IU-2014-8777). Paratypes: Madagascar, RV "Vauban", stn CH 72, 25°11.2′S 47°14.7′E, 85-90 m, 3 Mar. 1973, 1 male 23.0 mm cl (MNHN-Pa-587). ATIMO VATAE, stn CP3511, Secteur Fort-Dauphin, 25°15.0′S 47°14.5′E, 97-98 m, 29 Apr. 2010, 1 female 21.3 mm cl (MNHN-IU-2014-8778); stn CP3547, South Cap Andavaka, 25°18.0′S 46°40.3′E, 69-70 m, 4 May 2010, 1 male 19.2 mm cl (MNHN-IU-2014-8779); stn CP3572, 25°11.7′S 47°12.5′E, 75-77 m, 8 May 2010, 2 males 13.1-13.9 mm cl (MNHN-IU-2014-8782); stn CP3603, SW Cap Sainte Marie, 25°57.1′S 44°47.6′E, 62-63 m, 13 May 2010, 1 ovigerous female 19.3 mm cl (MNHN-IU-2014-8780); stn CP3624, South-eastern Faux-Cap, 25°38.1′S 45°57.0′E, 63 m, 15 May 2010, 2 females 21.3-23.0 mm cl (MNHN-IU-2014-8781).

Description.— Rostral tooth well developed and sharply pointed (fig. 1A). Pregastric tooth absent. Gastric tooth truncated, moderately elevated and about 1.5-2.1 times as high as rostral tooth (fig. 1C), followed by 7-8 transverse rows of squamae, each row with 3-6 fused and flatten squamae. Anterior submedian ridges not straight, composed of 8-10 irregular squamae. Cardiac tooth deeply bifurcate but blunt, not elevated, followed by 6 or 7 transverse rows of 2-4 fused flattened squamae. Posterior postrostral ridge flanked with patch of 10-12 squamae. Five to six distinct intermediate tubercles present. One or two tubercles present between intermediate tubercles and posterior branchial ridge. Cervical groove shallow and narrow, with anterior and posterior branchial carinae relatively close. Anterior branchial ridge bearing 2 large teeth anteriorly. Posterior branchial ridge with anterior end forming an acute tooth, followed posteriorly by double row of 7-9 squamiform tubercles. Postorbital ridge not sharp. Intercervical ridge composed of 2-4 small squamae. Anterolateral, mediolateral and posterolateral teeth only with anteriormost one as strong tooth, others reduced to squamiform tubercles. Posterior marginal groove deep but narrow. Posterior margin of carapace medially incised (fig. 1A).

Abdomen with arborescent sculpture and lacking median dorsal carina. Articulate surfaces of tergite I smooth, but those of tergite II to V each with 2 distinct

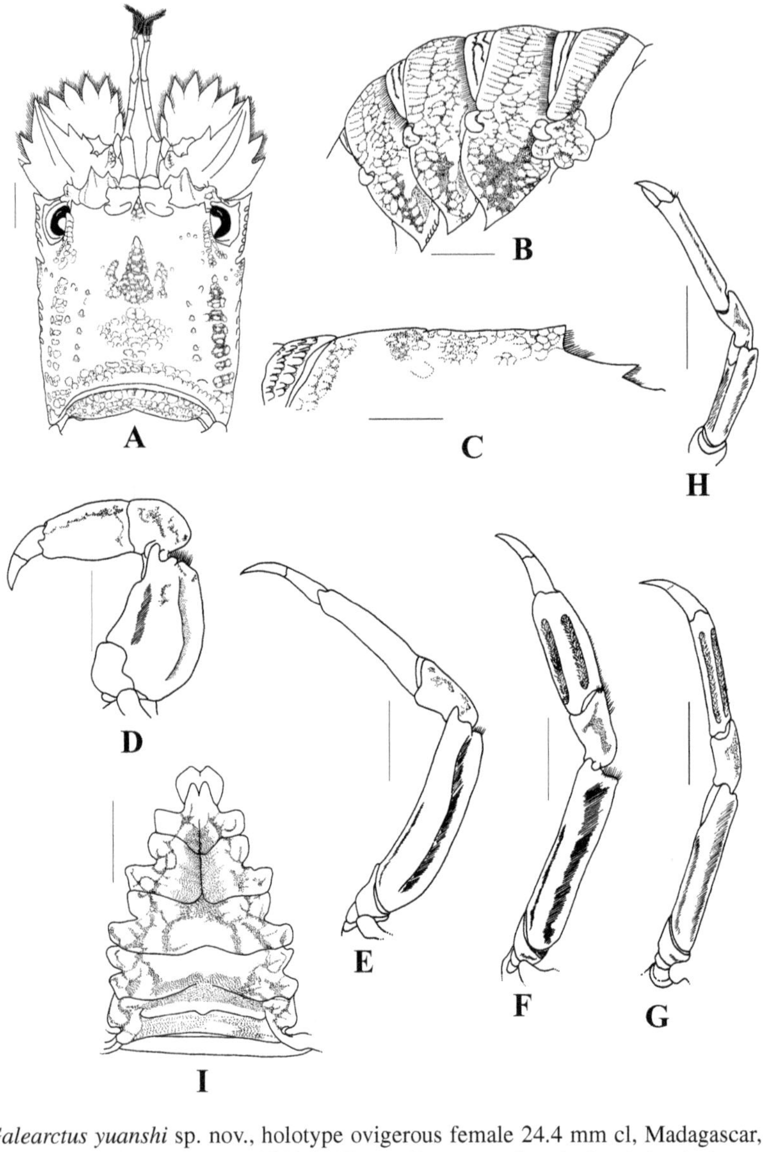

Fig. 1. *Galearctus yuanshi* sp. nov., holotype ovigerous female 24.4 mm cl, Madagascar, "ATIMO VATAE" stn. CP3583, MNHN-IU-2014-8777. A, Carapace, dorsal; B, abdominal somites I-IV, lateral; C, dorsal carapace, lateral; D, right pereiopod I, mesial; E, right pereiopod II, mesial; F, right pereiopod III, mesial; G, right pereiopod IV, mesial; H, right pereiopod V, mesial; I, thoracic sternum, ventral. Scale bars: 5 mm.

transverse grooves (one sometimes slightly crenulated and branched, fig. 1B). Posterior margins of tergites I to III deeply incised medially. Pleura II to IV sharply pointed posteroventrally, posterior margins smooth (fig. 1B). Posterior margin of telson bearing 2 pairs of similar size teeth.

Anterior margin of antennal segment VI with 7 teeth, outer 6 large and wide while innermost 1 obtuse and short. Antennal segment IV bearing 2 teeth on anterior margin, inner tooth small and sometimes indistinct (fig. 1B); outer margin armed with 2 large teeth; dorsal surface with only 1 oblique median carina. Fused antennal segments II and III bearing 2 teeth on anterior margin, with outer tooth much larger (fig. 1B).

Pereiopod I stout and shorter than other pereiopods; dorsal surface on propodus with 1 distinct groove, that on carpus with several large pits; that on merus with a wide groove; ventrolateral surface of merus also with a long depression longitudinally (fig. 1D). Propodus of pereiopod II slender (fig. 1E). Propodus of pereiopod III much wider and flatter than that of pereiopod II, bearing 2 long (about 0.8 time as long as propodus) and parallel hairy longitudinal grooves on mesial surface (fig. 1F). Pereiopod IV with 1 or 2 hairy grooves on mesial surface of propodus (fig. 1G). Mesial surfaces of carpus and merus of pereiopod II to V each with several large pits and 2 hairy grooves, respectively (fig. 1E-H).

Anterior part of thoracic sternum produced forwards, obtusely bifurcate; median part of anterior protrusion covered with a "M"-shaped sculpture (fig. 1I). Thoracic sternum with median part deeply sunken and with distinct median longitudinal groove extending from posterior half of sternite IV (position of pereiopod I) to entire sternite V (position of pereiopod II); posterolateral angle unarmed in females but developed into small tooth in males.

Coloration.— Entire body includes antennular flagella generally yellowish-brown. Eyes dark brown. Pereiopods yellowish-tan and distinctly covered with purple bands. Abdominal somite I with a large orange-brown median spot flanked by two smaller orange-brown spots. Abdominal somite VI and non-calcified parts of tail fan whitish-tan (fig. 3A).

Etymology.— For his eminence in Chinese crustacean research, Prof. J.Y. Liu was conferred the title of Academician of the Chinese Academy of Sciences. The new species name is derived from "Yuàn Shì", the Chinese transliteration for Academician, in honor of Prof. Liu. The name is used as a noun.

Distribution.— Currently only known from South Madagascar, at depths of 62-302 m.

Remarks.— The present new species closely resembles *G. timidus* but differs in: (1) cardiac tooth very low and almost leveled in *G. yuanshi* sp. nov. (fig. 1C), but highly elevated in *G. timidus* (fig. 2A); (2) gastric tooth only moderately high in *G. yuanshi* sp. nov. and being 1.5-2.1 times as high as rostral tooth (fig. 1C), but extraordinary high in *G. timidus* and being 3.5-4.7 times higher than the rostral tooth (fig. 2A); (3) the two transverse grooves on the articulate surfaces of abdominal tergite II to V with one of them sometimes crenulated and branched

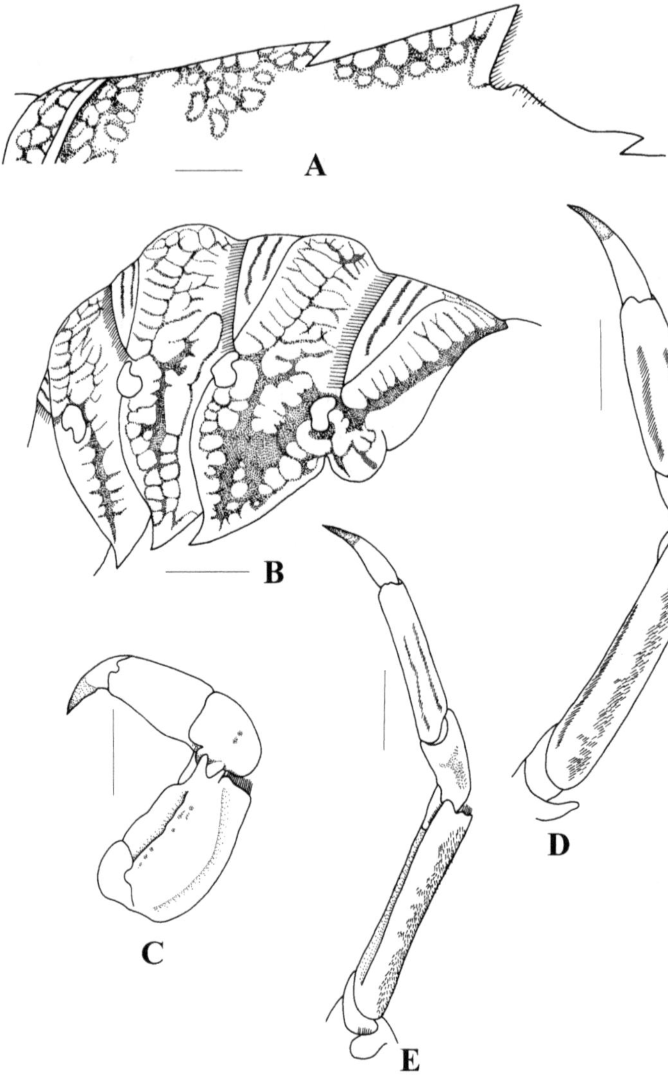

Fig. 2. *Galearctus timidus* (Holthuis, 1960), ovigerous female 27.7 mm cl, Philippines, Balicasag Island, NTOU M02237. A, Dorsal carapace, lateral; B, abdominal somites I-IV, lateral; C, right pereiopod I, mesial; D, right pereiopod III, mesial; E, right pereiopod IV, mesial.

in *G. yuanshi* sp. nov. (fig. 1B), but always straight and parallel (though one of them occasionally indistinct and interrupted) in *G. timidus* (fig. 2B); (4) median parts of abdominal tergite II to III nearly flatten in *G. yuanshi* sp. nov. (fig. 1B) but slightly elevated in *G. timidus* (fig. 2B); (5) dorsal surface of propodus and carpus in pereiopod I with distinct groove or pits in *G. yuanshi* sp. nov. (fig. 1D), but smooth or indistinctly pitted in *G. timidus* (fig. 2C); (6) the two parallel grooves on the propodus of pereiopods III to IV being very wide and longer (about 80%

Fig. 3. A, *Galearctus yuanshi* sp. nov., paratype male 13.9 mm cl, Madagascar, "ATIMO VATAE", CP3572, MNHN-IU-2014-8782; B, *Galearctus timidus* (Holthuis, 1960), female 21.6 mm cl, Philippines, "PANGLAO 2004", NTOU M02239.

propodus length) in *G. yuanshi* sp. nov. (fig. 1F-G), but much thinner and shorter (about 70% propodus length) in *G. timidus* (fig. 2D, E). Moreover, the coloration is rather different between the Philippines and Madagascan materials (fig. 3). Apart from the Madagascan form being distinctly lighter in color (body yellowish brown with banding on pereiopods purple, versus body reddish brown with banding on pereiopods dark purple), the median spot on the abdominal somite I is larger in the Philippines material. There is also a distinct red spot on the lateral part of the fourth antennal segment in the Philippines material, but this spot is lacking in the Madagascar form. Comparisons of the barcoding genes COI (657 bp) and 16S rRNA (543 bp) amongst the Madagascar (MNHN-IU-2014-8777; 8778; 8780; 8782) and Philippine materials (NTOU M00982; M02241; ovigerous female in ZRC 2001 0331) reveal as high as 16.3% sequence divergence in COI gene (accession Nos. MT364924 and JN701673) and 7.8-8.0 % divergence in 16S rRNA (accession Nos. MT365034-MT365039, JN701721). On the other hand, there is only 0.0-0.2% sequence divergence in 16S rRNA amongst either the Philippines (accession Nos. JN701721, MT365038-MT365039) or Madagascan specimens (accession Nos. MT365034-MT365037). Such high genetic divergences in both the mitochondrial COI (see Matzen da Silva et al., 2011; Yang et al., 2011; Meyer et al., 2013) and 16S rRNA (see Chan et al., 2008; Cabezas et al., 2009; Mazancourt et al., 2019) genes support a distinct specific status of the Madagascan form from the typotypic Philippines material of *G. timidus*.

Holthuis (2002) recorded *G. timidus* from a very wide range in the Indo-West Pacific including South Africa, Madagascar, Japan, Taiwan, Philippines, New Caledonia and Vanuatu. The holotype of *G. timidus* (RMNH.CRUS.D.14507) is almost identical to some of the recently collected specimens from the Philippines in this study. The Madagascan specimen (MNHN-Pa 587) recorded by Holthuis (2002) was re-examined and belongs to the present new species. Whether those *G. timidus* materials reported from other localities except the Philippines truly belong to that species will need to be re-examined, though the coloration of the Vanuatu and Taiwanese specimens reported in Holthuis (2002, fig. 67C-E) are more similar to *G. timidus* than to *G. yuanshi* sp. nov. The 99.8-100% sequence similarities in the 16S rRNA gene between a recently collected specimen from Taiwan (NMMBA CD5592) and the Philippines material of *G. timidus* confirmed the presence of *G. timidus* in Taiwan.

ACKNOWLEDGEMENTS

The ATIMO VATAE expedition was part of a cluster of Mozambique-Madagascar expeditions (La Planète Revistée/Our Planet Reviewed) funded by the Total, Prince Albert II of Monaco, and Stavros Niarchos foundations, and conducted by the Muséum national d'Histoire naturelle, Paris (MNHN, P. Bouchet principal investigator) and Pro-Natura International (PNI). The expedition was undertaken in partnership with Institut d'Halieutique et des Sciences Marines, University of Toliara, and the Madagascar bureau of the Wildlife Conservation Society. The PANGLAO 2004 Marine Biodiversity Project was a collaboration between the MNHN (P. Bouchet principal investigator) and the University of San Carlos, Cebu City (D. Largo principal investigator), and supported by the Total Foundation for Biodiversity and the Sea, the French Ministry of Foreign Affairs, and ASEAN Regional Center for Biodiversity Conservation. The Philippines Bureau of Fisheries and Aquatic Resources is gratefully acknowledged for issuing a research permit on the material collected by the PANGLAO 2004 expedition. Sincere thanks are extended to L. Corbari and P. Martin-Lefevre of MNHN for loaning the Madagascan material; P.K.L. Ng and J.C. Mendoza of the Lee Kong Chian Natural History Museum, Singapore for loaning part of the Philippines material, P.K.L. Ng also for the suggestion on the Etymology of the new species; C.H.J.M. Fransen of the Naturalis Biodiversity Center, Leiden for providing us with detailed photographs of the holotype of *G. timidus*; and C.-W. Lin of the National Museum of Marine Biology and Aquarium for loaning the Taiwanese specimen of *G. timidus*. This work was supported by grants from the Ministry of Science and Technology, Taiwan, R.O.C. and the Center of Excellence for the Oceans (National Taiwan Ocean University), which is financially supported from The Featured Areas Research Center Program within the framework of the Higher Education Sprout Project by the Ministry of Education in Taiwan, R.O.C.

REFERENCES

CABEZAS, P., E. MACPHERSON & A. MACHORDOM, 2009. Morphological and molecular description of new species of squat lobster (Crustacea: Decapoda: Galatheidae) from the Solomon and Fiji Islands (South-West Pacific). Zoological Journal of the Linnean Society, **156**: 465-493.

CHAN, T.-Y., J. TONG, Y. K. TAM & K. H. CHU, 2008. Phylogenetic relationships among the genera of the Penaeidae (Crustacea: Decapoda) revealed by mitochondrial 16S rRNA gene sequences. Zootaxa, **1694**: 38-50.

CHAN, T.-Y., S. T. AHYONG & C.-H. YANG, 2013. Priority of the slipper lobster genus *Crenarctus* Holthuis, 2002, over *Antipodarctus* Holthuis, 2002 (Crustacea, Decapoda, Scyllaridae). Zootaxa, **3701**: 471-472.

HOLTHUIS, L. B., 1946. Biological results of the Snellius Expedition XIV. The Decapoda Macrura of the Snellius Expedition. I. The Stenopodidae, Nephropsidae, Scyllaridae and Palinuridae. Temminckia, **7**: 1-178, plates 1-11.

HOLTHUIS, L. B., 1960. Preliminary descriptions of one new genus, twelve new species and three new subspecies of scyllarid lobsters (Crustacea Decapoda Macrura). Proceedings of the Biological Society of Washington, **73**: 147-154.

HOLTHUIS, L. B., 2002. The Indo-Pacific scyllarine lobsters (Crustacea, Decapoda, Scyllaridae). Zoosystema, **24**: 499-683.

MATZEN DA SILVA, J., S. CREER, A. DOS SANTOS, A. C. COSTA, M. R. CUNHA, F. O. COSTA & G. R. CARVALHO, 2011. Systematic and evolutionary insights derived from mtDNA COI barcode diversity in the Decapoda (Crustacea: Malacostraca). PLoS ONE, **6**(5): e19449.

MAZANCOURT, V., G. MARQUET & P. KEITH, 2019. Revision of freshwater shrimps belonging to *Caridina weberi* complex (Crustacea: Decapoda: Atyidae) from Polynesia with discussion on their biogeography. Journal of Natural History, **53**(13-14): 815-847.

MEYER, M., A. WEIS & R. R. MELZER, 2013. Decapoda of southern Chile: DNA barcoding and integrative taxonomy with focus on the genera *Acanthocyclus* and *Eurypodius*. Systematics and Biodiversity, **11**: 389-404.

YANG, C.-H. & T.-Y. CHAN, 2010. A new slipper lobster of the genus *Galearctus* Holthuis, 2002 (Decapoda: Scyllaridae) from Taiwan and Japan. In: C. H. J. M. FRANSEN, S. DE GRAVE & P. K. L. NG (eds.), Studies on Malacostraca: Lipke Bijdeley Holthuis memorial volume. Crustaceana Monographs, **14**: 735-745.

YANG, C.-H., I. S. CHEN & T.-Y. CHAN, 2011. A new slipper lobster of the genus *Galearctus* Holthuis, 2002 (Crustacea, Decapoda, Scyllaridae) from New Caledonia. Zoosystema, **33**: 207-217.

YANG, C.-H., H. D. BRACKEN-GRISSOM, D. KIM, K. A. CRANDALL & T.-Y. CHAN, 2012. Phylogenetic relationships, character evolution, and taxonomic implications within the slipper lobsters (Crustacea: Decapoda: Scyllaridae). Molecular Phylogenetics and Evolution, **62**: 237-250.

First received 18 November 2019.
Final version accepted 19 December 2019.

Stomatopoda

[When citing this volume, please refer to *Crustaceana* 93 (2020), 1233-1546]

Prof. Ruiyu Liu Memorial Issue

THE MANTIS SHRIMP SUPERFAMILY EURYSQUILLOIDEA CONFIRMED FROM TAIWAN: *LIUSQUILLA TAIWANICA* GEN. ET SP. NOV.

BY

SHANE T. AHYONG[1,2,5]) and CHIA-WEI LIN[3,4,6])

[1]) Marine Invertebrates, Australian Museum, 1 William St., Sydney, NSW 2010, Australia
[2]) School of Biological, Earth and Environmental Sciences, University of New South Wales, Kensington, NSW 2052, Australia
[3]) Department of Exhibition, National Museum of Marine Biology and Aquarium, 2 Houwan Road, Checheng, Pingtung, 944, Taiwan, R.O.C.
[4]) Institute of Marine Biology, National Dong Hwa University, Hualien 974, Taiwan, R.O.C.

ABSTRACT

Prior to the present study, the mantis shrimp superfamily Eurysquilloidea Manning, 1977, with the single family Eurysquillidae Manning, 1977, contained six genera and 32 species, most of which occur in the Indo-West Pacific. A new genus and species of eurysquilloid, *Liusquilla taiwanica*, recently discovered off southeastern Taiwan, is described herein, constituting the first record of the superfamily from the island. Sixty-eight species in six of the seven stomatopod superfamilies are now known from Taiwan. *Liusquilla* differs from other eurysquillids in the combination of four teeth on the dactylus of the raptorial claw and form of the eye and ocular scale. *Liusquilla* is possibly most closely related to *Sinosquillla* Liu & Wang, 1978, based on eye form and in the similarly ornamented, densely spinose telson and abdominal somite 6. A key to the eurysquillid genera is provided.

Key words. — Stomatopoda, mantis shrimp, Eurysquilloidea, Indo-West Pacific

RÉSUMÉ

Avant la présente étude, la superfamille de Stomatopodes Eurysquilloidea Manning, 1977, avec l'unique famille des Eurysquillidae Manning, 1977, comprenait six genres et 32 espèces, la plupart d'entre elles étant présentes dans l'Indo-Ouest Pacifique. Un genre nouveau et une espèce nouvelle d' Eurysquillidae, *Liusquilla taiwanica*, récemment découverts au large du sud-est de Taiwan, sont décrits ici, ce qui constitue le premier signalement de la superfamille de l'île. Soixante-six espèces appartenant à six des sept superfamilles de Stomatopodes sont maintenant connues de Taiwan. *Liusquilla* diffère des autres Eurysquillidae par la combinaison des quatre dents sur le dactyle de la patte ravisseuse, de la forme de l'œil et de l'écaille oculaire. *Liusquilla* est vraisemblablement très proche de *Sinosquillla* Liu & Wang, 1978, en considérant la forme de l'œil, le telson pourvu d'épines denses, ornementé de façon similaire et le 6ème somite abdominal. Une clé des genres d'Eurysquillidae est fournie.

[5]) Corresponding author; e-mail: shane.ahyong@austmus.gov.au
[6]) e-mail: linchiawei@nmmba.gov.tw

 | DOI:10.1163/9789004464353_016

INTRODUCTION

The mantis shrimp superfamily Eurysquilloidea Manning, 1977, includes the single family Eurysquillidae Manning, 1977. Eurysquillidae was long regarded as a member of the Gonodactyloidea Giesbrecht, 1910 (Manning, 1980, 1995; Ahyong, 1997a) on the basis of the shared ovate propodi of maxillipeds 3 and 4, and presence of a distinct median carina and two intermediate denticles on the telson. These features, however, are plesiomorphies, and Ahyong & Harling (2001) showed Eurysquillidae to instead be sister to Parasquillidae Manning, 1995, each respectively in separate superfamilies, Eurysquilloidea and Parasquilloidea. Eurysquilloidea and Parasquilloidea together form the sister clade to Squilloidea Latreille, 1802 (Ahyong & Harling, 2000; Porter et al., 2010; Van Der Wal et al., 2017), the most speciose stomatopod superfamily, with 49 genera and more than 180 species (Van Der Wal et al., 2019). In contrast, Eurysquilloidea is considerably smaller, containing seven genera and 33 species, most of which occur in the Indo-West Pacific (Ahyong, 2001, 2010; Lucatelli et al., 2013; Ahyong et al., 2020). Eurysquilloids are united by the unusual position of the intermediate and lateral denticles of the telson, of which at least the outermost intermediate denticle, if not the others, arise submarginally, rather than on the margin of the telson (Ahyong & Harling, 2000).

Knowledge of the stomatopod fauna of Taiwan has grown considerably in the past two decades, with 67 species now known, arrayed in five of the seven superfamilies (Manning & Chan, 1997; Ahyong et al., 1998, 2000, 2008; Huang & Hsueh, 2006; Ahyong & Chan, 2008; Yeh & Hsueh, 2010; Wang & Chiou, 2017). Recent field investigations by the second author collected the first representative of the Eurysquilloidea from Taiwan. Significantly, this Taiwanese eurysquilloid represents a new genus and species, and is formally described herein.

MATERIALS AND METHODS

Morphological terminology and size descriptors generally follow Ahyong (2001, 2012). Specimens are measured in millimetres (mm). Total length (TL) is measured along the dorsal midline from the apex of the rostral plate to the apices of the submedian teeth of the telson. Carapace length (CL) is measured along the dorsal midline and excludes the rostral plate. The corneal index (CI) is given as 100 CL/corneal width. The propodal index (PI) is given as 100 CL/propodus length. Material is deposited in the collections of the National Museum of Marine Biology and Aquarium, Kenting, Taiwan (NMMBA).

SYSTEMATICS

Superfamily EURYSQUILLOIDEA Manning, 1977
Family EURYSQUILLIDAE Manning, 1977
Liusquilla gen. nov.

Diagnosis.— Cornea bilobed, set transversely on stalk, with 6 midband rows of ommatidia. Rostral plate simple, apex rounded. Antennular somite not elongated, no longer than rostral plate. Mandibular palp present. Raptorial claw dactylus with 4 teeth on occlusal margin; propodus occlusal margin fully pectinate; carpus with 2 dorsal spines. Abdominal somite 6 densely covered with short spines. Telson wider than long; primary spines well-defined, prominent, acute, outline clearly visible in dorsal view; dorsal surface entirely covered with short spines; median carina indicated by line of short spines; submedian denticles absent; intermediate and lateral denticles without dorsal lobe. Uropodal protopod proximal dorsal surface covered with short spines; inner primary spine inner margin spinose, longer than outer spine.

Type species.— *Liusquilla taiwanica* sp. nov., by present designation.

Etymology.— Named for the late Prof. Ruiyu Liu, for his major contributions to crustacean research in China, and in particular, that of the Stomatopoda, which was one of his favourite groups. Gender: feminine.

Liusquilla taiwanica sp. nov.
(figs. 1, 2)

Type material.— Holotype: NMMB CD5591, female (TL 51 mm), Hojie, Pingtung County, SE Taiwan, burrow in sandy bottom, 26-28 m, coll. C. W. Lin, 25 March 2017.

Description.— Eye subtriangular, extending almost to end of antennular peduncle article 1; cornea strongly bilobed, set transversely on stalk; CI 296. Ophthalmic somite anterior margin triangular; ventral surface unarmed. Ocular scales triangular, separate, anterior margins concave, apices directed laterally.

Antennular peduncle 0.87 CL. Antennular somite dorsal processes directed anterolaterally, laterally compressed, apices blunt, rounded in lateral view. Antennal protopod with slender mesiodorsal spine, small ventrodistal spine and 1 ventral papilla. Antennal scale margins fully setose; length 0.43 CL.

Rostral plate broadly linguiform, wider than long, widest basally; lateral margins straight, convergent; apex rounded; low, indistinct median carina anteriorly. Carapace anterolateral angles rounded; anterior margins straight; carinae absent except for marginal carina, indicated posterolaterally.

Mandibular palp 3-segmented. Maxillipeds 1-5 with epipod. Maxillipeds 3-4 propodi ovate, rounded, without distoventral ribbing. Maxilliped 5 proximalmost

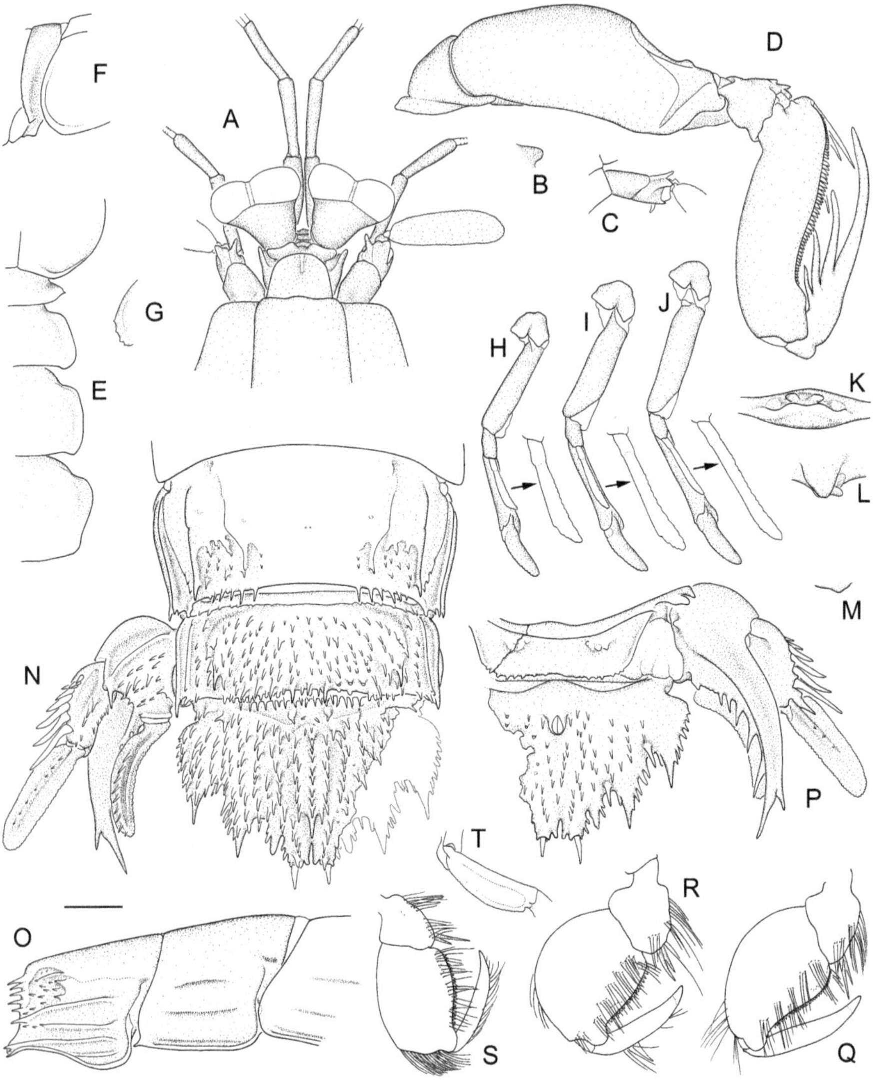

Fig. 1. *Liusquilla taiwanica* gen. et sp. nov., female holotype, TL 51 mm, Hojie, NMMB CD5591. A, Anterior cephalothorax, dorsal view; B, right dorsal process of antennular somite, lateral view; C, right antennal protopod; D, right raptorial claw; E, posterolateral portion of carapace and thoracic somites 5-8, right dorsal view; F, posterolateral portion of carapace and thoracic somite 5, right lateral view; G, lateral process of thoracic somite 5, anterior view; H-J, right pereopods 1-3, posterior view; K, L, female gonopore, right lateral view and ventral view; M, thoracic somite 8, sternal 'keel'; N, abdominal somites 5-6, telson and left uropod, dorsal view; O, abdominal somites 3-5, right lateral view; P, abdominal somite 6, telson and left uropod, ventral view; Q-S, right maxilliped 3–5 dactylus-carpus; T, right maxilliped 5 ischium. A-J, M-P, T = 2.5 mm; K, L, Q-S = 1.25 mm.

Fig. 2. *Liusquilla taiwanica* gen. et sp. nov., female holotype, TL 51 mm, Hojie, NMMB CD5591.

article without ventrally directed spine; ischium mesial margin broadly rounded, without anterior lobe.

Raptorial claw dactylus with 4 teeth; outer margin very weakly sinuous on proximal half, curving distally; proximal margin with distinct basal notch. Propodus shorter than carapace, with 3 movable spines proximally on occlusal margin; distal margin unarmed; when folded, not extending posterior beyond merus; PI 111 (female). Carpus dorsal margin terminating in 2 short, ventrally directed spines, distal longer. Merus inferodistal margin rounded, unarmed. Ischium shorter than one-fourth merus length. Basis lateral surface unarmed.

Pereopods 1-3 proximal-most article with outer ventrolaterally directed triangular lobe; inner margin unarmed. Pereopodal endopod distal article slender, liguliform, tapering distally; outer and inner distal margins setose.

Thoracic somite 5 lateral process obsolete, forming low, blunt anteroposteriorly compressed lobe with dentate lower margin; with small angular, ventrally directed lobe. Thoracic somites 6-8 lateral process subtruncate, each with faintly indicated lateral carina. Thoracic somite 6 female gonopore with bilobed median papillae flanked by lower lateral papilla. Thoracic somite 8 sternal keel low, blunt, angular.

Abdominal somite articulation compact; somites 1-4 smooth dorsally; dorsal posterior margin unarmed; 3 shallow grooves laterally (upper two corresponding to intermediate and lateral carina) and marginal carina; somites 1-3 posterolaterally unarmed; somite 4 with posterolateral spine; ventral pleural margin straight to faintly concave.

Abdominal somite 5 slightly tapering posteriorly in dorsal view; central dorsal surface smooth, with pair of minute median pits; surface on posterior one-third mesial to intermediate carina irregular and spinose; posterior margin between central one-third and level of intermediate carina lined with 5 or 6 posteriorly directed spines; lower lateral surface with posteriorly armed intermediate, lateral and marginal carinae, and also with posteriorly armed carina and several small spines between intermediate and lateral carina, and short unarmed carina between lateral and marginal carinae; ventral pleural margin distinctly concave. Abdominal somite 6 dorsal surface entirely covered with short posteriorly directed spines; posterior margin lined with posteriorly directed spines, submedian spines larger, more prominent than adjacent spines; intermediate spine larger, more prominent than surrounding spines; lateral carina smooth, terminating in prominent spine, mesially with shorter, smooth carina; ventrolaterally with triangular spine and blunt lobe anterior to uropodal articulation; sternum posterior margin with 8 low to short blunt spines either side of midline; surface with two clusters of low, blunt tubercles.

Telson wider than long; dorsal surface and margins of primary teeth densely covered with short, posteriorly curved spines; median carina indicated by longitudinal row of short, posteriorly directed spines of similar size to adjacent spines on telson surface, with prominent posterior spine; submedian teeth short, lateral margins spinose, movable apices slender, slightly curved dorsad, separated by smooth, narrow cleft; submedian denticles absent; with single spiniform, submarginal intermediate and lateral denticles, dorsal lobe absent. Intermediate and lateral teeth prominent, distally spiniform, of similar size; margins spinose; outer proximal margin of submedian teeth each with blunt, posteriorly directed lobe. Ventral surface covered with short spines; postanal carina absent.

Uropodal protopod dorsal surface between mid-dorsal carina and endopod articulation covered with short spines; anterior margin convex, unarmed; inner primary spine ventrally carinate, distinctly longer than outer spine; inner margin with 5 graded spines; outer margin unarmed. Uropodal exopod proximal article dorsal surface with row of short spines along midline and along upper mesial margin; inner margin slightly concave, unarmed; slender, curved distoventral spine; outer margin with 8 or 9 graded movable spines, distalmost spine not reaching midlength of distal article. Exopod distal article longer than proximal article; dorsal midrib with 7 minute spines; ventral midrib with 3 minute spines. Endopod elongate, reniform, apex bluntly rounded; length about 5 × width; dorsolateral surface with row of 9 or 10 short spines.

Colour in life.— Cornea metallic green. Body translucent white overall with diffuse brown and white speckling, and with several black brown spots on carapace and axial row of black brown spots, largest on abdominal somites 1 and 5. Telson with transverse row of 4 black spots along dorsal anterior margin and black spot on proximal notch of median carina. Thorax and abdomen appearing pale orange as a result of underlying gonads. Antennules and antennae translucent, with dark brown speckling; articles distally brown; antennular flagella reddish; antennal flagellum translucent white. Eyes and antennal protopod with brown speckling. Raptorial claw and pereopods 1–3 translucent white with diffuse brown speckling. Uropodal exopod distal article and endopod black-brown with dark-red marginal setae.

Measurements.— TL 51 mm, CL 9.30 mm, cornea width 3.14 mm, antennular peduncle length 8.09 mm, antennal scale length 3.96 mm, raptorial claw propodus length 8.35 mm.

Etymology.— Named *taiwanica*, after the island of Taiwan, alluding to the known geographic distribution of the species; the name thus is an adjective agreeing in gender with the (feminine) generic name.

Remarks.— With discovery of *Liusquilla taiwanica* gen. et sp. nov., eight genera are now recognized in Eurysquillidae. *Liusquilla*, together with *Coronidopsis* Hansen, 1926 and *Manningia* Serène, 1962, differs from all other eurysquillids in having four teeth on the dactylus of the raptorial claw. *Liusquilla* is readily separated from *Coronidopsis* and *Manningia* by the much more symmetrical eye-shape (fig. 1A) (versus strongly asymmetrical with outer margin much longer than inner margin, e.g., Ahyong, 1997b: fig. 1A) with six instead of two rows of corneal midband ommatidia, and prominent ocular scales that are visible in dorsal view (fig. 1A) (versus small, fully concealed by rostral plate, e.g., Ahyong, 1997b: fig. 1A). Apart from the armature of the raptorial claw, *Liusquilla* is further separated from *Eurysquilla* Manning, 1963, *Eurysquilloides* Manning, 1963 and *Raysquilla* Ahyong, 2000, by the densely spinose dorsal surface of abdominal somite 6, the telson and the uropods (figs. 1N, 2) (versus smooth or with only sparse spines or tubercles). In all species of the aforementioned genera, the median carina of the telson is dorsally smooth, contrasting with that of *L. taiwanica* in which the median carina is indicated by a line of spines (figs. 1N, 2). In the dense dorsal tailfan spination, *Liusquilla* superficially resembles *Sinosquilla* Liu & Wang, 1978, especially *S. hispida* Liu & Wang, 1978 in which the median carina of the telson is also indicated by a row of spines (that of *S. sinica* Liu & Wang, 1978 being dorsally unarmed). *Liusquilla*, however, appears to be most closely related to *Ankersquilla* Ahyong, Porter & Caldwell, 2020, also from the Indo-West Pacific (Ahyong et al., 2020). The two genera share remarkably similar uropod structure, eye shape, ocular scales and rostral plate form in addition to the densely spinose posterior abdominal somites and telson. *Liusquilla* is most readily distinguished from *Ankersquilla* by having four instead of three teeth on the dactylus of the raptorial claw and prominent rather than reduced primary teeth of the telson.

The holotype of *Liusquilla taiwanica* is unfortunately incomplete, with part of the right side of the telson damaged during collection (figs. 1N, P, 2). The specimen was collected at 26-28 m depth from a burrow that opened under a rock and continued downwards into the sediment.

Distribution.— At present known only from southeastern Taiwan.

KEY TO GENERA OF THE EURYSQUILLIDAE

1. A1 somite greatly elongate, extending anteriorly well beyond apex of rostral plate .... *Eurysquilloides*
– A1 somite short, covered by rostral plate 2
2. Abdominal somite 6 without spines or tubercles on dorsal surface or posterior margin (except short posterolateral spine, present or absent). Uropodal protopod with outer terminal spine longer than inner .... *Raysquilla*
– Abdominal somite 6 with 6 or more spines along dorsal posterior margin (including posterolateral spines). Uropodal protopod with outer terminal spine shorter than inner ... 3
3. Ocular scales and dorsal processes of antennular somite not visible in dorsal view, entirely covered by rostral plate 4
– Ocular scales and dorsal processes of antennular somite visible in dorsal view 5
4. Rostral plate bispinous anteriorly. Raptorial claw carpus with single dorsal spine .... *Coronidopsis*
– Rostral plate unarmed or with a median spine. Raptorial claw carpus with two dorsal spines .... *Manningia*
5. Rostral plate forming a long slender spine .... *Sinosquilla*
– Rostral plate subtriangular to subpentagonal, at most with a short anterior spine 6
6. Abdominal somite 6 and telson dorsal surface smooth or with scattered tubercles, but not entirely covered with short spines; median carina distinct, smooth .... *Eurysquilla*
– Abdominal somite 6 and telson dorsal surface entirely covered with short spines. Telson dorsal surface densely covered with short spines, obscuring median carina; median carina itself indicated by line of short spines similar to surrounding dorsal spines 7
7. Raptorial claw dactylus with 3 teeth. Telson primary teeth obscured in dorsal view .... *Ankersquilla*
– Raptorial claw dactylus with 4 teeth. Telson primary teeth clearly visible in dorsal view .... ***Liusquilla***

ACKNOWLEDGEMENTS

We thank Wenliang Liu, Wei Jiang and Xinzheng Li for inviting the present contribution, and two anonymous reviewers for their constructive comments on the manuscript.

REFERENCES

AHYONG, S. T., 1997a. A phylogenetic analysis of the Stomatopoda (Crustacea: Malacostraca). Journal of Crustacean Biology, **17**: 695-715.

AHYONG, S. T., 1997b. A new species of *Manningia* (Crustacea: Stomatopoda) from Irian Jaya, Indonesia, with remarks on the genus. Raffles Bulletin of Zoology, **45**: 327-333.

AHYONG, S. T., 2000. *Raysquilla manningi*, a new genus and species of stomatopod from the Australian Northwest Shelf. Journal of Crustacean Biology, **20**: 37-41.

AHYONG, S. T., 2001. Revision of the Australian stomatopod Crustacea. Records of the Australian Museum, Supplement **26**: 1-326.

AHYONG, S. T., 2010. Stomatopoda from the Philippines collected by the AURORA 2007 Expedition (Eurysquilloidea, Lysiosquilloidea and Gonodactyloidea). In: C. H. J. M. FRANSEN, S. DE GRAVE & P. K. L. NG (eds.), Studies on Malacostraca: Lipke Bijdeley Holthuis memorial volume. Crustaceana Monographs, **14**: 101-113.

AHYONG, S. T., 2012. The marine fauna of New Zealand: mantis shrimps (Crustacea: Stomatopoda). NIWA Biodiversity Memoir, **125**: 1-111.

AHYONG, S. T., T.-Y. CHAN & Y. J. LIAO, 1998. A new stomatopod (Crustacea: Malacostraca) of the genus *Harpiosquilla* Holthuis, 1964, from Taiwan and Australia. Proceedings of the Biological Society of Washington, **111**: 929-935.

AHYONG, S. T., T.-Y. CHAN & Y. J. LIAO, 2000. *Oratosquillina manningi*, a new species of stomatopod from Taiwan and Australia. Journal of Crustacean Biology, **20**: 42-47.

AHYONG, S. T., T.-Y. CHAN & Y.-C. LIAO, 2008. A catalog of the mantis shrimps (Stomatopoda) of Taiwan: 191 pp. (National Science Council, Taiwan, R.O.C., Taipei).

AHYONG, S. T. & C. HARLING, 2000. The phylogeny of the stomatopod Crustacea. Australian Journal of Zoology, **48**: 607-642.

AHYONG, S. T., M. L. PORTER & R. L. CALDWELL, 2020. The leopard mantis shrimp, a new genus and species of eurysquillid from Indo-West Pacific coral reefs. Records of the Australian Museum, **72**(1): 1-8.

GIESBRECHT, W., 1910. Stomatopoden, Erster Theil. Fauna und Flora des Golfes von Neapel, **33**: i-vii, 1-239, pl. I-XI.

HANSEN, H. J., 1926. The Stomatopoda of the Siboga Expedition. Siboga-Expeditie Monograph, **35**: 1-48, pls. I, II.

HUANG, J.-F. & P.-W. HSUEH, 2006. A new record of *Haptosquilla glyptocercus* (Wood-Mason, 1875) (Stomatopoda: Protosquillidae) from Taiwan. Endemic Species Research, **8**: 97-101.

LATREILLE, P. A., 1802. Histoire naturelle, générale et particulière, des Crustacés et des Insectes **3**: 467 pp. (F. Dufart, Paris).

LIU, J. Y. & Y. WANG, 1978. Description of a new genus and two new species of stomatopod Crustacea from the South China Sea. Oceanologia et Limnologia Sinica, **9**: 89-93, pl. 81.

LUCATELLI, D., S. T. AHYONG, L. E. A. BEZERRA & P. J. P. SANTOS, 2013. A new western Atlantic species of *Eurysquilla* (Crustacea: Stomatopoda: Eurysquilloidea) with a key to the world species. Journal of the Marine Biological Association of the United Kingdom, **93**: 1329-1333.

MANNING, R. B., 1963. Preliminary revision of the genera *Pseudosquilla* and *Lysiosquilla* with descriptions of six new genera. Bulletin of Marine Science of the Gulf and Caribbean, **13**: 308-328.

MANNING, R. B., 1977. A monograph of the west African stomatopod Crustacea. Atlantide Report, **12**: 25-181.

MANNING, R. B., 1980. The superfamilies, families, and genera of Recent stomatopod Crustacea, with diagnoses of six new families. Proceedings of the Biological Society of Washington, **93**: 362-372.

MANNING, R. B., 1995. Stomatopod Crustacea of Vietnam: the legacy of Raoul Serène. Crustacean Research, Special Number, **4**: 1-339.

MANNING, R. B. & T.-Y. CHAN, 1997. The genus *Faughnia* from Taiwan, with the description of a new species (Stomatopoda: Parasquillidae). Journal of Crustacean Biology, **17**: 546-554.

PORTER, M., Y. ZHANG, S. DESAI, R. L. CALDWELL & T. W. CRONIN, 2010. Evolution of anatomical and physiological specialization in the compound eyes of stomatopod crustaceans. Journal of Experimental Biology, **213**: 3473-3486.

SERÈNE, R., 1962. Révision du genre *Pseudosquilla* (Stomatopoda) et définition de genres nouveaux. Bulletin de l'Institut Océanographique, Monaco, **1241**: 1-27.

VAN DER WAL, C., S. T. AHYONG, S. Y. W. HO, L. S. F. LINS & N. LO, 2019. Combining morphological and molecular data resolves the phylogeny of Squilloidea (Crustacea: Malacostraca). Invertebrate Systematics, **33**: 89-100.

VAN DER WAL, C., S. T. AHYONG, S. Y. W. HO & N. LO, 2017. The evolutionary history of Stomatopoda (Crustacea: Malacostraca) inferred from molecular data. PeerJ, **5**: e3844.

WANG, J.-W. & T.-H. CHIOU, 2017. Three new records of Nannosquillidae from Taiwan with notes on their ecology (Crustacea, Stomatopoda, Lysiosquilloidea). ZooKeys, **721**: 33-43.

YEH, T.-C. & P.-W. HSUEH, 2010. *Taku spinosocarinatus* (Fukuda, 1909): first record of a takuid stomatopod from Taiwan. Crustaceana, **83**: 369-373.

First received 19 March 201.
Final version accepted 26 August 201.

PERACARIDA

[When citing this volume, please refer to *Crustaceana* 93 (2020), 1233-1546]

Prof. Ruiyu Liu Memorial Issue

GAMMARUS LIURUIYUI SP. NOV. (AMPHIPODA, GAMMARIDAE), A NEW BLIND SPECIES FROM TIAN SHAN, CHINA

BY

YAMI ZHENG[1,2]), ZHONGE HOU[1,3]) and SHUQIANG LI[1])

[1]) Key Laboratory of Zoological Systematics and Evolution, Institute of Zoology, Chinese Academy of Sciences, Beijing 100101, China

[2]) University of Chinese Academy of Sciences, Beijing 100049, China

ABSTRACT

A new species of the genus *Gammarus* Fabricius, 1775 is described and illustrated from the Tian Shan, China. *Gammarus liuruiyui* sp. nov. is characterized by eyes being absent; pereopods V-VII bases not expanded; uropod III with scale-like inner ramus, second article of outer ramus vestigial; and posterior margins of epimeral plates nearly straight. Detailed morphological comparisons with related species are discussed. Genetic distances of the new species and related species are provided as proof of the species delimitation.

Key words. — Morphology, genetic distance, taxonomy

RÉSUMÉ

Une nouvelle espèce du genre *Gammarus* Fabricius, 1775 est décrite et illustrée de Tian Shan, Chine. *Gammarus liuruiyui* sp. nov. est caractérisée par l'absence d'yeux ; les bases des péréiopodes V-VII non élargies ; l'uropode III avec une rame interne en forme d'écaille ; le second article de la rame externe vestigial ; et les bords postérieurs des plaques épimérales presque droites. Des comparaisons morphologiques détaillées avec des espèces apparentées sont discutées. Les distances génétiques entre la nouvelle espèce et les espèces apparentées sont fournies comme preuves de la délimitation des espèces.

Mots clés. — Morphologie, distance génétique, taxonomie

INTRODUCTION

The Tian Shan is a large system of mountain ranges in Central Asia and stretches some 2900 kilometres westward from Xinjiang in China. The geological history of the Tian Shan was influenced by the uplift of the Tibetan Plateau and regression of

[3]) Corresponding author; e-mail: houze@ioz.ac.cn

 | DOI:10.1163/9789004464353_017

the Tethys (Caves et al., 2017), which defined its biodiversity features, sustaining ancient Tethyan relicts (Hou et al., 2011). Thus, the discovery of species diversity will contribute to the understanding of the regional palaeogeographical history and the evolution of the region's fauna (Li & Quan, 2017).

The genus *Gammarus* Fabricius, 1775 contains more than 200 freshwater, brackish, and marine species in the northern hemisphere (Väinölä et al., 2008). It has been suggested that *Gammarus* originated from the ancient Tethys, then diversified in Eurasia (Hou et al., 2011). Up to now, four species have been recorded along the Tian Shan in China, including *Gammarus tianshan* Zhao, Meng & Hou, 2017, *G. takesensis* Hou, Li & Platvoet, 2004, *G. brevipodus* Hou, Li & Platvoet, and *G. decorosus* Meng, Hou & Li, 2003. In the current paper, a blind species attributable to the genus *Gammarus* was described and illustrated from a spring of the Tian Shan, China. A distribution map and a key to the species of the genus *Gammarus* known from Tian Shan in China are presented. Moreover, partial fragments of the mitochondrial cytochrome oxidase subunit I (COI) and nuclear 28S rRNA were obtained. The genetic distances between these species are calculated to confirm the species delimitation.

MATERIAL AND METHODS

Morphological observation

The specimens were collected in a spring with a fine-meshed hand net. Samples were preserved in 95% ethanol in the field, then deposited at −20 °C freezer for long-term preservation. The body length was recorded by holding the specimen straight and measuring the distance along the dorsal side of the body from the base of the first antenna to the base of the telson. All dissected appendages were mounted on slides according to the methods described by Holsinger (1967). Appendages were drawn using a Leica DM2500 compound microscope equipped with a drawing tube. Terminology and taxonomic descriptions follow the literature (Zheng et al., 2017), especially the terms "spines" and "setae" are used to distinguish between thin or fine and more robust setal structures. All types and other materials are maintained in the Institute of Zoology, Chinese Academy of Sciences (IZCAS), Beijing.

Molecular methods

Partial fragments of mitochondrial COI and nuclear 28S were amplified to confirm identifications and validate species delimitation (Costa et al., 2007; Hou et al., 2009). The primers used are LCO1490 (5′-GGTCAACAAATCATAAAGATAT TGG-3′) and HCO2198 (5′-TAAACTTCAGGGTGACCAAAAAATCA-3′) for

TABLE I
GenBank accession numbers and uncorrected pairwise distance of the COI and 28S partial sequences between species in this text

	Species	COI	28S				
			1	2	3	4	5
			MK455898	EF582971	JF965809	JF965681	EF582949
1	*Gammarus liuruiyui* sp. nov.	MK455899	–	0.056	0.070	0.049	0.056
2	*G. tianshan* Zhao, Meng & Hou, 2017	EF570327	0.216	–	0.045	0.042	0.016
3	*G. takesensis* Hou, Li & Platvoet, 2004	JF965979	0.271	0.216	–	0.056	0.043
4	*G. brevipodus* Hou, Li & Platvoet, 2004	–	–	–	–	–	0.039
5	*G. decorosus* Meng, Hou & Li, 2003	EF570303	0.229	0.210	0.232	–	–

COI (Folmer et al., 1994), and 28F (5′-TTAGTAGGGGCGACCGAACAGGGAT-3′) and 28R (5′-GTCTTTCGCCCCTATGCCCAACTGA-3′) for 28S (Hou et al., 2007). Genomic DNA extraction, amplification and sequencing procedures were performed as in Hou et al. (2007). Uncorrected pairwise distances between species were calculated using MEGA 7.0.16 (Kumar et al., 2016). The new sequences were deposited in GenBank and the accession numbers are provided in Table 1.

TAXONOMY

Family GAMMARIDAE Leach, 1814
Genus *Gammarus* Fabricius, 1775
Gammarus liuruiyui sp. nov.
(figs. 1-9)

Material examined.— Holotype male, 11.0 mm (IZCAS-I-A1634-1): Sumutash Town (78.19°E, 40.88°N), Aheqi County, Kizilsu Kirghiz Autonomous Prefecture, Xinjiang Uygur Autonomous Region, altitude 2170 m, 25 July, 2014, collected by Kaibaryer Meng. Paratype female, 9.0 mm (IZCAS-I-A1634-2): same data as holotype. Other material: six males, five females, and two juveniles, same data as holotype.

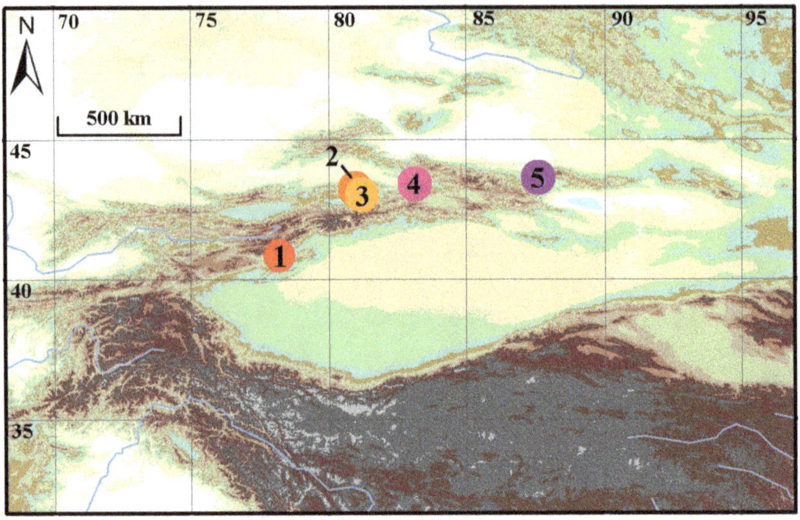

Fig. 1. Type localities of *Gammarus* species from Tian Shan, China. 1, *Gammarus liuruiyui* sp. nov.; 2, *Gammarus tianshan* Zhao, Meng & Hou, 2017; 3, *Gammarus takesensis* Hou, Li & Platvoet, 2004; 4, *Gammarus brevipodus* Hou, Li & Platvoet, 2004; 5, *Gammarus decorosus* Meng, Hou & Li, 2003.

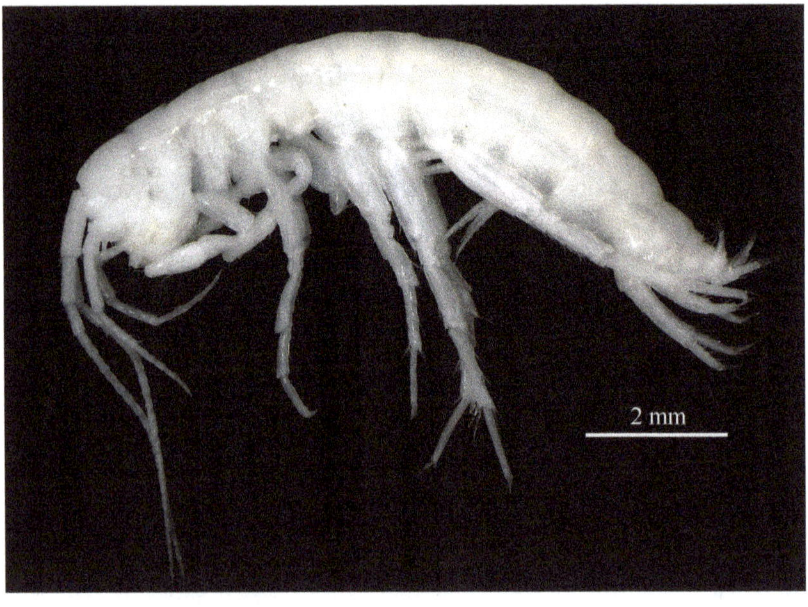

Fig. 2. *Gammarus liuruiyui* sp. nov. male holotype.

Etymology.— The species name honours Prof. Liu Ruiyu for his contributions to crustacean research; the name is a noun in the genitive singular.

Diagnosis.— Eyes absent; pereopod IV coxal plate weakly lobate; pereopods V–VII bases similar, not expanded; epimeral plates nearly straight on posterior margins; urosomites I-III with spines on dorsal margins; uropod III with scale-like inner ramus, second article of outer ramus vestigial; telson deeply cleft, with three distal spines.

Description of holotype male (IZCAS-I-A1634-1).— Body length 11.0 mm.

Head (fig. 3A): eyes absent, inferior antennal sinus shallow.

Antenna I (fig. 3B): longer than antenna II, peduncle articles 1-3 in length ratio 1.0 : 0.6 : 0.4, with distal setae; flagellum with 23 articles; accessory flagellum with two articles; both primary and accessory flagella with short distal setae.

Antenna II (fig. 3C): peduncle articles 3-5 in length ratio 1.0 : 2.5 : 2.2, peduncle article 3 with a seta on posterior margin, three setae on anterior margin, articles 4-5 nearly same length, each with stiff setae along anterior and posterior margins; flagellum with eight articles, each article with distal setae; calceoli absent.

Upper lip (fig. 3D): ventral margin convex, bearing short minute setae.

Mandible (fig. 3E): asymmetrical, right mandible incisor with four teeth; lacinia mobilis with three teeth, setal row with eight plumose setae; palp composed of three articles, third article with one long seta and four short setae on apical margin.

Lower lip and maxilla I: broken, similar to those of female (fig. 6G-I).

Maxilla II (fig. 3F): inner plate with 11 lateral setae; outer plate with 12 apical setae.

Maxilliped (fig. 3G): inner plate with ten apical setae; outer plate with eight peg-like spines and three setae; palp with four articles, second article stout with a row of long simple setae on inner margin and four setae on outer margin, third article with setae on both margins, terminal article hooked, nail small.

Pereon. Gnathopod I (fig. 4A, B): coxal plate sub-rectangular, with three setae on anterior margin, one seta on inner face, three setae on posterior margin; basis sub-linear, bearing long setae on posterior margin; merus with five long setae on posterior margin; carpus with setae on anterior and posterior margins, with tapered distolateral lobe; propodus twice as long as wide, oval, palm with six spines on posterodistal corner, medial spine absent; dactylus reaching approx. 60% length of propodus.

Gnathopod II (fig. 4C, D): coxal plate with three setae on anterior margin and two setae on posterior margin; basis with three long setae on anterior margin and one seta on distal margin, posterior margin with a row of long setae; merus with five setae on posterior margin; carpus with long setae on posterior margin; propodus 1.7 times as long as wide, sub-rectangular, palm with one medial spine and three spines on posterior margin; dactylus reaching approx. 67% length of propodus.

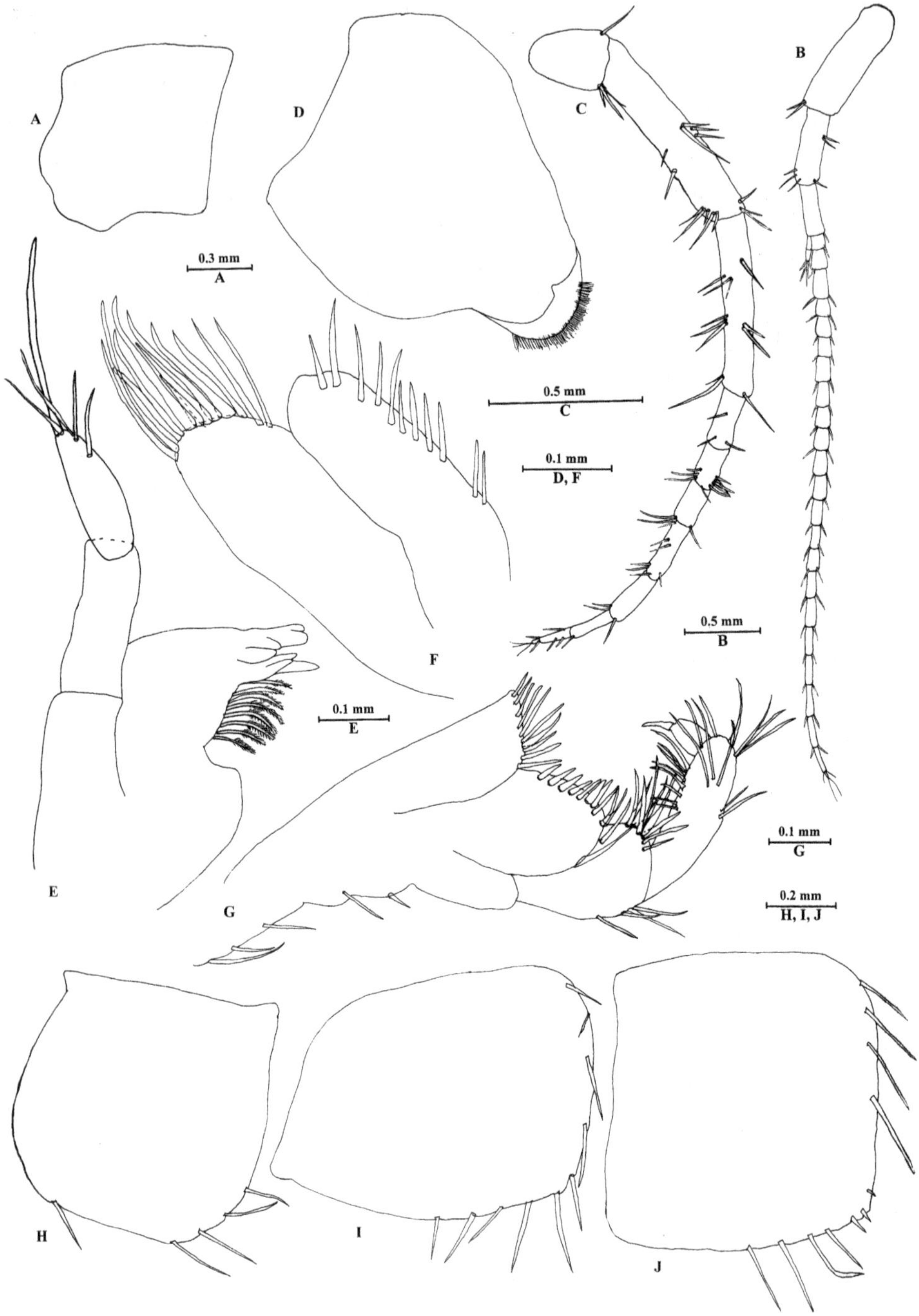

Fig. 3. *Gammarus liuruiyui* sp. nov. male holotype. A, Head; B, antenna I; C, antenna II; D, upper lip; E, right mandible; F, maxilla II; G, maxilliped; H, epimeral plate I; I, epimeral plate II; J, epimeral plate III.

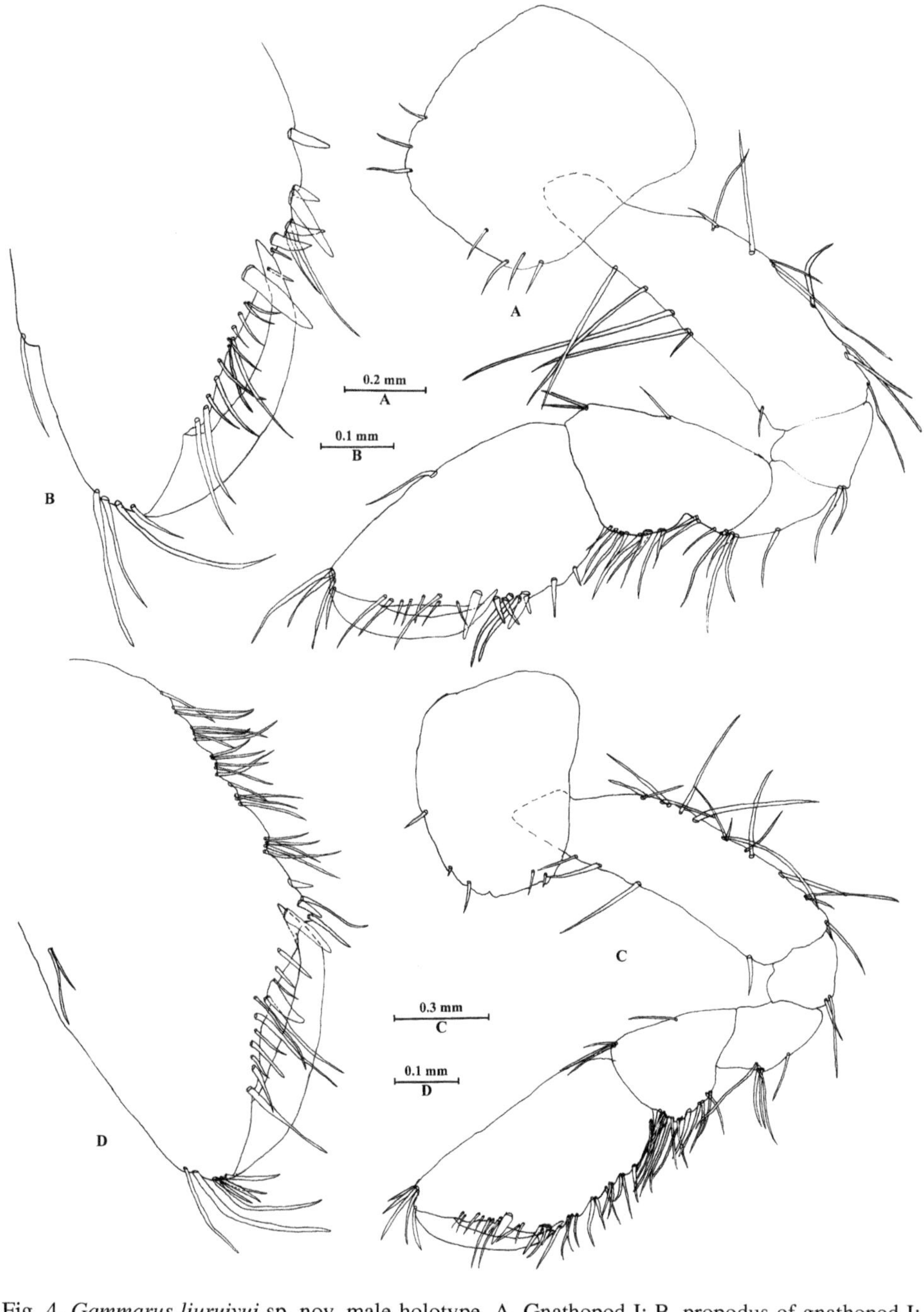

Fig. 4. *Gammarus liuruiyui* sp. nov. male holotype. A, Gnathopod I; B, propodus of gnathopod I; C, gnathopod II; D, propodus of gnathopod II.

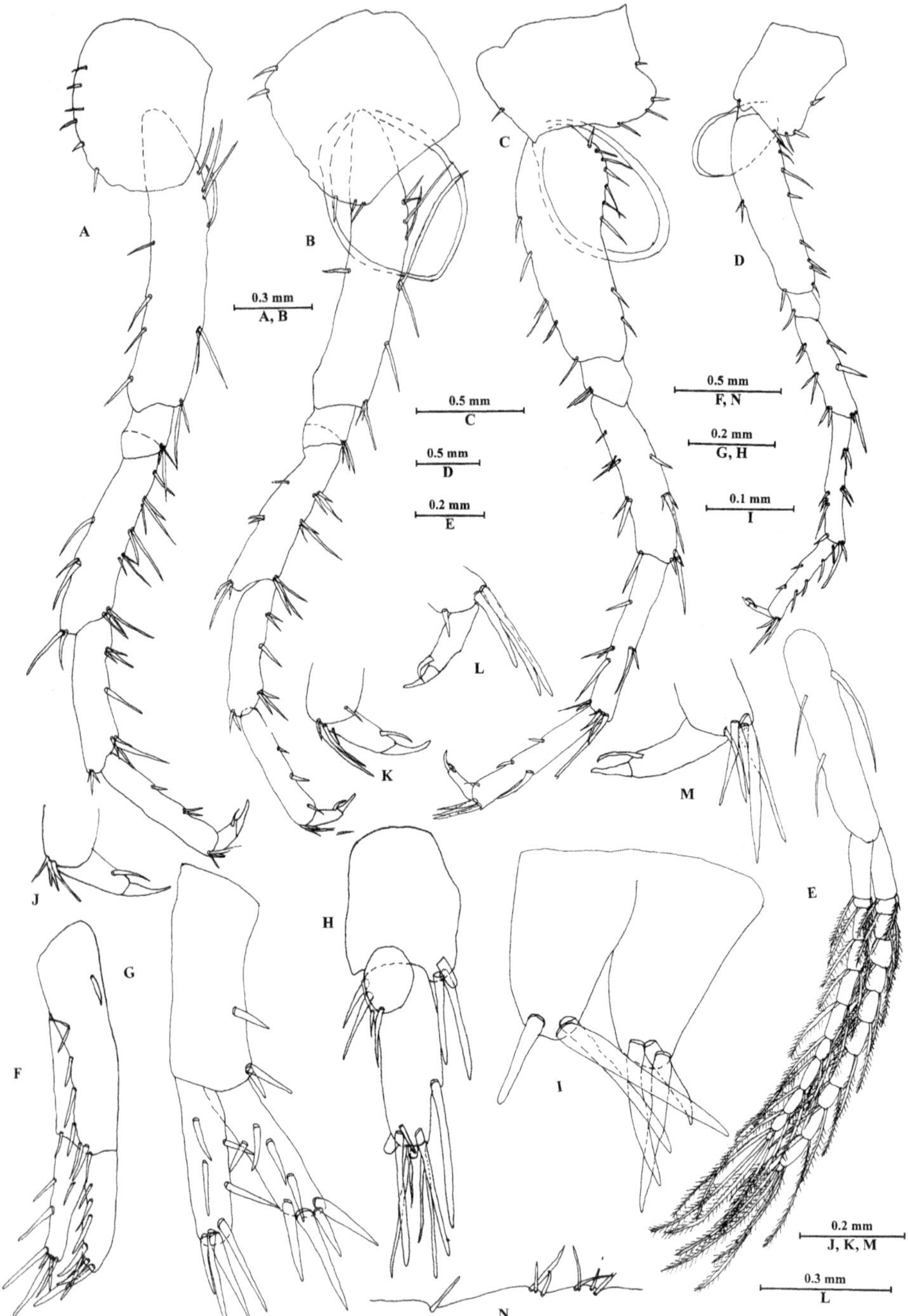

Fig. 5. *Gammarus liuruiyui* sp. nov. male holotype. A, Pereopod III; B, pereopod IV; C, pereopod VI; D, pereopod VII; E, pleopod I; F, uropod I; G, uropod II; H, uropod III; I, telson; J, dactylus of pereopod III; K, dactylus of pereopod IV; L, dactylus of pereopod VI; M, dactylus of pereopod VII; N, urosomites (dorsal view).

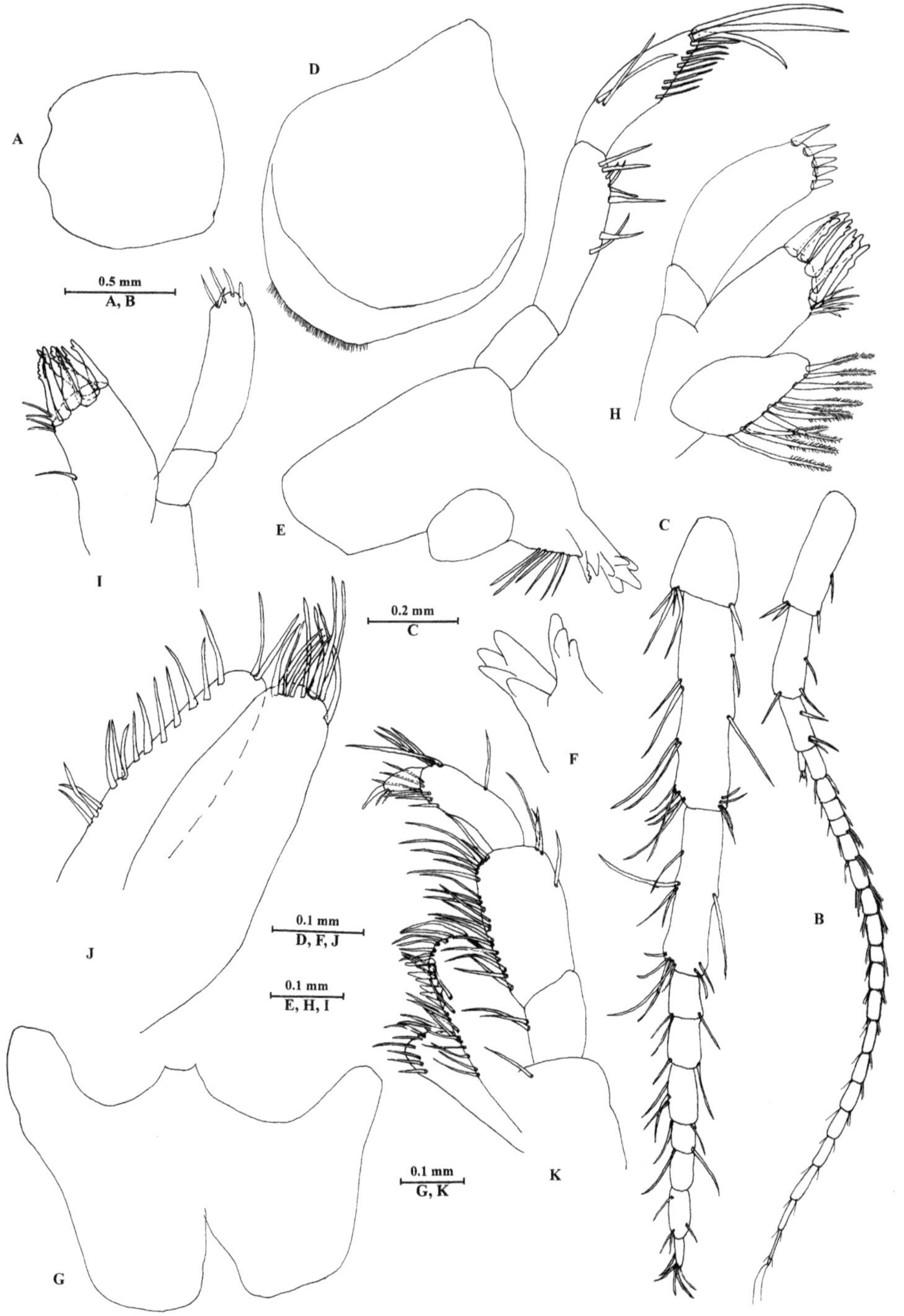

Fig. 6. *Gammarus liuruiyui* sp. nov. female paratype. A, Head; B, antenna I; C, antenna II; D, upper lip; E, right mandible; F, left mandible; G, lower lip; H, maxilla I (right); I, maxilla I (left); J, maxilla II; K, maxilliped.

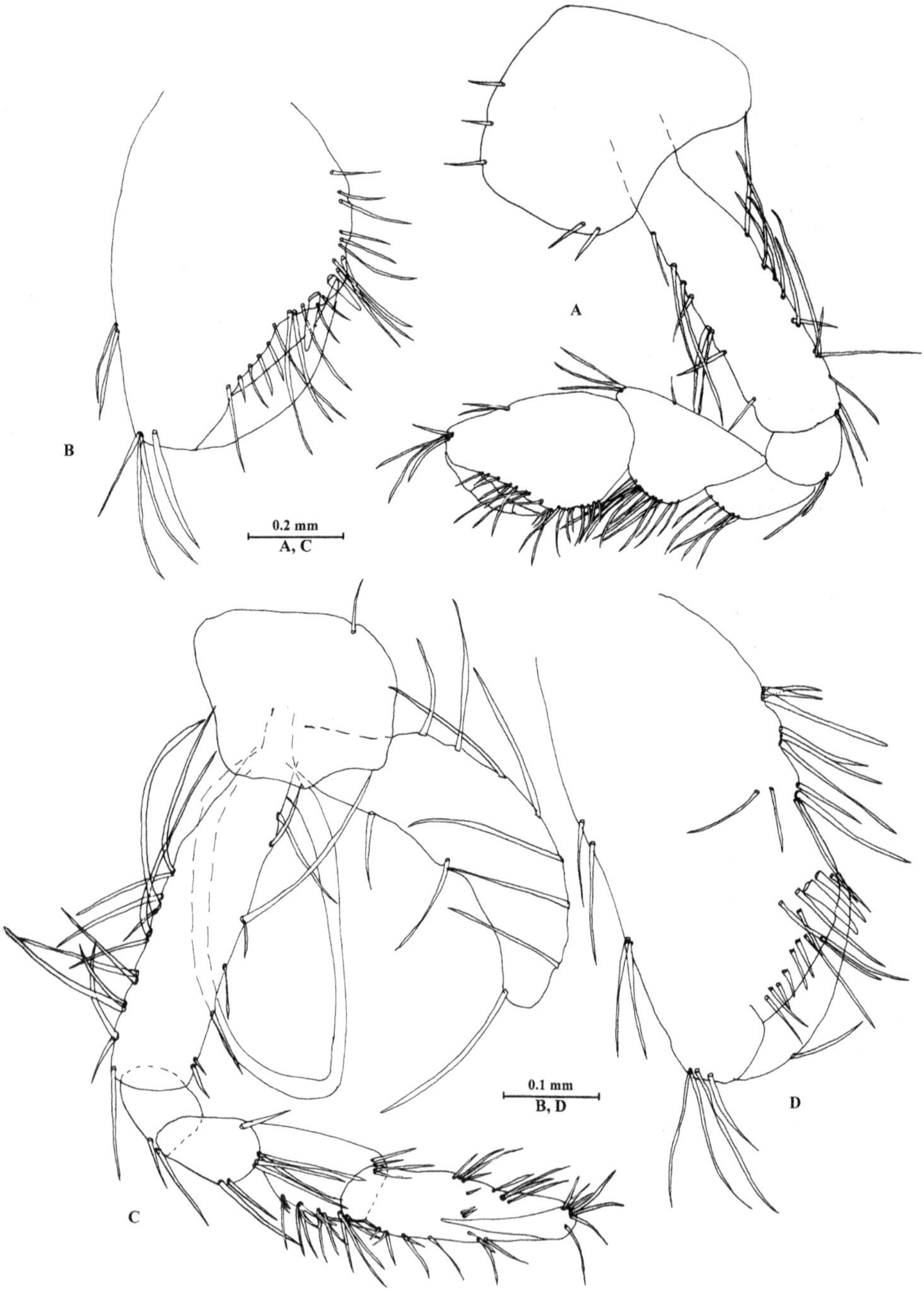

Fig. 7. *Gammarus liuruiyui* sp. nov. female paratype. A, Gnathopod I; B, propodus of gnathopod I; C, gnathopod II; D, propodus of gnathopod II.

Pereopods III-IV (fig. 5A-B, J-K): similar to each other, coxal plate III with six setae on anterior margin; coxal plate IV weakly concave, with two setae on anterior margin; bases sub-linear, bearing one to four long setae and short spines on both margins; merus to propodus with spines accompanied by short setae along anterior and posterior margins; dactylus with one seta at hinge of unguis.

Pereopod V: missing.

Pereopod VI (fig. 5C, L): coxal plate weakly bilobate, with four short setae on posterior margin; basis with six spines on anterior margin and nine spines on posterior margin; merus to propodus with spines and setae along anterior and posterior margins; dactylus with one seta at hinge of unguis.

Pereopod VII (fig. 5D, M): coxal plate small, irregular, with four setae on posterior margin; basis linear, with three short spines on anterior margin and seven spines on posterior margin; merus to propodus with spines and setae along anterior and posterior margins; dactylus with one seta at hinge of unguis.

Coxal gills (fig. 5B-D): present on gnathopod II and pereopods III-VII, sac-like.

Pleon. Epimeral plates (fig. 3H-J): plate I ventrally rounded, with one seta on ventral margin and four setae on posterior margin; plates II-III nearly straight on posterodistal corners, with long setae along ventral and posterior margins.

Pleopods I–III (fig. 5E): similar to each other, peduncle with some marginal setae; inner and outer rami nearly the same length, both rami fringed with plumose setae.

Urosome. Urosomites I-III (fig. 5N): urosomite I with one spine on dorsal margin; urosomite II with three spines on dorsal margin; urosomite III with four spines on dorsal margin.

Uropod I (fig. 5F): peduncle longer than rami, with one basofacial spine, one and six spines on inner and outer margins, respectively; inner ramus with two spines on inner margin; outer ramus with three spines on outer margin; both rami with five terminal spines.

Uropod II (fig. 5G): peduncle with three spines on inner margin; inner ramus with three spines on inner margin, one spine on outer margin; outer ramus with two spines on inner and outer margins, respectively; both rami with four terminal spines.

Uropod III (fig. 5H): shorter than uropods I-II, peduncle shorter than outer ramus, with three spines on distal margin; inner ramus with one distal seta, scale-like; outer ramus with three spines on inner margin, one spine on outer margin, and five terminal spines, second article very small.

Telson (fig. 5I): deeply cleft, each lobe with three distal spines.

Description of paratype female (IZCAS-I-A1634-2).— Body length 9.0 mm.

Head. Antenna I (fig. 6B): peduncle articles 1-3 in length ratio 1.0 : 0.7 : 0.4, with distal setae; flagellum with 20 articles; accessory flagellum with two articles.

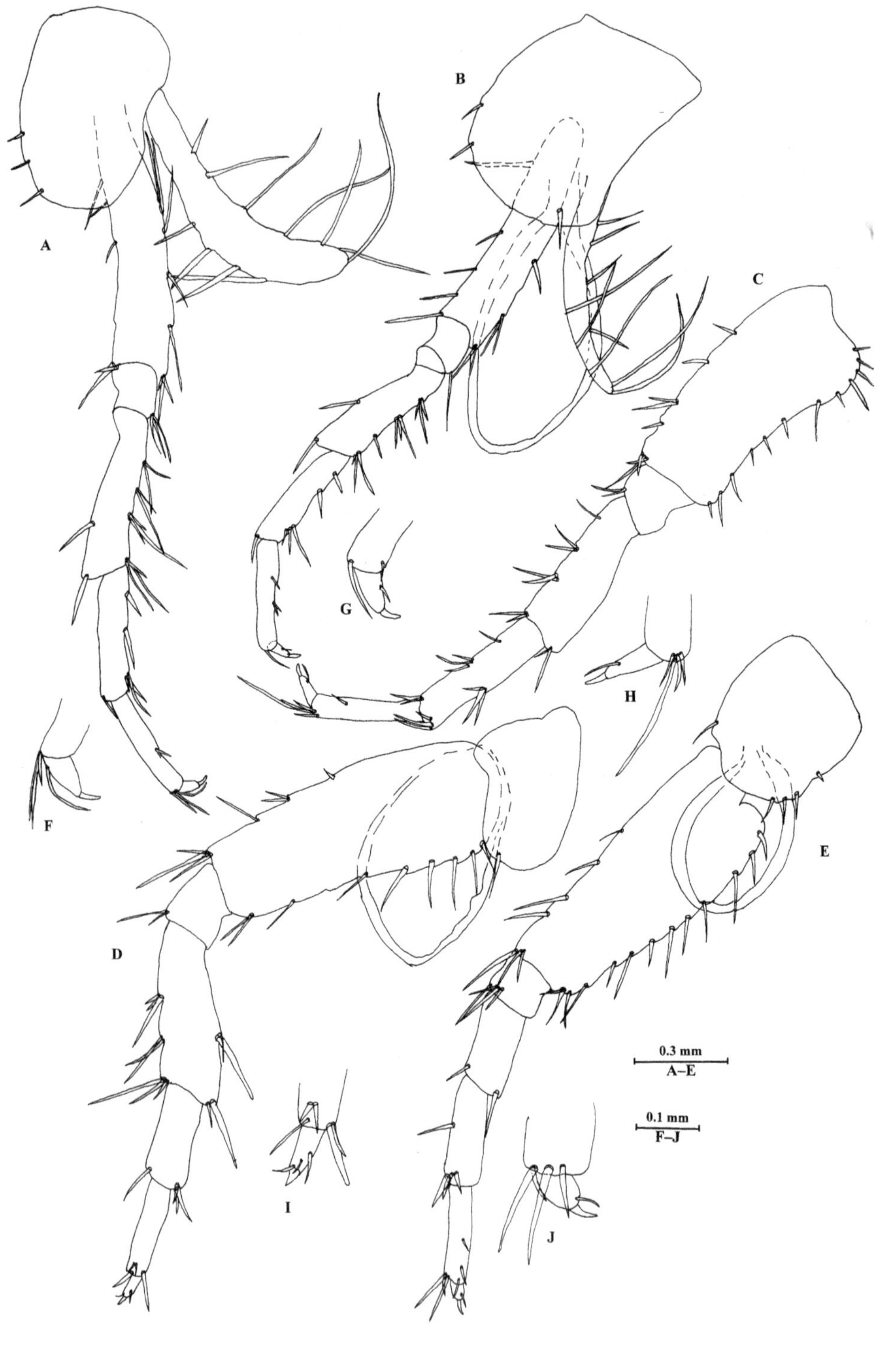
A
B
C
D
E
F
G
H
I
J
0.3 mm
A–E
0.1 mm
F–J

Antenna II (fig. 6C): peduncle articles 3-5 with setae along anterior and posterior margins; flagellum with seven articles, both with distal setae.

Upper lip (fig. 6D). convex, with minute setae.

Mandible (fig. 6E-F): asymmetrical, right incisor with four teeth; lacinia mobilis with three teeth; spine row with six setae; second article of palp with eight setae on inner margin, third article with two setae on outer face, a row of eight setae on inner margin, three long setae on apical margin. Incisor of left mandible with four teeth; lacinia mobilis with three teeth.

Lower lip (fig. 6G): inner lobes lacking.

Maxilla I (fig. 6H-I): inner plate of right maxilla I with ten long plumose setae; outer plate with seven pectinate robust spines; article 2 of palp bearing five spines apically; left palp with five stiff setae.

Maxilla II (fig. 6J): outer plate broader than inner plate, both with several setae.

Maxilliped (fig. 6K): inner plate and outer plate with spines and setae on both margins; second article of palp expanded, with a row of setae on inner margin, four setae on outer margin, third article with marginal setae, fourth article claw-shaped.

Pereon. Gnathopod I (fig. 7A-B): basis sub-linear bearing several long setae on both margins; propodus oval, palm with two spines on posterodistal corner.

Gnathopod II (fig. 7C-D): merus and carpus with setae; propodus twice as long as wide, palm with two spines on posterodistal corner.

Pereopods III-VII (fig. 8A-J): similar to those of male. Bases of pereopods V-VII slender.

Oostegites present on gnathopod II and pereopods III-V.

Pleon. Epimeral plates I-III (fig. 9A-C): plates I-III with three, four and four setae on posterior margins, respectively; plates II-III with one and two setae on ventral margins, respectively.

Pleopods I-III (fig. 9D): both rami fringed with plumose setae.

Urosome. Urosomites I-III (fig. 9I): with three spines on dorsal margins, respectively.

Uropod I (fig. 9E): peduncle with two and five spines on outer and inner margins, respectively; inner ramus with three spines on inner margin; outer ramus with one and two spines on inner and outer margins, respectively; both rami with five terminal spines.

Uropod II (fig. 9F): peduncle with three spines on outer margin and one spine on inner margin; outer ramus slightly shorter than inner ramus, with three spines

---

Fig. 8. *Gammarus liuruiyui* sp. nov. female paratype. A, Pereopod III; B, pereopod IV; C, pereopod V; D, pereopod VI; E, pereopod VII; F, dactylus of pereopod III; G, dactylus of pereopod IV; H, dactylus of pereopod V; I, dactylus of pereopod VI; J, dactylus of pereopod VII.

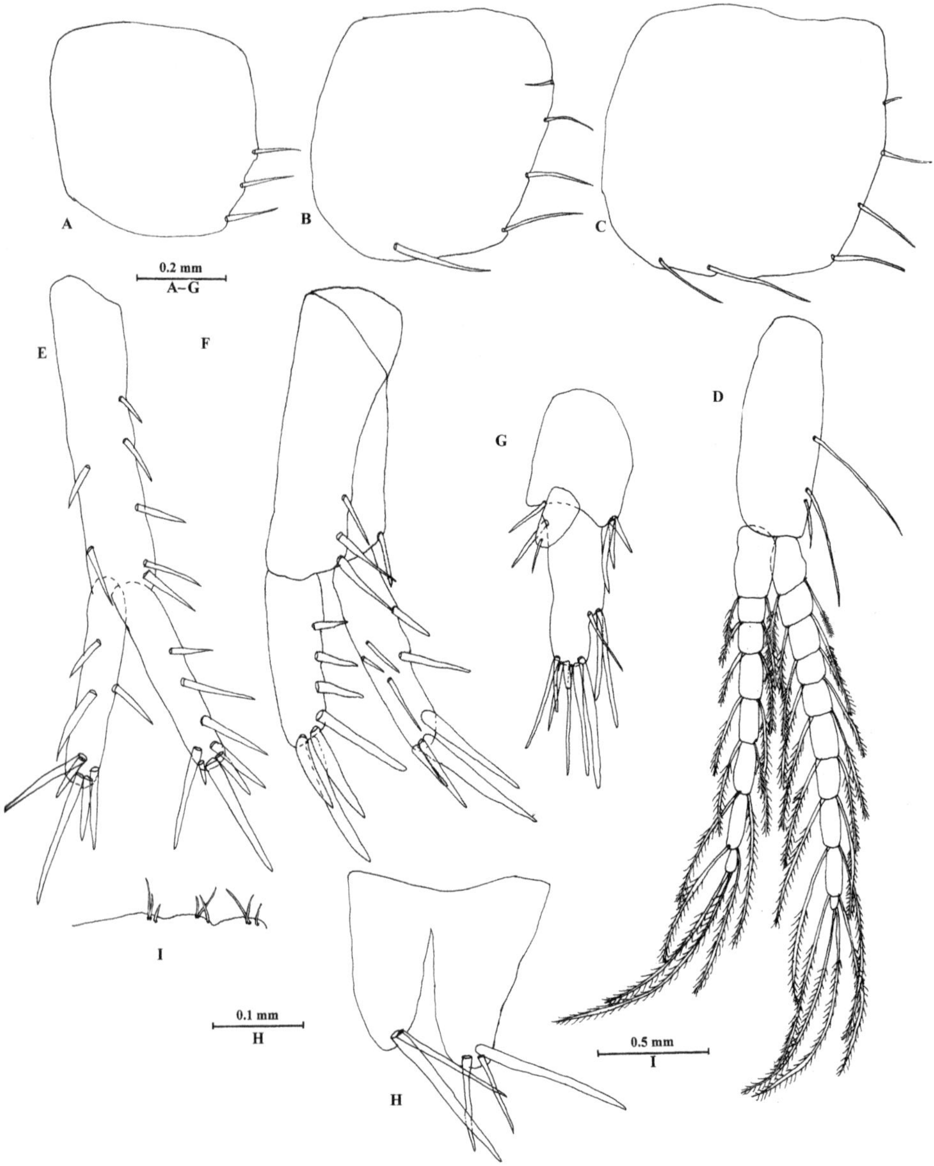

Fig. 9. *Gammarus liuruiyui* sp. nov. female paratype. A, Epimeral plate I; B, epimeral plate II; C, epimeral plate III; D, pleopod I; E, uropod I; F, uropod II; G, uropod III; H, telson; I, urosomites (dorsal view).

on inner margin; inner ramus with two and three spines on inner and outer margins, respectively; both rami with five terminal spines.

Uropod III (fig. 9G): peduncle shorter than outer ramus, with one and three spines on anterior and posterior margins, respectively; inner ramus scale-like; outer ramus with three spines on outer margin and several terminal spines, second article vestigial.

Telson (fig. 9H): deeply cleft, each lobe with two to three distal spines.

Habitat.— Fifteen individuals were collected from a spring (area 0.2 m^2), which is connected to a stream flowing from Tian Shan two kilometers away. The spring is influenced by the groundwater in the deep strata. When the snow melts, the river flows through the desert and forms subterranean water. A spring will erupt from the shallow layer. All specimens were found under rubble in the bottom of the spring, with eyes totally absent without subcutaneous ocelli. The blind species were considering as isolated populations from a species widespread in surface water that have adapted to subterranean life (Simon, 2012). However, epigean gammarids have not been seen around the type locality.

Remarks.— Freshwater *Gammarus* are characterized by direct development of fertilized eggs in a marsupium and no free-swimming larvae to be transported passively by currents (Hughes & Ahyong, 2016). Their weak dispersal potential means that they are easily influenced by geological barriers, such as mountain ranges (Hou et al., 2014). Previous studies indicated that most are narrowly endemic species (Copilaş-Ciocianu et al., 2018). We have made comparisons between *Gammarus liuruiyui* sp. nov. and the species of *Gammarus* from Central Asia, finding that the new species can be easily distinguished by the scale-like inner ramus of uropod III (Sidorov et al., 2018). Therefore, we focused our discussion on the differences between the new species and the species of *Gammarus* known from Tian Shan, China.

Gammarus liuruiyui sp. nov. is most similar to *Gammarus brevipodus* in that the calceoli are absent in antenna II and that pereopods III and IV have a few short setae. The new species differs from *G. brevipodus* in that the eyes are absent in *Gammarus liuruiyui*, but eyes of *G. brevipodus* are present; bases of pereopods V-VII are slender, but bases of *G. brevipodus* are slightly expanded on medial margin; uropod III has the scale-like inner ramus, but inner ramus of *G. brevipodus* is about one third the length of the outer ramus.

Gammarus liuruiyui sp. nov. resembles *Gammarus takesensis* in that pereopods III-VII have a few short setae, the peduncle of uropod I has a basofacial spine, and the telson is deeply cleft. It can be distinguished from *G. takesensis* by the eyes being absent in *Gammarus liuruiyui* while present in *G. takesensis*; the coxal plate of pereopod IV is weakly concave, while it is concave in *G. takesensis*; epimeral plates II and III are nearly straight on posterodistal corners, while subacute in *G. takesensis*; uropod III has a scale-like inner ramus, while the inner ramus of *G. takesensis* is about three-fourth the length of the outer ramus.

We downloaded COI and 28S sequences of the genus *Gammarus* from Tian Shan, China. Molecular analyses showed high interspecific divergences. The uncorrected pairwise distances between *Gammarus liuruiyui* sp. nov. and *G. tianshan*, *G. takesensis*, *G. decorosus* range from 21.6-27.1% for COI, and 4.9-7.0% for 28S. These COI distance values are larger than the 16% divergence recommended by Lefebure et al. (2006) as a threshold for crustacean species delimitation. The 28S differentiation is comparable to interspecific distances calculated for several species of *Gammarus* well delineated in morphological terms (Hou & Li, 2010; Copilaş-Ciocianu & Petrusek, 2016). Therefore, morphological and molecular data support *G. liuruiyui* sp. nov. being a new species.

The comparison between *Gammarus* species from Tian Shan, China is presented in the following key.

Key to *Gammarus* Species from the Tian Shan, China

1. Eyes present, inner ramus of uropod III longer than one-third of outer ramus 2
– Eyes absent, inner ramus of uropod III scale-like .... *Gammarus liuruiyui* sp. nov.

2. Pereopods III-IV and uropod III bearing long setae 3
– Pereopods III-IV and uropod III with a few short setae 4

3. Uropod III inner ramus about half the length of outer ramus, both rami with simple setae .... *Gammarus tianshan*
– Uropod III inner ramus about three-fourth of outer ramus, both rami with simple and plumose setae .... *Gammarus decorosus*

4. Calceoli of antenna II present .... *Gammarus takesensis*
– Calceoli of antenna II absent .... *Gammarus brevipodus*

ACKNOWLEDGEMENTS

We thank Kaibaryer Meng for their help in the field collection. This study was supported by the Strategic Priority Research Program of Chinese Academy of Sciences (XDA2005020102, XDB31000000), the National Natural Sciences Foundation of China (NSFC-31772417), and the Southeast Asia Biodiversity Research Institute, Chinese Academy of Sciences (Y4ZK111B01).

REFERENCES

Caves, J. K., B. U. Bayshashov, A. Zhamangara, A. J. Ritch, D. E. Ibarra, D. J. Sjostrom, H. T. Mix, M. J. Winnick & C. P. Chamberlain, 2017. Late Miocene uplift of the Tian Shan and Altai and reorganization of Central Asia climate. GSA Today, **17**: 19-26.

Copilaş-Ciocianu, D. & A. Petrusek, 2016. Phylogeography of a freshwater crustacean species complex reflects a long-gone archipelago. J. Biogeogr., **44**: 421-432.

Copilaş-Ciocianu, D., A. Zimţa & A. Petrusek, 2018. Integrative taxonomy reveals a new *Gammarus* species (Crustacea, Amphipoda) surviving in a previously unknown southeast European glacial refugium. Zool. Syst. Evol. Res., **57**: 272-297.

COSTA, F. O., J. R. DEWAARD, J. BOUTILLIER, S. RATNASINGHAM, R. T. DOOH, M. HAJIBABAEI & P. D. N. HEBERT, 2007. Biological identifications through DNA barcodes: the case of the Crustacea. Can. J. Fish. Aquat. Sci., **64**: 272-295.

FABRICIUS, J. C., 1775. Systema Entomologiae, Sistens Insectorum Classes, Ordines, Genera Species Adiectis Synonymis, Locis, Descritionibus, Observationibus: 832 pp. (Flensbergi et Lipsiae).

FOLMER, O., M. BLACK, W. HOEH, R. LUTZ & R. VRIJENHOEK, 1994. DNA primers for amplification of mitochondrial cytochrome *c* oxidase subunit I from diverse metazoan invertebrates. Mol. Mar. Biol. Biotechnol., **3**: 294-299.

HOLSINGER, J. R., 1967. Systematics, speciation, and distribution of the subterranean amphipod genus *Stygonectes* (Gammaridae). Bull. U.S. Natn. Mus., **259**: 1-176.

HOU, Z. & S. LI, 2010. Intraspecific or interspecific variation: delimitation of species boundaries within the genus *Gammarus* (Crustacea, Amphipoda, Gammaridae), with description of four new species. Zool. J. Linn. Soc. Lond, **260**: 215-253.

HOU, Z., B. SKET, C. FISER & S. LI, 2011. Eocene habitat shift from saline to freshwater promoted Tethyan amphipod diversification. Proc. Natl Acad. Sci. USA, **108**: 14533-14538.

HOU, Z., J. FU & S. LI, 2007. A molecular phylogeny of the genus *Gammarus* (Crustacea: Amphipoda) based on mitochondrial and nuclear gene sequences. Mol. Phylogenet. Evol., **45**: 596-611.

HOU, Z., S. LI & D. PLATVOET, 2004. Three new species of the genus *Gammarus* from Ili River, China. Rev. Suisse Zool., **111**: 257-284.

HOU, Z., Z. LI & S. LI, 2009. Identifying Chinese species of *Gammarus* (Crustacea: Amphipoda) using DNA barcoding. Curr. Zool., **55**: 158-164.

HOU, Z., J. LI & S. LI, 2014. Diversification of low dispersal crustaceans through mountain uplift: a case study of *Gammarus* (Amphipoda: Gammaridae) with descriptions of four novel species. Zool. J. Linn. Soc., **170**: 591-633.

HUGEHES, L. E. & S. T. AHYONG, 2016. Collecting and processing amphipods. J. Crust. Biol., **36**: 584-588.

KUMAR, S., G. STECHER & K. TAMURA, 2016. MEGA7: molecular evolutionary genetics analysis version 7.0 for bigger datasets. Mol. Phylogenet. Evol., **33**: 1870-1874.

LEACH, W. E., 1814. Crustaceology. The Edinburgh Encyclopaedia, **7**: 402-403.

LEFÉBURE, T., C. J. DOUADY, M. GOUY & J. GIBERT, 2006. Relationship between morphological taxonomy and molecular divergence within Crustacea: proposal of a molecular threshold to help species delimitation. Mol. Phylogenet. Evol., **40**: 435-447.

LI, S. & R. QUAN, 2017. Taxonomy is the cornerstone of biodiversity conservation — SEABRI reports on biological surveys in Southeast Asia. Zool. Res., **38**: 213-214.

MENG, K., Z. HOU & S. LI, 2003. A new species from Xinjiang, China (Crustacea: Amphipoda: Gammaridae). Acta Zootaxonomica Sinica, **28**: 621-628.

SIDOROV, D., Z. HOU & B. SKET, 2018. Three new remarkable amphipod species (Crustacea: Gammaridae) from springs and subterranean waters of Central Asia. Zootaxa, **444**: 437-461.

SIMON, K., 2012. Cave ecosystems. In: W. B. WHITE & D. C. CULVER (eds.), Encyclopedia of caves (2nd ed.): 99-102. (Academic Press, Elsevier, Amsterdam).

VÄINÖLÄ, R., J. D. S. WITT, M. GRABOWSKI, J. H. BRADBURY, K. JAZDZEWSKI & B. SKET, 2008. Global diversity of amphipods (Amphipoda; Crustacea) in freshwater. Hydrobiologia, **595**: 241-255.

ZHAO, S., K. MENG & Z. HOU, 2017. Two new *Gammarus* species and a new name (Crustacea: Amphipoda: Gammaridae) from Northwest China. Zootaxa, **4273**: 195-215.

ZHENG, Y. & Z. HOU, 2017. *Myanmarorchestia victoria* sp. nov. (Crustacea, Amphipoda, Talitridae), a new species of landhopper from the high altitude forests in Myanmar. Zool. Res., **38**: 1-11.

First received 19 January 2019.
Final version accepted 29 August 2019.

[When citing this volume, please refer to *Crustaceana* 93 (2020), 1233-1546]

Prof. Ruiyu Liu Memorial Issue

A NEW SPECIES OF THE GENUS *PARIONINELLA* (EPICARIDEA, BOPYRIDAE, PSEUDIONINAE) FROM AUSTRALIA

BY

JIANMEI AN[1,3]), QIANQIAN XI[1]) and GUSTAV PAULAY[2])

[1]) School of Life Science, Shanxi Normal University, Linfen, 041000, P. R. China

[2]) Florida Museum of Natural History, University of Florida, Gainesville, FL 32611-7800, USA

ABSTRACT

Two new records of the rare parasitic isopod genus *Parioninella* are described. *Parioninella liuruiyui* n. sp. represents the first record of the genus from Australia and the first bopyrid described from the porcelain crab *Pachycheles pisoides* (Heller, 1865). The female of the new species differs from other described species in lacking eyes, produced into pair of anterolateral flaps, the brood pouch completely covered by oostegites, while the male lacks pigmentation. *Parioninella astridae* Nierstrasz & Brender à Brandis, 1930 is recorded from Mindoro, Philippines, in *Pachycheles sculptus* (H. Milne Edwards, 1837). An identification key to the three *Parioninella* species is provided.

Key words. — Epicaridea, Bopyridae, *Parioninella*, new species, new records

RÉSUMÉ

Deux nouvelles observations du parasite, rare, *Parioninella* sont décrites. *Parioninella liuruiyui* n. sp. représente le premier signalement du genre en Australie et le premier bopyre décrit sur le crabe porcelaine *Pachycheles pisoides* (Heller, 1865). La femelle de cette nouvelle espèce diffère des autres espèces décrites par son absence d'yeux, transformés en une paire de languettes antérolatérales, une poche incubatrice complètement couverte par les oostégites, tandis que le mâle n'a pas de pigmentation. *Parioninella astridae* Nierstrasz & Brender à Brandis, 1930 a été observé à Mindoro, Philippines, chez *Pachycheles sculptus* (H. Milne Edwards, 1837). Une clé d'identification des trois espèces de *Parioninella* est donnée.

Mots clés. — Epicaride, Bopyridé, *Parioninella*, nouvelle espèce, nouveau signalement

INTRODUCTION

Nierstrasz & Brender à Brandis (1930) erected *Parioninella* for their new species *Parioninella astridae* found infesting the porcelain crab *Pachycheles sculptus* (H. Milne Edwards, 1837) from Indonesia. Bourdon (1976) reexamined the

[3]) Corresponding author; e-mail: anjianmei@hotmail.com

 | DOI:10.1163/9789004464353_018

type specimens and recorded *P. astridae* from the Philippines in the same host species. Shiino (1942) described *Aporobopyrosa pacifica* from the Palau Islands, also infesting *P. sculptus*. Bourdon (1976) transferred Shiino's species to *Parioninella*. Shiino (1958) described *Parioninella obovata* infesting *Munida japonica* Stimpson, 1858 from Japan. Boyko (2004) noted that some key characters, such as the longer body, well-developed lateral plates on the pleon, and the absence of uropods warranted the erection of a new genus, *Ovoionella* for *O. obovata*. Thus, *Parioninella* presently includes two valid species. In this paper we provide two new records for the genus: *P. astridae* infesting *Pachycheles sculptus* from the Philippines, and *P. liuruiyui* n. sp. infesting *Pachycheles pisoides* (Heller, 1865) from Australia. This is the first record of the genus from Australia, and the first bopyrid recorded from *P. pisoides*. A key to the three species of *Parioninella* is presented.

MATERIAL AND METHODS

The specimens reported were found infesting decapods in the collections of the Florida Museum of Natural History, University of Florida (UF). Animals were viewed and drawn using a Nikon SMZ1500 microscope. Males were studied with scanning electron microscopy. They were fixed in 2.5% glutaraldehyde in Millonig's phosphate buffer at pH 7.4 for 1.5 hours and postfixed in 1% osmium tetroxide in Millonig's buffer for 1 hour. The specimens were then dehydrated through a graded ethanol series, critical-point dried, sputter coated with colloidal gold, and examined with a HITACHI S-3400N scanning electron microscope.

SYSTEMATIC ACCOUNT

Order ISOPODA Latreille, 1817

Family BOPYRIDAE Rafinesque-Schmaltz, 1815

Subfamily PSEUDIONINAE R. Codreanu, 1967

Parioninella Nierstrasz & Brender à Brandis, 1930

Type species.— *Parioninella astridae* Nierstrasz & Brender à Brandis, 1930, by monotypy.

Parioninella liuruiyui n. sp.
(figs. 1, 2)

Material examined.— UF42274, holotype, ♀, allotype, ♂, Australia, Western Australia, Ningaloo Reef, -22.7691°S 113.7046°E, 10 m depth, host in dead *Pocillopora* coral head, collected by Julian Caley, Laetitia Plaisance, May 2009, infesting *Pachycheles pisoides* (Heller, 1865) (UF22517).

Description of holotype female.— Body oval, length 2.33 mm, maximal width 1.90 mm across fourth pereomere, head length 0.70 mm, head width 0.93 mm. Body distorted about 13°. All body segments distinct (fig. 1A, B).

Head trapezoid, with well-developed frontal lamina (fig. 1A). Eyes absent. Antennae only visible in ventral view, first of three, second of five articles respectively, ending in tuft of terminal setae (fig. 1C). Barbula (fig. 1D) with blunt, bilobed projections on each side. Maxilliped (fig. 1E) lacking palp, nearly

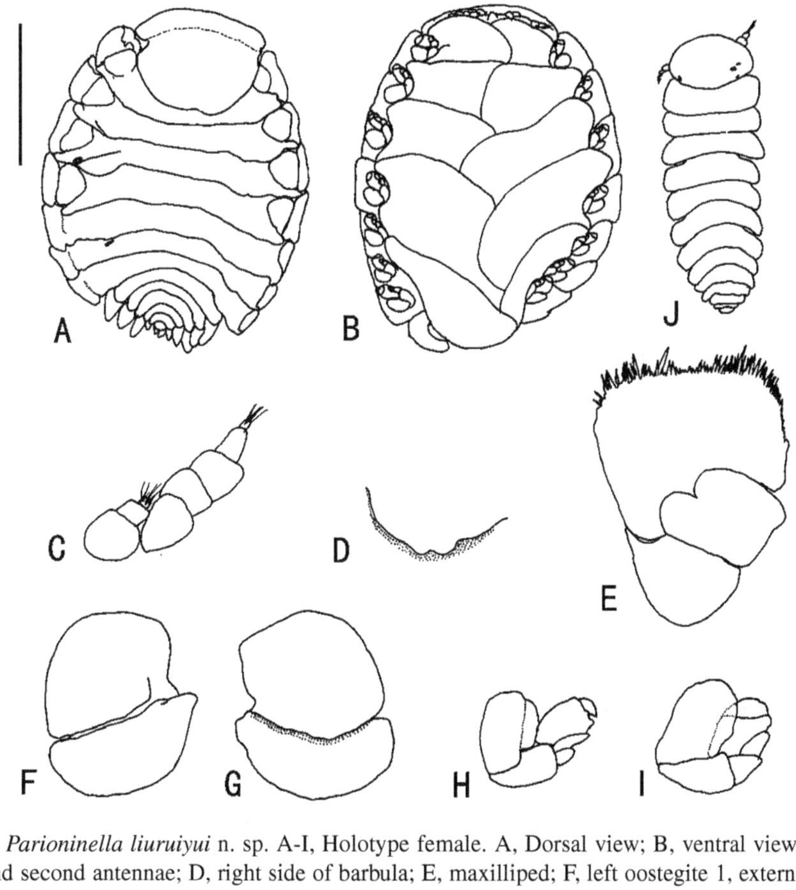

Fig. 1. *Parioninella liuruiyui* n. sp. A-I, Holotype female. A, Dorsal view; B, ventral view; C, left first and second antennae; D, right side of barbula; E, maxilliped; F, left oostegite 1, external view; G, left oostegite 1, internal view; H, left pereopod 1; I, left pereopod 7. J, Allotype male, dorsal view. Scale bars: A, B = 1 mm; C, E, H, I = 0.35 mm; D, J = 0.81 mm; F, G = 0.64 mm.

triangular, anterior part of maxilliped much larger than posterior part, anterior margin with densely setose fringe.

Pereomeres distinct, fourth broadest. All pereomeres with conspicuous coxal plates, those on left side longer than those on right. Dorsolateral bosses on first four pereomeres. Brood pouch highly vaulted, completely covered by oostegites. First oostegite (fig. 1F, G) with two subequal articles, without posterolateral point, with smooth internal ridge. Fifth pair of oostegites extending over and covering pleon in ventral view. Pereopods (fig. 1H, I) of similar structure, increasing slightly in size toward posterior, dactyli blunt, meri of last four pereopods larger than of anterior ones.

Pleon small, with six distinct pleomeres, reduced lateral plates, terminating in minute (sixth) pleotelson. Five pairs of biramous pleopods, with endopodites smaller than exopodites, and with small, uniramous uropods, similar in size to fifth pleopods.

Description of allotype male.— Length 1.98 mm, maximal width (across fifth pleomere) 0.68 mm, head length 0.38 mm, head width 0.60 mm, pleonal length 0.45 mm. All regions and segments distinctly separated (figs. 1J, 2A).

Head semicircular, with dark eyes near posterolateral corners (fig. 1J). Antennae visible from dorsal view, first of three, second of five articles (fig. 2B), second much longer than first (fig. 1J).

Seven distinct pereomeres of similar width, all lacking midventral projections (fig. 2A). Pereopods subequal (fig. 2A), first three with sharp dactyli (fig. 2C), last four with blunt dactyli (fig. 2D).

Pleon of six segments, lacking midventral projections, terminating in pleotelson bearing a ventral anal cone (fig. 2G). First five pleomeres bearing cushion-like, ovate pleopods (fig. 2E), uropods reduced to two or three setae emerging from pleotelson (fig. 2G). All pereomeres and pleomeres with dense clusters of setae along their lateral margins (fig. 2F).

Etymology.— The specific name, *liuruiyui*, refers to Prof. Ruiyu Liu to commemorate the contributions he has made to carcinology. The name is a noun in the genitive singular.

Remarks.—The round body, small pleon, well-developed coxal plates on all pereomeres, and reduced lateral plates on the pleon indicate that this species belongs to *Parioninella*. The new species can be distinguished from the two previously described species in the genus by the following characters: 1. Maxilliped without palp and with dense setae along anterior margin. In contrast the maxilliped of *P. astridae* bears a long palp, while that of *P. pacifica* lacks a palp but the anterior margin in both is barren. 2. First oostegite without a posterolateral point, and with a smooth internal ridge. Both *P. astridae* and *P. pacifica* have a first oostegite

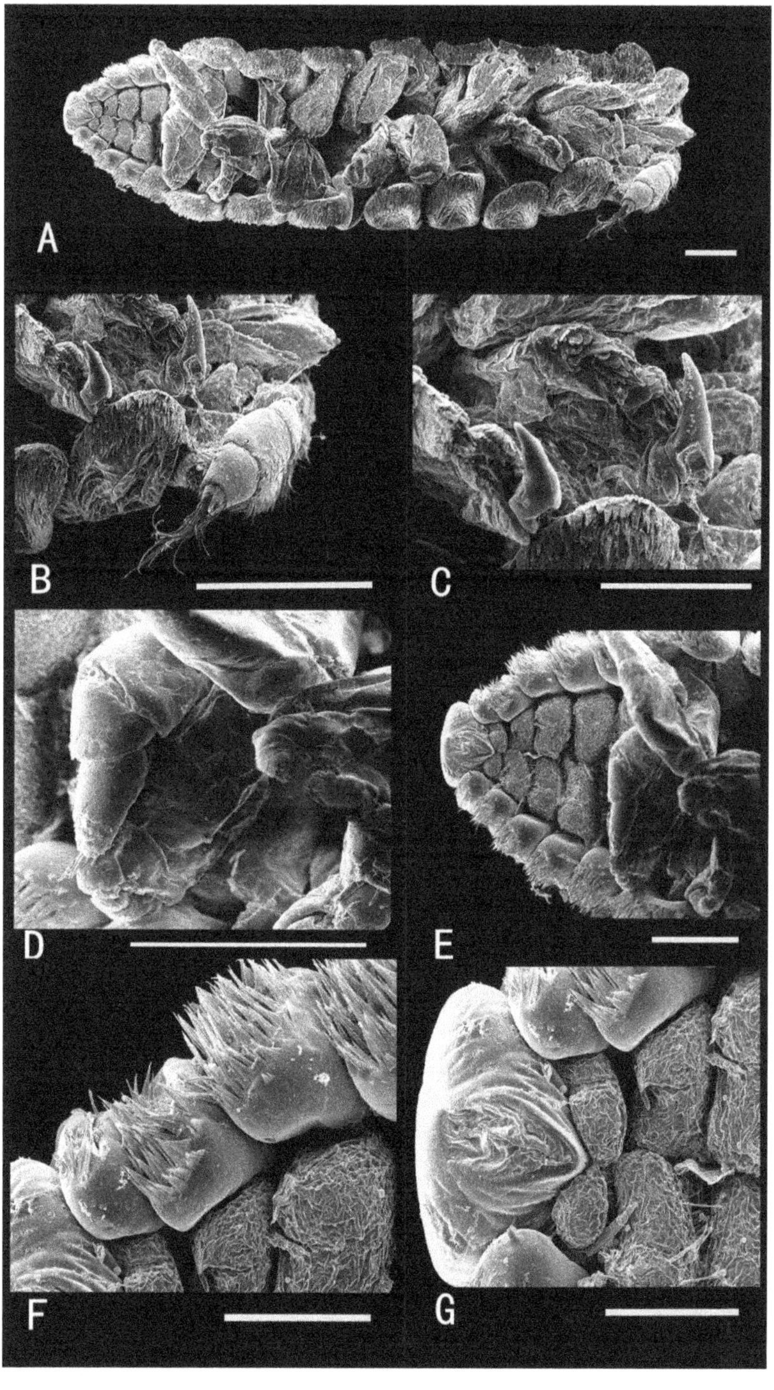

Fig. 2. *Parioninella liuruiyui* n. sp. Allotype male, scanning electron micrographs. A, Ventral view; B, left antennae; C, right pereopods 1, 2; D, right pereopod 7; E, ventral view of pleon; F, lateral margin of the right side of pleon; G, terminal segments of the pleon. Scale bar: 50 μm.

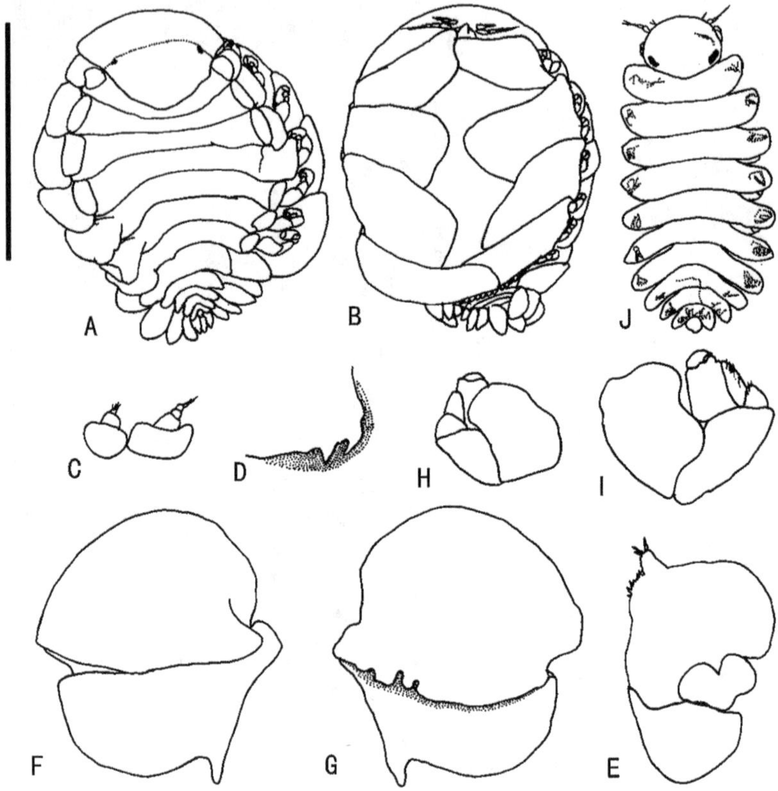

Fig. 3. *Parioninella astridae* Nierstrasz & Brender à Brandis, 1930 (UF42810). A-I, Female. A, Dorsal view; B, ventral view; C, left antennae; D, left side of barbula; E, left maxilliped, external view; F, left oostegite 1, internal view; G, left oostegite 1, external view; H, left pereopod 2; I, left pereopod 7. J, Male, dorsal view. Scale bars: A, B = 1 mm; C, D, E, H, I = 0.29 mm; F, G, J = 0.43 mm.

with a sharp posterolateral point, and with a finely tuberculate internal ridge. 3. Pleon of male with six segments. Males of *P. astridae* have 4-5 pleomeres, while those of *P. pacifica* possess 3 pleomeres.

Parioninella astridae Nierstrasz & Brender à Brandis, 1930
(figs. 3, 4)

Parioninella astridae Nierstrasz & Brender à Brandis, 1930: 14-15, figs. 17-20 [type locality: Indonesia; infesing *Pachycheles sculptus*]; Shiino, 1952: 39; 1958: 45; Bourdon, 1967: 111; 1976: 166, 214, 215, 221-224, 240, tab.V, figs. 34-35; Markham, 1982: 326-329, 385, figs. 1, 2, tab. 1; 1990: 555-556.

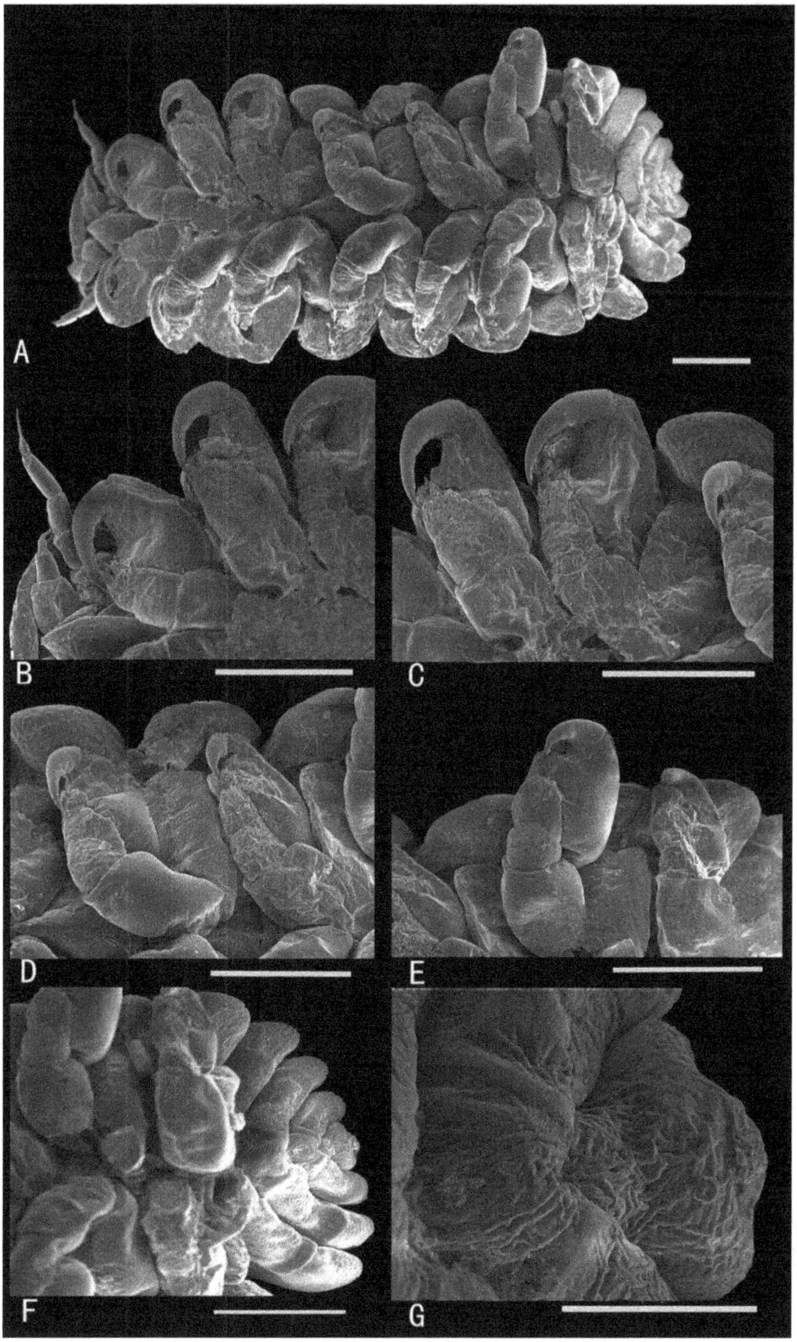

Fig. 4. *Parioninella astridae* Nierstrasz & Brender à Brandis, 1930 (UF42810). Male, scanning electron micrographs. A, Ventral view; B, left first and second antennae; C, left pereopods 2, 3; D, left pereopods 4, 5; E, left pereopods 6, 7; F, ventral view of pleon; G, pleotelson. Scale bar: 50 μm.

Material examined.— UF42810, 1♀, 1♂, Philippines, Mindoro, Puerto Galera, off Sabang Beach, 13.52207°N 120.97522°E, 4-6 m depth, collected by G. Paulay, infesting *Pachycheles sculptus*, collected at night (UF 43125).

Remarks.— *Parioninella astridae* has been recorded from Indonesia, the Philippines and Hong Kong, thus the present specimen does not represent a significant range extension. Nierstrasz & Brender à Brandis (1930), Bourdon (1976) and Markham (1982, 1986) described this species thoroughly. The species has been recorded from the porcelain crabs *Pachycheles sculptus* and *Polyonyx* sp. (Markham, 1990). The reference female is similar to Bourdon's Philippine specimen, with black eyes (fig. 3A), round body (fig. 3A, B), well-developed coxal plates on all pereomeres, barbula with two projections on each side (fig. 3D), and oostegite 1 with pointed posterolateral point (fig. 3G, H). However, the maxilliped palp of the present female (fig. 3E) is setose as in the specimen from Hong Kong reported by Markham (1982), and not smooth as in Bourdon's (1976) Philippine sample. The present male is most similar to the allotype male, with eyes, antennae visible from dorsal view, and pleon with four segments (Nierstrasz & Brender à Brandis, 1930), and differs only in having some pigment on all body segments (fig. 3J).

KEY TO THE SPECIES OF THE GENUS *PARIONINELLA*

1. Maxilliped without palp 2
– Maxilliped with palp .... *P. astridae*

2. First oostegite produced backward in a round lobe .... *P. liuruiyui* n. sp.
– First oostegite produced backward in a small acuminate lobe .... *P. pacifica*

ACKNOWLEDGEMENTS

This work was supported by the National Natural Science Foundation of China (No. 31471970) and the Fund Program for the Scientific Activities of Selected Returned Overseas Professionals in Shanxi Province (2016). This project was also supported by a Program of Ministry of Science and Technology of the People's Republic of China (2015FY210300). We are indebted to all collectors of specimens in the Florida Museum of Natural History. We also wish to thank Mandy Bemis and John Slapcinsky of the Florida Museum of Natural History for all the help during the first author's visit. We thank Alexandra Hiller and Bernd Werding for identifying the hosts of the two species treated herein.

REFERENCES

ADKISON, D. L. & R. W. HEARD, 1978. Description of a new genus and species of Pseudioninae (Isopoda: Bopyridae) parasite of hermit crab *Pagurus annulipes* (Stimpson) from North Carolina. Proceedings of the Biological Society of Washington, **91**: 408-417.

AN, J. M., 2012. Taxonomy of parasitic isopods from China: 163 pp. (China Scientific Books, Hong Kong).

AN, J. M., H. Y. YU & J. C. MARKHAM, 2009. First record of the genus *Gigantione* (Epicaridea: Bopyridae: Pseudioninae) from Chinese waters, with description of three new species. Journal of Natural History, **43**: 335-353.

BOURDON, R., 1976. Les bopyres des porcellanes. Bulletin du Muséum national d'Histoire naturelle, (3) Zoologie, **252**: 165-245.

BOYKO, C. B., 2004. The Bopyridae (Crustacea: Isopoda) of Taiwan. Zoological Studies, **43**: 677-703.

MARKHAM, J. C., 1982. Bopyrid isopods parasitic on decapod crustaceans in Hong Kong and southern China. In: B. S. MORTON & C. K. TSENG (eds.), Proceedings of the first international marine biological workshop: the marine flora and fauna of Hong Kong and southern China, **1**: 325-391. (Hong Kong University Press, Hong Kong).

MARKHAM, J. C., 1990. Further notes on the Isopoda Bopyridae of Hong Kong. In: B. S. MORTON (ed.), Proceedings of the second international marine biological workshop: the marine flora and fauna of Hong Kong and southern China, Hong Kong, 1986, **2**: 555-568. (Hong Kong University Press, Hong Kong).

NIERSTRASZ, H. F. & G. A. BRENDER A BRANDIS, 1930. Isopoda Epicaridea. Résultats scientifiques du voyage aux Indes Orientales Néerlandaises de L.L.A.A.R.R. le Prince et la Princesse Léopold de Belgique. Mémoires du Muséum Royale d'Histoire Naturelle de Belgique, (Hors serie) **3**(1): 12-17.

RICHARDSON, H., 1904. Contributions to the natural history of the Isopoda. Proceedings of the United States National Museum, **27**: 1-89.

SHIINO, S. M., 1942. Bopyrids from the South Sea Islands with description of a hyperparasitic cryptoniscid. Palao Tropical Biological Station Studies, **2**: 437-458.

SHIINO, S. M., 1958. Note on the bopyrid fauna of Japan. Report of the Faculty of Fisheries, Prefectural University of Mi, **3**: 29-73.

First received 11 March 201.
Final version accepted 29 August 201.

CIRRIPEDIA

[When citing this volume, please refer to *Crustaceana* 93 (2020), 1233-1546]

Prof. Ruiyu Liu Memorial Issue

A NEW SPECIES OF SPONGE-INHABITING BARNACLE *NEOACASTA* (ARCHAEOBALANIDAE, ACASTINAE) IN KYUSHU, JAPAN

BY

MENG-CHEN YU[1,2]), TAKEFUMI YORISUE[3,4]), GREGORY A. KOLBASOV[5]) and BENNY K. K. CHAN[2,6])

[1]) Doctoral Degree Program in Marine Biotechnology, National Sun Yat-sen University and Academia Sinica, Kaohsiung 80424, Taiwan

[2]) Biodiversity Research Center, Academia Sinica, Taipei 11529, Taiwan

[3]) Institute of Natural and Environmental Sciences, University of Hyogo, 6 Yayoigaoka, Sanda, Hyogo, 669-1546, Japan

[4]) Museum of Nature and Human Activities, Hyogo, 6 Yayoigaoka, Sanda, Hyogo, 669-1546, Japan

[5]) White Sea Biological Station, Biological Faculty, Moscow State University, 119991, Moscow, Russia

ABSTRACT

This paper describes a new species of sponge-inhabiting barnacle, *Neoacasta liui* sp. nov., collected in Fukuoka, Kyushu, Japan. In the present study we used both the morphological examination and DNA barcode (cytochrome c oxidase I gene) to distinguish this new species from acastine species including its congeners from *Neoacasta* Kolbasov, 1993. *Neoacasta liui* sp. nov. is similar to *Neoacasta glans* (Lamarck, 1818) in shell morphology, but the divergence in COI gene between them reached inter-specific differences. *N. liui* sp. nov. is characterized by externally latticed, narrow scutum, beak-shaped tergum with wider spur, and lack of six inwardly prominent teeth on margin of basis. *N. glans* having an Indo-West Pacific distribution may represent a complex of cryptic species.

Key words. — Acastinae, sponge-inhabiting barnacles, taxonomy, new species, DNA barcode

RÉSUMÉ

Cet article décrit une nouvelle espèce de balane *Neoacasta liui* sp. nov., habitant une éponge, récoltée à Fukuoka, Kyushu, Japon. Dans cette étude nous avons utilisé à la fois l'examen morphologique et le code-barre ADN (gène cytochrome c oxydase I) pour distinguer cette nouvelle espèce des espèces acastines incluant ses congénères de *Neoacasta* Kolbasov, 1993. *Neoacasta liui* sp. nov. est semblable à *Neoacasta glans* (Lamarck, 1818) dans la morphologie de sa carapace, mais la divergence dans les gènes COI entre elles parvient à des différences interspécifiques. *N. liui* sp. nov. est caractérisée par un scutum étroit et grillagé, un tergum en forme de bec avec un aiguillon

[6]) Corresponding author; e-mail: chankk@gate.sinica.edu.tw

 | DOI:10.1163/9789004464353_019

plus large et l'absence de six dents proéminentes vers l'intérieur sur le bord de la base. *N. glans* ayant une distribution Indo-Ouest-Pacifique pourrait représenter un complexe d'espèces cryptiques.

Mots clés. — Acastinae, balane habitant une éponge, taxonomie, nouvelle espèce, Code-barre ADNs

INTRODUCTION

Leach (1817) erected the genus *Acasta* for barnacles which are living in sponges, having six shell plates, separated opercular valves and cup-shaped basis. The genus *Acasta* included about 80 species worldwide and exhibited great diversity in morphological forms until Kolbasov (1993) conducted a detailed revision of *Acasta*. Four new genera: *Archiacasta*, *Neoacasta*, *Euacasta* and *Pectinoacasta* were erected and grouped together with *Acasta* into the subfamily Acastinae. In addition to Acastinae, sponge-inhabiting barnacles also include the genus *Membranobalanus* (subfamily Archaeobalaninae) and subfamily Bryozobiinae in Archaeobalanidae, and *Pyrgospongia* (subfamily Pyrgomatinae) in Pyrgomatidae (Kolbasov, 1993; Achituv & Simon-Blecher, 2006; van Syoc & Newman, 2010; van Syoc et al., 2015; Yu et al., 2016, 2017a, b, 2019, and in press).

One-quarter of species known as sponge-inhabiting barnacles had been collected from Japan, including 20 species in 6 genera (table I; Pilsbry, 1911; Broch, 1922; Nilsson-Cantell, 1931; Hiro, 1931, 1933, 1936, 1937; Utinomi, 1953, 1959, 1962, 1967; Newman & Ross, 1976; Newman, 1981; Ise et al., 2004; Achituv & Simon-Blecher, 2006; Kim, 2011). Among those records, 12 new species of sponge-inhabiting barnacles had been reported by Pilsbry, Nilsson-Cantell, and Utinomi (previously as Hiro) (Pilsbry, 1911; Nilsson-Cantell, 1931; Hiro, 1931, 1933, 1936, 1937; Utinomi, 1953, 1959, 1962, 1967). In the present study we examined the acastine specimens collected from Fukuoka in Kyushu of Japan. From a combined molecular (mitochondrial marker COI) and morphological (characters of the shell parts and somatic body) approach, these specimens represent a new species, which is described herein as one of the cryptic forms of *N. glans*.

MATERIAL AND METHODS

Specimens were collected in Fukuoka, Kyushu, Japan in 2013. Separated barnacles and their host sponge were preserved in 95% ethanol. The morphological characters of barnacle shell (basis, wall plates, scutum, and tergum) and somatic bodies (six pairs of cirri, penis, and mouth parts) were examined. DNA was extracted from muscle tissues of barnacles and amplified through polymerase chain reaction (PCR) with specific primers to obtain partial sequences of mitochondrial gene cytochrome c oxidase subunit I (COI) (GenBank accession Nos: MN653641-MN653645). Sequences were aligned in Geneious (Drummond et al.,

TABLE I
Distributions of sponge-inhabiting barnacles in Japan

Species	First description			Distribution in Japan
	Author	Year	Locality	
Subfamily Acastinae				
Acasta alcyonicola	Utinomi	1953	Kusimoto and Tanabe Bays, Kii Peninsula, Japan	Kusimoto and Tanabe Bays, Honsyu (Utinimi, 1959)
Acasta denticulata	Hiro	1931	Misaki, Sagami. Japan	Misaki (Hiro, 1931)
Acasta echinata	Hiro	1937	Tomioka, Amakrusa	Tomioka (Utinomi,1959,1962)
Acasta fenestrata	Darwin	1854	Philippines	Vicinity of Seto, Japan (Hiro, 1939); Kannonzuka-dasi, Amadaiba (Utinomi, 1958); Japan (Rosell, 1972)
Acasta flexuosa	Nilsson-Cantell	1931	Northern Japan	Tomioka, Amakrusa (Hiro, 1931; Utinomi, 1949)
Acasta gregaria	Utinomi	1959	Tanabe Bay, Kii Peninsula, Japan	Tanabe Bay (Utinomi, 1959)
Acasta japonica	Pilsbry	1911	Kagosima Gulf, Japan	Toyama Bay and Kagosima Gulf, Sagami Bay (Hiro, 1939); Tomioka (Utinomi, 1962); Sagami Sea (Broch, 1922)
Acasta semota	Hiro	1933	Kasima-ura, Ibaragi, Japan	Ibaragi (Hiro, 1933)
Acasta spongites	Poli	1791	Not given	Between Kannonzuka-dasi and Maruyama-dasi (Utinomi, 1958); Japan (Weltner, 1897)
Acasta sulcata	Lamarck	1818	Australia	Kagoshima (Weltner, 1897); Sagami bay and adjacent waters (Utinomi, 1958); Kagoshima, Tokyo and Tanbe Bays (Hiro,1937)
Acasta umitosaka	Utinomi	1962	Nomosaki, Kyusyu, Japan	Nomosaki (Utinomi, 1962)

TABLE I
(Continued)

Species	First description			Distribution in Japan
	Author	Year	Locality	
Archiacasta spinifera	Utinomi	1967	Off Mera, Tiba-ken, mouth of Tokyo Bay, Japan	Tokyo Bay (Utinomi, 1967)
Archiacasta tulipa	Hiro	1933	Osumi Strait, off Satamisaki, Kagoshima, Japan	Kagoshima (Hiro, 1933)
Euacasta dofleini	Krüger	1911	Tropical sea	Seto, Kii Peninsula; Toba, Shima; Misaki, Sagami (Hiro, 1931); Tomioka, intertidal (Utinomi, 1962); Japan (Rosell, 1972)
Neoacasta coriobasis	Broch	1947	Indo-Chinese shallow waters	Nomi-sima, Hiroshima Bay (Utinomi,1953)
Pectinoacasta cancellorum	Hiro	1931	Seto, Kii Peninsula, Japan	Seto, Kii Peninsula (Hiro1931)
Pectinoacasta pectinipes	Pilsbry	1912	Malavatuan island, Philippines	Seto, Kii Peninsula (Hiro1931); Tomioka, Amakrusa (Utinomi, 1949); Nomosaki, intertidal (Utinomi, 1962)
Subfamily Archaeobalaninae				
Membranobalanus cuneiformis	Hiro	1936	Port Darwin, Australia	Vicinity of Seto (Hiro, 1939)
Membranobalanus longirostrum	Hoek	1913	Celebes and dangar besar islands, Indonesia	Hateruma-jima, Ishigaki-jima, Iriomote-jima & Kuro-shima, Ryukyu island (Ise et al., 2004)
Subfamily Pyrgomatinae				
Pyrgospongia stellula	Rosell	1973	Bauang Dakula Reef, Philippines	Japan (Achituv & Simon-Blecher, 2006)

2013) with the default parameter settings and then checked by eye. Genealogical relationships of the species of acastines were constructed through the maximum-likelihood (ML), maximum-parsimony (MP), and neighbor-joining (NJ) methods by MEGA version 6.06 (Tamura et al., 2013). *Galkinius indica* was selected as the outgroup; *Neoacasta glans* (new sequence), *Acasta aspera*, *Acasta huangi*, *Acasta radenta*, *Acasta sulcata*, *Acasta undulaterga* (Yu et al., 2017b), and *Neoacasta laevigata* (Gray, 1825) (GenBank accession no.: KU986746; downloaded from GenBank, National Center for Biotechnology Information Search database, NCBI) were selected for analyses of species differentiation. We used jModel Test 2.1.5 (Darriba et al., 2012) to determine the best-fit nucleotide substitution model of the COI dataset. Evolutionary distances (number of base differences per site) among sequences of acastines were calculated with uncorrected p-distance and Kimura-2-parameter (K2P) models by using MEGA. A ML tree search was conducted with Nearest-Neighbor-Interchange (Nei & Kumar, 2000) in MEGA. Node support values of the most parsimonious tree were estimated, with 1000 bootstrap replicates. A MP tree search was conducted with Subtree-Pruning-Regrafting (Nei & Kumar, 2000), starting from a tree generated through 10 random additions of sequences in MEGA. Node support values of the most parsimonious tree were estimated with 1000 bootstrap replicates. The NJ tree searches were conducted using the K2P models, with a uniform rate among sites and a homogeneous pattern among lineages in MEGA. Node support was estimated with 1000 bootstrap replicates. The methods of collecting and species identification (morphology and molecular) are described in Yu et al. (2019). The morphology of the setae is described in accordance with Chan et al. (2008). The holotype was deposited at the National Museum of Nature and Science (NSMT), Tokyo and paratypes were deposited at Biodiversity Research Museum, Academia Sinica, Taiwan (ASIZCR).

TAXONOMY

Subclass CIRRIPEDIA Burmeister, 1834
Superorder THORACICA Darwin, 1854
Order SESSILIA Lamarck, 1818
Suborder BALANOMORPHA Pilsbry, 1916
Superfamily BALANOIDEA Leach, 1817
Family ARCHAEOBALANIDAE Newman & Ross, 1976
Subfamily ACASTINAE Kolbasov, 1993
Genus *Neoacasta* Kolbasov, 1993

Neoacasta liui sp. nov.
(figs. 1-5)

Material examined.— Holotype: NSMT-Cr 28259, Katsuma, Shikanoshima, Fukuoka, Kyushu, Japan (N33 41.266 E130 17.767), sponge stranded on the sandy beach, November 2013, coll. Benny

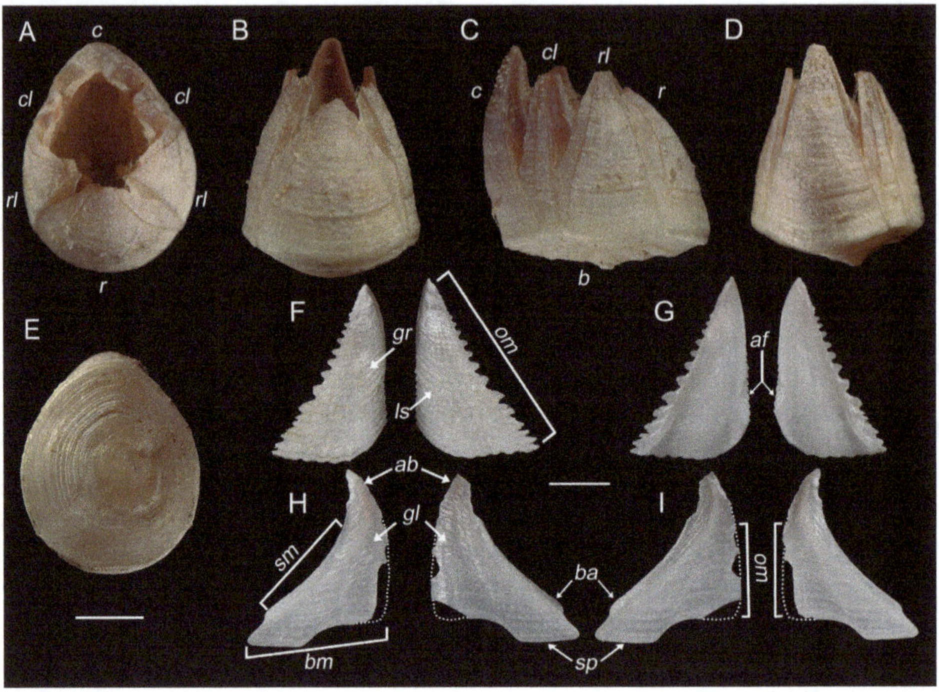

Fig. 1. *Neoacasta liui* sp. nov. Complete shell (NSMT-Cr 28259), holotype (A-I). A-E, General view of shell from orifice, rostrum, lateral, carina and basis, respectively; F and G, external and internal side of scuta, respectively; H and I, external and internal side of terga (occludent margin broken), respectively. Abbreviations: *ab*, apex beaked; *af*, articular furrow; *b*, basis; *ba*, basiscutal angle; *bm*, basal margin; *c*, carina; *cl*, carinolateral; *gl*, growth lines; *gr*, growth ridges; *ls*, longitudinal striations; *om*, occludent margin; *r*, rostrum; *rl*, rostrolateral; *sp*, spur of tergum; *sm*, scutal margin. Scale bars = 1 mm.

K. K. Chan, in host sponge *Spongia* sp. Paratype: ASIZCR000452 (empty shell) and ASIZCR000453 same data as for holotype. Complete shell: NSMT-Cr 28259. Disassembled Shell: ASIZCR000453. The holotype is deposited in the National Museum of Nature and Science, Tokyo and paratypes are stored at the Biodiversity Research Museum, Academia Sinica, Taiwan.

Diagnosis.— Shell thick and white, tinged yellowish brown towards apex. Basis flat or saucer-shaped with smooth margin. Scutum narrow with external surface latticed. Tergum with sharp beaked apex and tinged with yellowish brown on apex; tergal spur approx. 1/2 of basal margin. Cirri II to VI with sharp denticles or ctenes on anterior rami. Basidorsal point of penis vestigial.

Description.— Shell thick and white, tinged yellowish brown towards apex (figs. 1A-D, 2A-F). Basis flat or saucer-shaped, with 6 indistinct, shallow external radial furrows extending from center to smooth margin (figs. 1B-E, 2G). Parietes with few small external projections and fine growth lines (figs. 1A-D, 2A-F). Inner surface of wall plates with wide horizontal ridges and two marginal ribs reaching to basis (fig. 2A-F). Radii almost reaching to basis, external surface with inclined

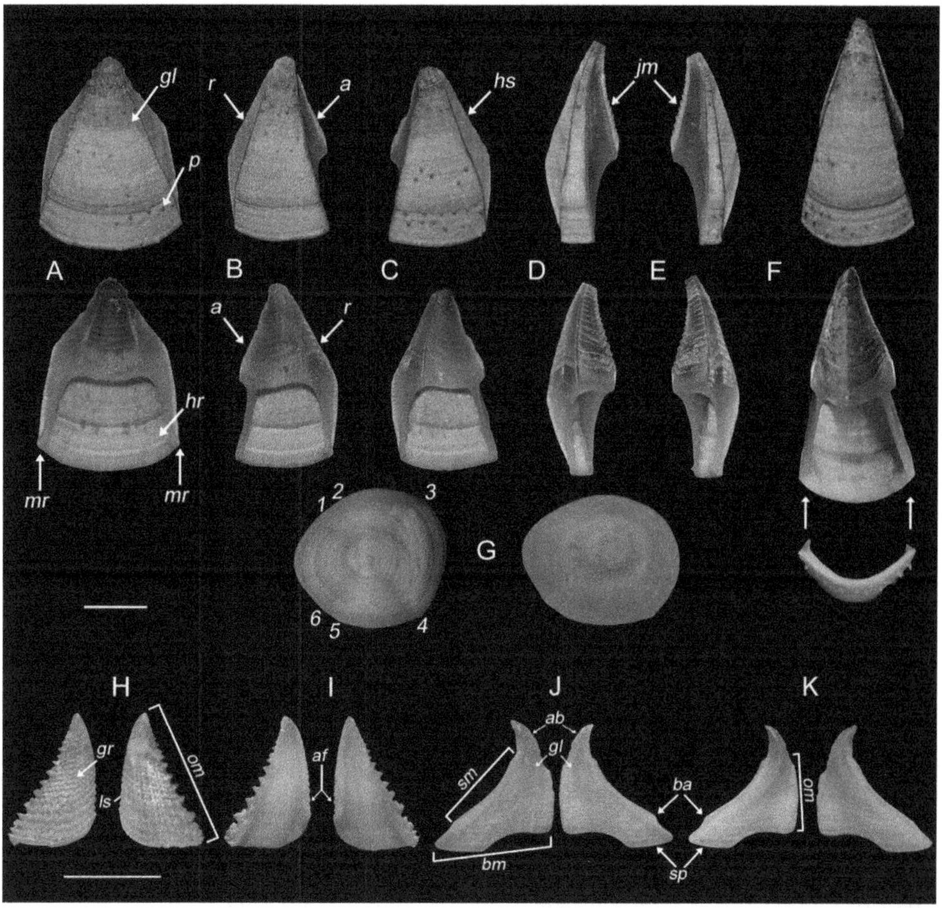

Fig. 2. *Neoacasta liui* sp. nov. Disassembled shell after bleach treatment showing separated plates and basis, paratype (ASIZCR000453). A, External and internal side of rostrum; B and C, external and internal side of rostrolaterals; D and E, external and internal side of carinolaterals; F, external, internal side and basal view of carina, basal view showing rising marginal ridges indicated by arrows; G, external and internal side of basis showing six radial furrows (1-6); H and I, external and internal side of scuta; J and K, external and internal side of terga. Abbreviations: *a*, ala; *ab*, apex beaked; *af*, articular furrow; *b*, basis; *ba*, basiscutal angle; *bm*, basal margin; *gl*, growth lines; *gr*, growth ridge; *hr*, horizontal ridge/fold of parietes; *hs*, horizontal striation; *jm*, jagged margin; *ls*, longitudinal striations; *mr*, marginal ridge; *om*, occludent margin; *p*, projection of wall plate; *r*, radius; *sm*, scutal margin; *sp*, spur of tergum. Scale bars = 1 mm.

and horizontal striations; alae tinged pinkish, with strong horizontal ridges. Upper outer edges of alae and lower inner edges of radii white and thickened; sheath with horizontal ridges. Alae of carinolaterals and carina with jagged margin (fig. 2A-F).

Scutum longer than wide; externally latticed by prominent growth ridges and longitudinal striations, with teeth on occludent margin. Internal surface smooth, articular furrow developed, articular ridge feeble, adductor and depressor muscle pits feeble (figs. 1F-G, 2H-I).

Tergum with sharp beaked apex, tinged with yellowish brown on apex, growth lines on external surface, carinal margin perpendicular to basal margin, spur truncated, width approx. 1/2 of basal margin, narrowly separated from basiscutal angle, spur furrow wide and shallow, internally with scutal margin raised, articular ridge feeble, crests of depressor muscle absent (figs. 1H-I, 2J-K).

Labrum bilobed, with V-shaped notch, 2 or 3 conspicuous, sharp denticles/teeth on each crest (fig. 3A). Mandibular palp truncated with dense, long plumose setae, outer margin concave with dense, serrulate and serrate setae (fig. 3B). Mandible with 5 teeth (fig. 3C), second to fourth teeth bifid (fig. 3D, E), inferior angle with denticles and stout, serrulate setae, lower margin with simple setae (fig. 3D). Maxillule cutting margin straight, without notch, with 7 sharp setae between two pairs of upper and lower large, cuspidate setae (fig. 3F), upper and lower margins with simple setae (fig. 3F-H). Maxilla bilobed, oblong, distal lobe with dense serrulate and few simple setae, outer edge without setae, inner edge straight, basal lobe with serrulate setae (fig. 3H).

Cirrus I protopod with plumose setae on proximal posterior margin. Rami unequal (fig. 4A); anterior ramus (17 annuli) twice as long as posterior ramus (8 annuli); anterior surfaces of both rami with dense setae; serrulate and serrate setae on distal ends of terminal annuli of anterior and posterior rami (fig. 4A, B). Cirrus II with rami unequal; anterior ramus (11 annuli) longer than posterior ramus (9 annuli); anterior and posterior margins of protopod with serrulate and plumose setae, respectively (fig. 4C); distal annuli of rami with serrulate and serrate setae; anterior and posterior rami with small ctenes, without denticles (fig. 4C-E).

Cirrus III protopod anterior and posterior margins with serrulate and plumose setae, respectively (fig. 4F); anterior ramus (12 annuli) slightly longer than posterior (10 annuli); anterior and posterior rami with sharp denticles and ctenes on intermediate annuli in anterior distal margin; distal annuli of both rami with serrate and serrulate setae (fig. 4F-H). Cirrus IV, anterior ramus broken, with 17 annuli on remaining part, posterior ramus with 15 annuli; anterior margin of protopod with simple setae (fig. 5A). Annuli of anterior ramus with 3 pairs of long, middle and short, serrulate setae on anterior edge; intermediate annuli of posterior ramus with 2 pairs of long and short anterior setae (fig. 5A-C); proximal annuli of both rami with 1-4 denticles on anterior margin, row of ctenoid scales along upper margins (fig. 5C); terminal annulus of posterior ramus with serrate and simple setae (fig. 5A-C). Cirrus V with 19 and 21 annuli in anterior and posterior rami; anterior margin of protopod with serrulate setae (fig. 5D). Intermediate annuli of both rami with 3 pairs of long, middle and short serrulate setae on anterior edge; proximal annuli of both rami with 1-2 denticles on anterior margin; row of ctenoid

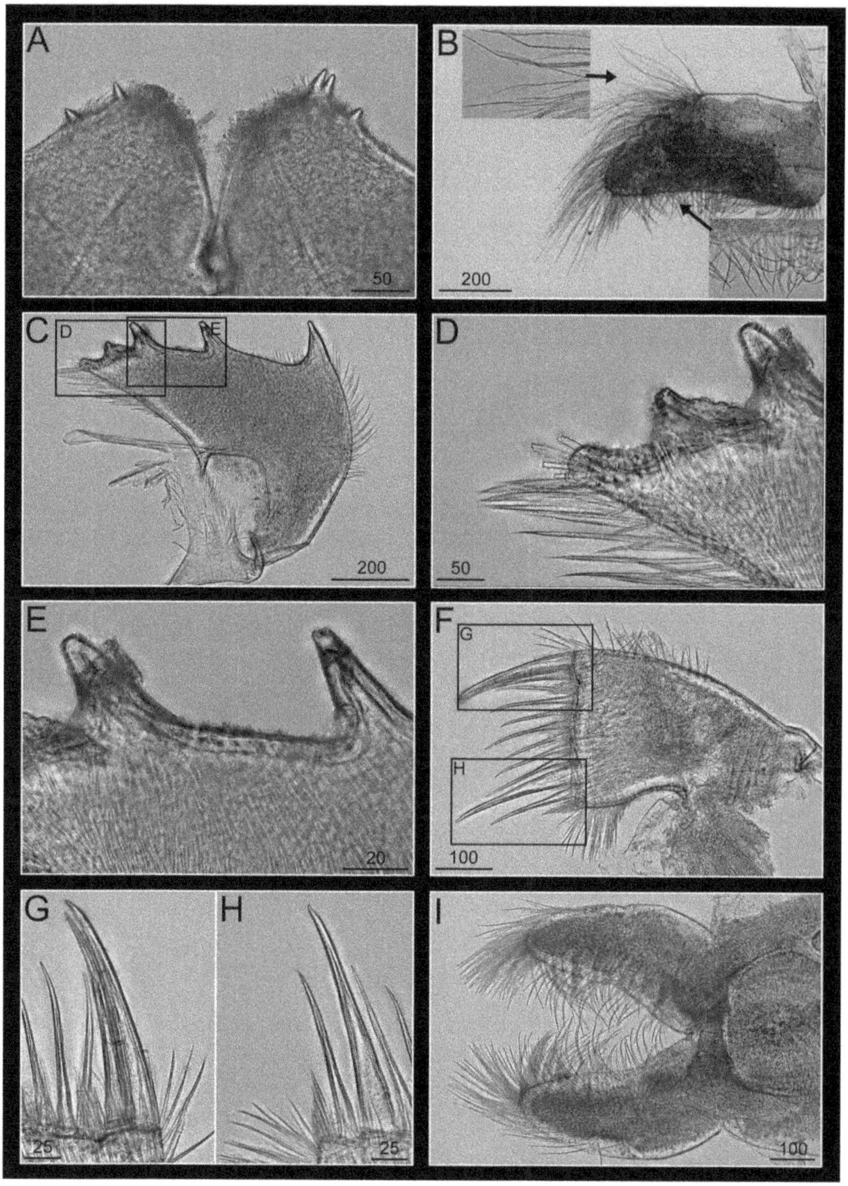

Fig. 3. *Neoacasta liui* sp. nov., mouth parts (holotype). A, Crests of labrum; B, mandibular palp, plumose setae on distal lobe enlarged in left-upper rectangle area; serrulate and serrate setae on outer margin enlarged in right-lower rectangle area; C, mandible; D, lower margin and inferior angle of mandible; E, bifid second and third teeth; F, maxillule, straight cutting edge with 11 cuspidate setae; G and H, large cuspidate setae on upper and lower parts, respectively; I, maxillae, bilobed with dense serrulate setae on distal lobe. Scale bars in μm.

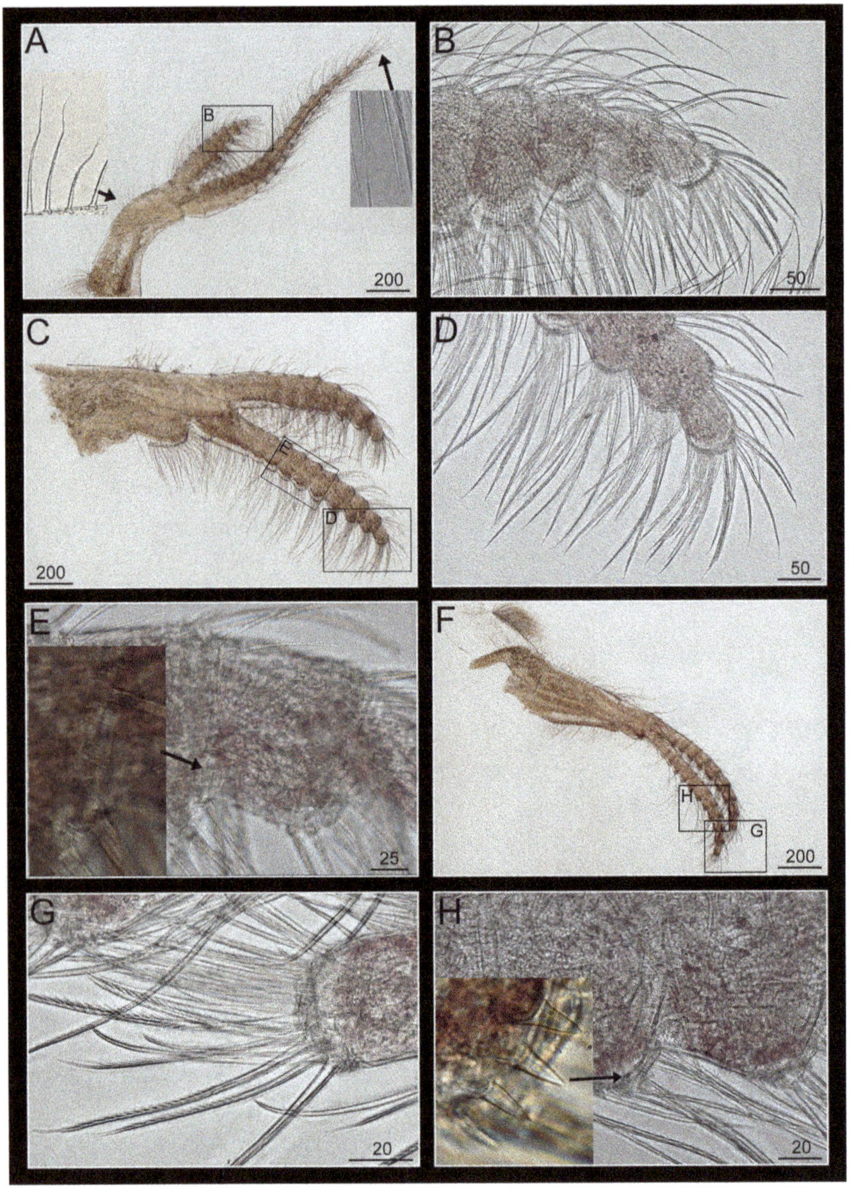

Fig. 4. *Neoacasta liui* sp. nov. Cirri I (A, B), II (C, D) and III (E-H) (holotype). A, Cirrus I, plumose setae in posterior distal margin of coxa and distal setae of anterior ramus enlarged in rectangle areas; B, distal annuli of posterior ramus; C, cirrus II, with plumose and serrulate setae on protopod; D, distal annuli of anterior ramus; E, small ctenes on anterior ramus enlarged in rectangle area; F, cirrus III; G terminal annulus of posterior ramus; H, intermediate annuli of anterior ramus with small, sharp denticles in anterior distal angle enlarged in rectangle area. Scale bars in μm.

scales along upper margins; distal annuli of both rami with serrate and simple setae (fig. 5D, E). Cirrus VI with 21 and 19 annuli in anterior and posterior rami; anterior margin of protopod with serrulate setae. Intermediate annuli of both rami with 3 pairs of long, middle and short serrulate setae on anterior edge; proximal annuli of both rami with row of ctenoid scales along upper margins; distal annuli of both rami with serrate and simple setae (fig. 5F).

Penis finely annulated, gradually tapering to tip, tip with setae (fig. 5G), basidorsal point vestigial (fig. 5H).

Remarks.— Currently, the genus *Neoacasta* is composed of 6 species: *N. coriobasis* (Broch, 1947), *N. fossata* (Barnard, 1924), *N. glans* (Lamarck, 1818), *N. laevigata* (Gray, 1825), *N. planibasis* Kolbasov, 1993, and *N. scuticosta* (Weltner, 1887). The new species *Neoacasta liui* sp. nov. is similar to its congeners in having two marginal ribs of each shell extending the base and basis with a smooth margin and 6 radial furrows. There are some diagnostic features of *N. liui* sp. nov. to distinguish it from other species in *Neoacasta*. The basis of *N. coriobasis*is is deeply cup-shaped, whilst that of *N. liui* sp. nov. is flat and saucer-shaped. In *N. coriobasis*, the surface of compartments is delicately striped horizontally; shell plates are loosely connected with each other. *N. liui* sp. nov. is further differing from *N. coriobasis* in having six indistinct radiating furrows on its basis. The scutum of *N. scuticosta* is smooth on the external surface, while the scutum of *N. liui* sp. nov. is latticed externally. The margin of the basis of *N. glans* has six bifid teeth, while the margin of the basis of *N. liui* sp. nov. is smooth; the tergal spur is about half the width of the basal margin of the tergum in *N. liui* sp. nov, while a quarter of the width of the basal margin of the tergum in *N. glans*. The terga of both *N. laevigata* and *N. planibasis* are broad, not beak-shaped, while the tergum of *N. liui* sp. nov. is narrower and beak-shaped; the scutum in the new species is latticed externally, while the scuta are without radial striae in *N. laevigata* and *N. planibasis*. The crests of the depressor muscle of the tergum of *N. fossata* are distinct, while absent in *N. liui* sp. nov.

Etymology.— The species is named in honour of Prof. Liu Ruiyu, Institute of Oceanology, Chinese Academy of Science and Technology, China, in appreciation of his numerous contributions to our knowledge of the Cirripedia (Liu & Ren, 2007; see Xiang & Li, 2013). The name thus is a noun in the genitive singular.

RESULTS

Molecular analysis

The general time reversible model with a proportion of invariant sites and gamma-distributed rate heterogeneity (GTR + I + G) were selected under the

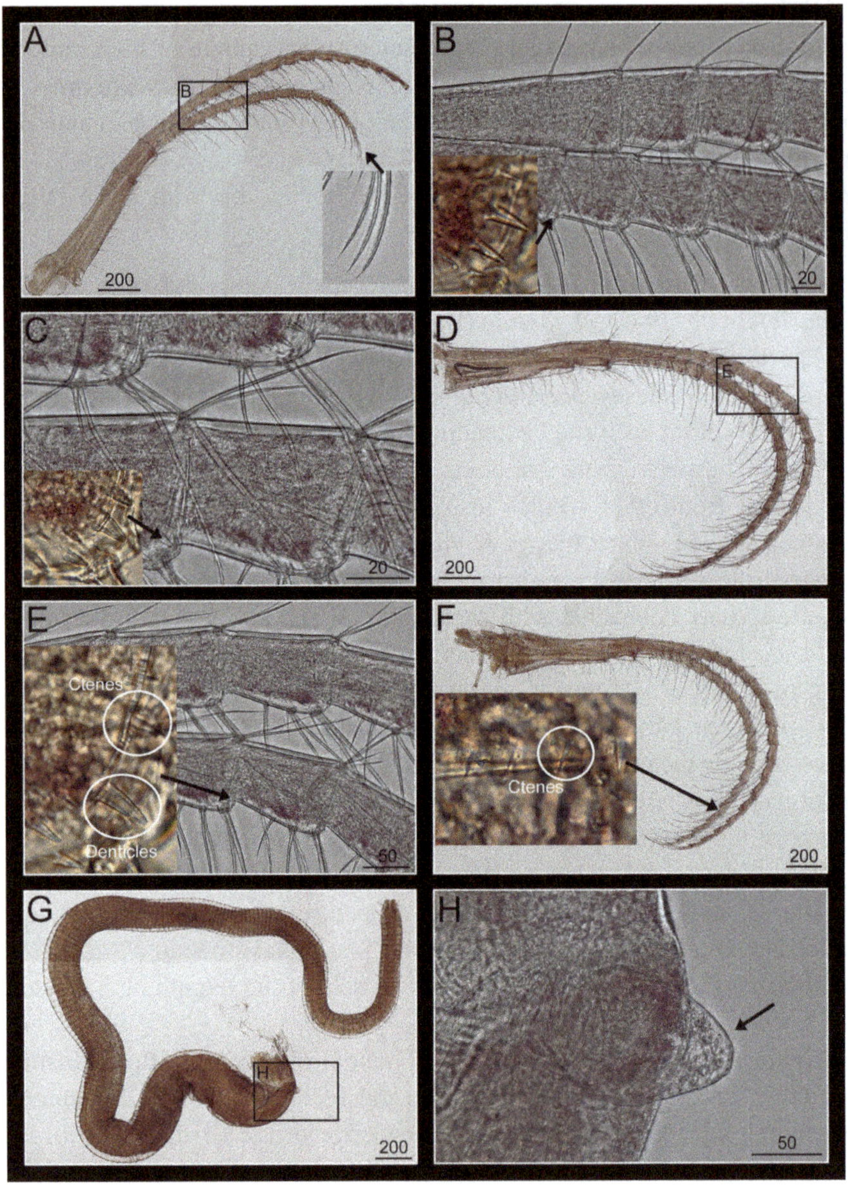

Fig. 5. *Neoacasta liui* sp. nov. Cirri IV (A-C), V (D, E), VI (F), and penis (G, H) (holotype). A, Cirrus IV, simple and serrate setae of distal annulus of anterior ramus enlarged in rectangle area; B and C, annuli of rami with three pairs serrulate setae and proximal annuli with sharp denticles enlarged in rectangle area; D, cirrus V; E, sharp denticles and ctenes on anterior rami enlarged in rectangle area; F, cirrus VI and row of ctenoid scales along upper margins of rami enlarged in rectangle area; G, penis; H, vestigial basidorsal point (indicated by arrow). Scale bars in μm.

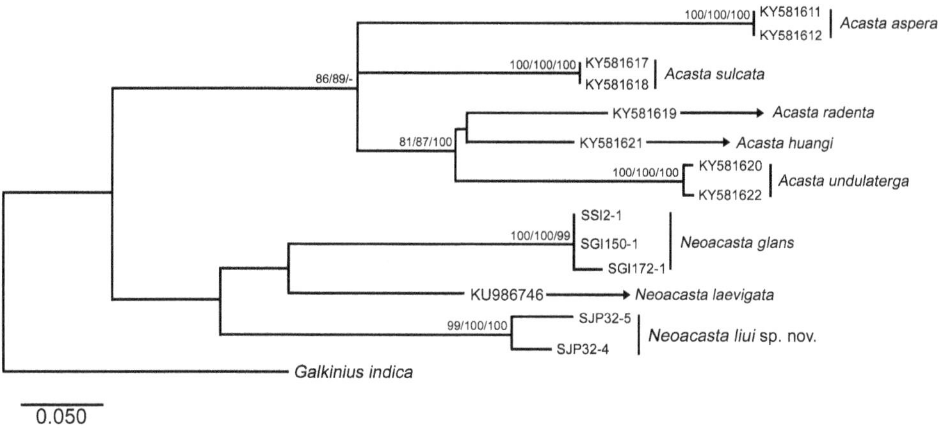

Fig. 6. The best scoring maximum-likelihood (ML) tree based on concatenated cytochrome c oxidase subunit I (COI) dataset. Node labels are supporting values (%) estimated from the ML methods, neighbor-joining (NJ), and maximum-parsimony (MP), respectively. Only the nodes with ⩾80% support values for all 3 methods are listed. Scale bars indicate the number of expected substitutions per site.

Akaike information criterion (Akaike, 1974) for the COI (390 base pairs) dataset. The phylogenetic trees of sponge barnacles, based on ML, NJ, and MP analyses, exhibited similar topologies (fig. 6). *Neoacasta liui* sp. nov., *N. glans*, *N. laevigata*, and *Acasta* species (*A. huangi*, *A. radenta*, *A. undulaterga*, and *A. sulcata*) were represented by distinct species-level clades. The intraspecific evolutionary distances of the new species were much lower than the interspecific distances for COI sequences (*p*-distance: 0.048; K2P: 0.046 versus *p*-distance: 0.128-0.164; K2P: 0.118-0.159). These findings further support *N. liui* sp. nov. to be a new species different from *N. glans*, *N. laevigata* and other sponge barnacles in our phylogenetic tree (fig. 6, table II).

DISCUSSION

The new species in the genus *Neoacasta*, *N. liui* sp. nov., was collected from Fukuoka, Kyushu, Japan. The genus *Neoacasta* is diagnosed from *Acasta* by having distinct six radial furrows extending from the center of the basis to the edge. However, these six furrows in *N. liui* sp. nov. are not very obvious when observed from the external surfaces of the bases. As a result, from the examination of external shell features only, *N. liui* sp. nov. could be erroneously identified as a species of *Acasta*. When dissected, the internal morphological characters of the shell plates of *N. liui* sp. nov. can be easily distinguished from those of *Acasta* by their smooth internal surface of parietes, two marginal ribs of wall plates, and six radial furrows extending from the center of the basis to the edge. From these

TABLE II

Estimates of average evolutionary distance (numbers of base differences per site) of COI sequences of *Acasta aspera*, *A. huangi*, *A. radenta*, *A. sulcata*, *A. undulaterga*, *Neoacasta glans*, *N. laevigata* and *N. liui* sp. nov.

	A. aspera	*A. huangi*	*A. radenta*	*A. sulcata*	*A. undulaterga*	*N. glans*	*N. laevigata*	*N. liui* sp. nov.
A. aspera	*0.000*	0.131	0.149	0.141	0.159	0.184	0.212	0.169
A. huangi	0.145	N/C	0.094	0.133	0.116	0.158	0.160	0.154
A. radenta	0.167	0.087	N/C	0.138	0.106	0.153	0.141	0.173
A. sulcata	0.158	0.121	0.126	*0.000*	0.133	0.155	0.166	0.162
A. undulaterga	0.180	0.106	0.116	0.148	*0.013*	0.157	0.189	0.164
N. glans	0.211	0.178	0.171	0.174	0.177	*0.013*	0.128	0.164
N. laevigata	0.185	0.144	0.128	0.149	0.167	0.118	N/C	0.142
N. liui sp. nov.	0.192	0.172	0.197	0.183	0.186	0.146	0.159	*0.046/0.048*

Upper-right: inter-specific distance based on uncorrected *p*-distance model; lower-left: inter-specific distance based on Kimura-2-parameter model; italics: intra-specific distance based on *p*-distance/Kimura-2-parameter models (only one number shown if the same). N/C, only one sequence available in this species therefore unable to calculate intra-specific distance.

distinct morphological characters, we confirm the present new species to belong to the genus *Neoacasta*.

Previously, only one species of the genus *Neoacasta*, *Neoacasta coriobasis*, was described off Japan (Broch, 1947, as *Acasta coriobasis*). The present study added a second *Neoacasta* species identified from Japanese waters.

Morphologically this new species is closely related to *Neoacasta glans*, but can be distinguished from this species in having a wider tergal spur (approx. 1/2 of basal margin, instead of approx. 1/4 in *N. glans*), the presence of sharp denticles in cirrus V and a vestigial basidorsal point of the penis. *N. glans* has a wide geographical distribution (South East Africa, Persian Gulf, Bay of Bengal, South West Australia, Indonesia, Philippines and South China Sea), and specimens from different locations and sponge species possess a variable morphology of shell, e.g. absence or presence of small slits between wall plates, cup-shaped or saucer-shaped basis, presence of absence of crests of tergal depressors (Kolbasov, 1993; pers. data). Thus, *N. glans* may represent a complex of cryptic species and the identification of *N. liui* sp. nov. based on morphological and molecular approaches may prove this supposition.

The molecular phylogenetic tree and evolutionary distances of *Acasta* and *Neoacasta* species also support that *N. liui* sp. nov. is a different species from *N. glans*, *N. laevigata* and belongs to the genus *Neoacasta*. Although the bootstrap value for the monophyly of the *Neoacasta* clade is not high, the presence of six rays on the base in *N. liui* sp. nov. supports its allocation in the genus *Neoacasta*. Based on morphological characteristics and molecular analyses, it was confirmed that *Neoacasta liui* sp. nov. is a new taxon within the acastine genus *Neoacasta* and constitutes a new record of a *Neoacasta* species from Japan. A taxonomic key for the identification of the *Neoacasta* species is presented herein.

KEY TO THE SPECIES OF THE GENUS *NEOACASTA*

1. Scutum latticed externally 2
   Scutum not latticed externally 3
2. Tergal spur ~ 1/4 of width of basal margin of tergum .... *N. glans*
   Tergal spur ~ 1/2 of width of basal margin of tergum .... *N. liui*
3. Scutum externally with radial striation 4
   Scutum externally without radial striation 5
4. Spur of tergum with deep furrow .... *N. scuticosta*
   Spur of tergum with rudimentary furrow .... *N. coriobasis*
5. Inner surface of parietes smooth 6
   Inner surface of parietes with developed longitudinal ribs .... *N. fossata*
6. Basis saucer-shaped, wall plates externally with dense, sharp denticles .... *N. planibasis*
   Basis cup-shaped, wall plated externally with or without rare small denticles .... *N. laevigata*

ACKNOWLEDGEMENTS

The authors would like to thank the Senior Investigator Award (Academia Sinica, Taiwan) for supporting the study. Thanks to the Doctoral Degree Program in Marine Biotechnology, National Sun Yat-sen University and Academia Sinica granted to MCY. BKKC thanks the Ministry of Science and Technology for Russian-Taiwan Bilateral funding (106-2923-B-001-002-MY3). GAK thanks the Russian Foundation for Basic Research (grants 17-54-52006 MNT_a, 18-04-00624 A).

REFERENCES

ACHITUV, Y. & N. SIMON-BLECHER, 2006. *Pyrgopsella* (Cirripedia: Balanomorpha: Pyrgomatidae) is not a sponge-inhabiting barnacle. Zootaxa, **1319**: 29-42.

AKAIKE, H., 1974. A new look at the statistical model identification. IEEE Transactions on Automatic Control, **AC-19**: 716-723. https://doi.org/10.1109/TAC.1974.1100705.

BARNARD, K. H., 1924. Contributions to the crustacean fauna of South Africa. 7. Cirripedia. Annals of the South African Museum, **20**: 1-103.

BROCH, H., 1922. Studies on Pacific cirripeds. Papers from Dr. Th. Mortensen's Pacific Expedition 1914-1916. No. X. Videnskabelige Meddelelser fra Dansk naturhistorisk Forening i Kjøbenhavn, **73**: 215-358.

BROCH, H., 1947. Cirripedes from Indo-Chinese shallow waters. Avhandlinger Utgitt av det Norske Videnskaps. Akademii Oslo I. Mathematisk-Naturvidenskapelig Klasse, **7**: 1-32.

BURMEISTER, H., 1834. Beiträge zur Naturgeschichte der Rankenfüsser (Cirripedia): i-viii, 1-60. (G. Reimer, Berlin).

CHAN, B. K. K., A. GARM & J. T. HØEG, 2008. Setal morphology and cirral setation of thoracican barnacle cirri: adaptations and implications for thoracican evolution. Journal of Zoology, **275**: 294-306. https://doi.org/10.1111/j.1469-7998.2008.00441.x.

DARRIBA, D., G. L. TABOADA, R. DOALLO & D. POSADA, 2012. jModelTest 2: more models, new heuristics and parallel computing. Nature Methods, **9**: 772. https://doi.org/10.1038/nmeth.2109.

DARWIN, C., 1854. A monograph on the sub-class Cirripedia, with figures of all the species. The Balanidae: i-viii, 1-684. (Ray Society, London).

DRUMMOND, A. J., B. ASHTON, S. BUSTON, M. CJEUNG, A. COOPER, C. DURAN, M. FIELD, J. HELED, M. KEARSE, S. MARKOWITZ, R. MOIR, S. STONES-HAVAS, S. STUROCK & T. THIERER, 2013. Geneious v7.0.5. website: www.geneious.com.

GRAY, J. E., 1825. A synopsis of the genera of Cirripedes arranged in natural families, with a description of some new species. Annals of Philosophy, **10**: 97-107.

HIRO, F., 1931. Notes on some new Cirripedia from Japan. Memoirs of the College of Science, Kyoto Imperial University, Series B, **7**(3): 143-158.

HIRO, F., 1933. Report of the Cirripedia collected by the surveying ships of the Imperial Fisheries Experimental Station on the continental shelf bordering Japan. Records of oceanographic works in Japan, **5**(1): 11-84.

HIRO, F., 1936. Report on the Cirripedia collected in the Malayan waters by the ship "Zuiho-maru". Japanese Journal of Zoology, **6**(19): 621-636.

HIRO, F., 1937. A new barnacle *Acasta echinata* n. sp. imbedded in the stalk of an alcyonarian from southern Japan. Zoological magazine, Tokyo, **49**(2): 70-71.

HIRO, F., 1939. Studies on the cirripedian fauna of Japan. III. Supplementary notes on the cirripeds found in vicinity of Seto. Memoirs of the College of Science, Kyoto Imperial University, Series B, **15**(2): 237-244.

HOEK, P. P. C., 1913. The Cirripedia of the Siboga Expedition. B. Cirripedia Sessilia. Siboga-Expeditie Monogr. XXXIb: i-xxv, 129-275.

ISE, Y., 2004. Psammobiontic Clionaidae (Demospongiae: *Hadromerida*) in lagoons of the Ryukyu Islands, southwestern Japan. Bollettino dei Musei e degli Istituti Biologici, **68**: 381-389.

KIM, I. H., 2011. Invertebrate fauna of Korea: Arthropoda: Crustacea: Cirripedia. Barnacles. Flora and Fauna of Korea, **21**(6): 1-144.

KOLBASOV, G. A., 1993. Revision of the genus *Acasta* Leach (Cirripedia: Balanoidea). Zoological Journal of the Linnean Society, **109**(4): 395-427.

KRÜGER, D. P., 1911. Beiträge zur Cirripedienfauna Ostasien. Beiträge zur Naturgeschichte Ostasiens herausgegeben von F. Doflein. Konglige Bayerische Akademie der Wissenschaften, Munich Mathematische-physikalische Klasse. Abhandlungen Supplement Band, **2**: 1-72.

LAMARCK, J. B. P., 1818. Histoire Naturelle des Animaux sans Vertèbres. Précédée d'une Introduction Offrant la Détermination des Caractères Essentiels de l'animal, sa Distinction du Végétal et des Autres Corps Naturels, Enfin, l'exposition des Principes Fondamentaux de la Zoologie: 1-612. (Verdière, Paris).

LEACH, W. E., 1817. Distribution, systematique de la class Cirripedes. Journal de Physique de Chimie et d'Histoire Naturelle, **85**: 67-69.

LIU, R. Y. & X. Q. REN, 2007. Fauna Sinica. Invertebrata. Vol. 42 Crustacea Cirripedia Thoracica: 1-633. (Science Press, Beijing). [In Chinese.]

NEI, M. & S. KIMAR, 2000. Molecular evolution and phylogenetics: 1-333. (Oxford University Press, Oxford).

NEWMAN, W. A. & A. ROSS, 1976. Revision of the Balanomorpha barnacles; including a catalog of the species. Memoirs of the San Diego Society of Natural History, **9**(1): 1-108. https://doi.org/10.1163/156854077X00999.

NEWMAN, W. A., 1981. Huzio Utinomi: 24 February 1910-19 June 1979. Crustaceana, **40**(1): 100-108.

NILSSON-CANTELL, C. A., 1931. Revision der Sammlung recenter Cirripedien des Naturhistorischen Museums in Basel. Verhandlungen der Naturforschenden Gesellschaft in Basel, **42**: 103-137.

PILSBRY, H. A., 1911. Barnacles of Japan and Bering Sea. Bulletin of the Bureau of Fisheries, **29**: 59-84.

PILSBRY, H. A., 1912. Diagnoses of new barnacles from the Philippine Archipelago and China Sea. Proceedings of the United States National Museum, **42**(1902): 291-294.

PILSBRY, H. A., 1916. The sessile barnacles (Cirripedia) contained in the collections of the U. S. National museum; including a monograph of the American species, by Henry A. Pilsbry: 1-358. (Washington Government Printing, Washington). https://doi.org/10.5962/bhl.title.1144.

POLI, G. S. & S. DELLE CHIAIE, 1791. Testacea Utriusque Siciliae Eorumque Historia et Anatome Tabulis Aeneis Illustrata. Regium Typographeio: 1-192. (Regium Typographeio, Parma).

ROSELL, N. C., 1972. Some barnacles (Cirripedia Thoracica) of Puerto Galera found in the vicinity of the UP Marine Biological Laboratory. Natural and Applied Science Bulletin, **24**: 143-285.

ROSELL, N. C., 1973. Some thoracic barnacles (Crustacea: Cirripedia) of Manila Bay. Kalikasan Philippine Journal of Biology, **2**: 69-95.

TAMURA, K., G. STECHER, D. PETERSON, A. FILIPSKI & S. KUMAR, 2013. MEGA 6: molecular evolutionary genetics analysis version 6.0. Molecular Biology and Evolution, **30**: 2725-2729. https://doi.org/10.1093/molbev/mst197.

UTINOMI, F., 1949. Studies on the cirripedian fauna of Japan. VI. Cirripeds from Kyusyu and Ryukyu Islands. Publications of the Seto Marine Biological Laboratory, **1**(2): 19-37.

UTINOMI, F., 1953. On two interesting species of epizoic barnacle *Acasta* from Japan. Memoirs of the College of Science, Kyoto Imperial University, Series B, **20**(3): 139-144.

UTINOMI, F., 1958. Studies on the cirripedian fauna of Japan. VII. Cirripeds from Sagami Bay. Publications of the Seto Marine Biological Laboratory, **6**(3): 281-311.

UTINOMI, F., 1959. A new gall-forming barnacle imbedded in the bark of a gorgonacean colony (*Acasta gregaria* n. sp.). Publications of the Seto Marine Biological Laboratory, **4**(3): 313-318.

UTINOMI, F., 1962. Studies on the cirripedian fauna of Japan. VIII. Thoracic cirripeds from western Kyusyu. Publications of the Seto Marine Biological Laboratory, **10**(2): 211-239.

UTINOMI, F., 1967. Comments on some new and already known cirripeds with emended taxa, with special reference to the parietal structure. Publications of the Seto Marine Biological Laboratory, **15**(3): 199-237.

VAN SYOC, R. J. & W. A. NEWMAN, 2010. Morphology and evolutionary ecology of a sponge-barnacle symbiosis: four new genera of barnacles (Archaeobalanidae, Bryozobiinae). Journal of Experimental Marine Biology and Ecology, **392**: 65-88. https://doi.org/10.1016/j.jembe.2010.04.011.

VAN SYOC, R. J., R. W. M. VAN SOEST, J. R. XAVIER & J. N. A. HOOPER, 2015. A phylogenetic overview of sponge-inhabiting barnacles and their host specificity (Crustacea, Cirripedia). Proceedings of the California Academy of Natural Sciences, Series 4, **62**: 331-357.

WELTER, W., 1887. Die von Dr. Sander 1883-85 gesammelten Cirripedien. Archiv für Naturgeschichte, Berlin, **53**: 98-117.

WELTER, W., 1897. Verzeichnis der bisher beschriebene recenten Cirripedienarten. Mit Angabe der im berliner Museum vorhandenen Species und ihrer Fundorte. Archiv für Naturgeschichte, Berlin, **63**(1): 227-280.

XIANG, J. & F. LI, 2013. Ruiyu Liu (used name J. Y. Liu) 4 November 1922-16 July 2012. Journal of Crustacean Biology, **33**(5): 744-750. https://doi.org/10.1163/1937240X-00002180.

YU, M.-C., B. K. K. CHAN, Y. ACHITUV & G. A. KOLBASOV, 2017a. Four new sponge-inhabiting barnacles of the genus *Acasta* (Thoracica: Archaeobalanidae: Acastinae) from the Indo-Pacific. Raffles Bulletin of Zoology, **65**: 585-615.

YU, M.-C., G. A. KOLBASOV & B. K. K. CHAN, 2016. A new species of sponge inhabiting barnacle *Bryozobia* (Archaeobalanidae, Bryozobiinae) in the West Pacific. ZooKeys, **571**: 1-20.

YU, M.-C., G. A. KOLBASOV, B. K. K. CHAN & M. GAMMANEE, (in press). Biodiversity and host usage of sponge-associated barnacles (Cirripedia: Thoracica) in Thailand. Journal of Crustacean Biology, [in press].

YU, M.-C., G. A. KOLBASOV, J. T. HØEG & B. K. K. CHAN, 2019. Crustacean-sponge symbiosis: collecting and maintaining sponge-inhabiting barnacles (Cirripedia: Thoracica: Acastinae) for studies on host specificity and larval biology. Journal of Crustacean Biology, **39**(4): 522-538. https://doi.org/10.1093/jcbiol/ruz025.

YU, M.-C., G. A. KOLBASOV, A. M. HOSIE, T. M. LEE & B. K. K. CHAN, 2017b. Descriptions of four new sponge-inhabiting barnacles (Thoracica: Archaeobalanidae: Acastinae). Zootaxa, **4277**(2): 151-198. https://doi.org/10.11646/zootaxa.4277.2.1.

First received 23 October 2019.
Final version accepted 19 September 2020.

COPEPODA

[When citing this volume, please refer to *Crustaceana* 93 (2020), 1233-1546]

Prof. Ruiyu Liu Memorial Issue

A NEW SPECIES OF *STYGIOPONTIUS* (COPEPODA, SIPHONOSTOMATOIDA, DIRIVULTIDAE) FROM A DEEP-SEA COLD SEEP IN THE SOUTH CHINA SEA

BY

LIN MA[1,2]), MINXIAO WANG[1,2,4]) and XINZHENG LI[1,2,3,4,5])

[1]) Institute of Oceanology, Chinese Academy of Sciences, Qingdao 266071, P.R. China

[2]) CAS Center for Ocean Mega-Science, Chinese Academy of Sciences, Qingdao, 266071, P.R. China

[3]) University of Chinese Academy of Sciences, Beijing 100039, P.R. China

[4]) Pilot National Laboratory for Marine Science and Technology, Qingdao 266237, P.R. China

ABSTRACT

A new species of the copepod genus *Stygiopontius* is described based on samples from a cold seep in northeastern South China Sea southwest to Taiwan, which were collected by the ROV FAXIAN with its mother vessel R/V KEXUE in September 2017. The copepods were obtained by washing *Shinkaia crosnieri* Baba & Williams, 1998 (Decapoda: Munidopsidae) caught at a depth of 1124 m. The new species can be distinguished from its congeners by the combination of the following characteristics: caudal rami about 3.5 times as long as wide; basis of maxilliped with plumose seta, endopodal claw of maxilliped stout, with a row of fine spinules on inner margin; coxa of leg 1 with one inner seta; basis of leg 1 with stout spine exceeding to end of first endopodal segment; second endopodal segment of female leg 4 with pointed process; third exopodal segment of leg 4 with three outer spines. This is the first record of a *Stygiopontius* species from a cold seep.

Key words. — Copepoda, Siphonostomatoida, Dirivultidae, new species, deep-sea, cold seep, South China Sea

RÉSUMÉ

Une nouvelle espèce du genre de copépode *Stygiopontius* est décrite à partir d'échantillons provenant d'un suintement froid situé au nord-est de la mer de Chine du Sud, au sud-ouest de Taiwan, collectés par le ROV "Faxian" avec son navire-mère R/V "Kexue" en septembre 2017. Les copépodes ont été obtenus par lavage de *Shinkaia crosnieri* Baba & Williams, 1998 (Decapoda : Munidopsidae) capturés à une profondeur de 1124 m. La nouvelle espèce peut être distinguée de ses congénères par la combinaison des caractéristiques suivantes : rames caudales environ 3,5 fois aussi longues que larges ; basis du maxillipède avec une soie plumeuse, croc endopodal du maxillipède fort avec une rangée de fines spinules au bord interne ; coxa de la patte 1 avec une soie interne ; basis de la patte 1 avec une forte épine dépassant l'extrémité du premier segment de l'endopodite ; second segment

[5]) Corresponding author; e-mail: lixzh@qdio.ac.cn

 | DOI:10.1163/9789004464353_020

de l'endopodite de la patte 4 femelle avec un processus pointu ; troisième segment de l'exopodite de la patte 4 avec trois soies externes. Cette espèce constitue le premier signalement d'une espèce de *Stygiopontius* d'un suintement froid.

Mots clés. — Copepoda, Siphonostomatoida, Dirivultidae, nouvelle espèce, eau profonde, suintement froid, mer de Chine du Sud

INTRODUCTION

The copepods of the order Siphonostomatoida are known from deep-sea hydrothermal vents and cold seeps (Humes, 1988a, b; Ivanenko & Defaye, 2004; Mahatma et al., 2008). Within the order Siphonostomatoida, the family Dirivultidae Humes & Dojiri, 1980 is one of the most diverse families in deep-sea hydrothermal vent habitats (Gollner et al., 2010; Ivanenko et al., 2011). In the family, *Stygiopontius* Humes, 1987 is the largest genus with 27 valid species, some species are associated with other invertebrates (Uyeno et al., 2018; Lee et al., 2020). Species of the genus *Stygiopontius* are known from the Indian Ocean, the Pacific Ocean and the Mid-Atlantic Ridge in the Atlantic Ocean (Humes, 1987, 1988, 1989, 1990, 1996, 1997; Ivanenko et al., 2006; Ivanenko & Ferrari, 2013; Gollner et al., 2016; Uyeno et al., 2018; Lee et al., 2020). All species of the genus were reported from and over diverse deep-sea hydrothermal vent fields (Ivanenko, 1998; Gollner et al., 2010; Ivanenko et al., 2007, 2011; Cuvelier et al., 2014).

Among dirivultid copepods of the genus *Stygiopontius*, only *S. pectinatus* Humes, 1987 and *S. cladarus* Humes, 1996 are found associating with deep-sea decapod crustaceans representing alvinocaridid shrimps (Ivanenko et al., 2006). The new species of *Stygiopontius* described here is the first record of the family Dirivultidae in a cold seep, as well as the first report of the family from the South China Sea.

MATERIAL AND METHODS

The copepods were collected during the cruise by the scientific research vessel (RV) KEXUE (Institute of Oceanology, Chinese academy of Sciences, China) on 17 September 2017. The sampling site was at a cold seep in the South China Sea (22°06′55″N, 119°17′08″E) (see Fig. 1 in Shen et al., 2016). Environmental conditions were reported by Hui in the year 2018 (Hui et al., 2018). The main surficial features of the cold seep system include authigenic carbonate crusts, chemosynthetic communities, and gas plumes (see Wang et al., 2018 for details). The entire area of this cold seep is colonized by the galatheid decapod *Shinkaia crosnieri* Baba & Williams, 1998 and mussels (fig. 1). All copepod material was obtained by washing the galatheid decapod *Shinkaia crosnieri* and were preserved

Fig. 1. Sampling site of the cold seep in the South China Sea, colonized by galatheid decapod *Shinkaia crosnieri* and mussels.

in 75% alcohol. Before dissection, the samples were cleared in lactic acid and the habitus was drawn and the whole-body length was measured. They were dissected in lactic acid, mounted on slides in lactophenol, covered by a cover glass, and subsequently sealed with nail polish (Humes & Gooding, 1964). The observations and drawings were made with a differential interference contrast microscope (Nikon Eclipse Ni), equipped with a drawing tube.

Body length was measured from the anterior margin of the prosome to the posterior margin of the caudal rami. All specimens including holotype and paratypes are deposited in the Marine Biological Museum, Chinese Academy of Sciences, Qingdao, China (MBMCAS).

SYSTEMATIC PART

Order SIPHONOSTOMATOIDA Burmeister, 1835
Family DIRIVULTIDAE Humes & Dojiri, 1980
Genus *Stygiopontius* Humes, 1987
Stygiopontius liui sp. nov.
(figs. 2-4)

Type material.— Holotype: adult ♀ (MBM 189264), dissected on six slides, depth 1124 m, associated with *Shinkaia crosnieri* Baba & Williams, 1998 (Decapoda: Munidopsidae), 17 September 2017, collected by Zheng Xu from Institute of Oceanology, Chinese Academy of Sciences. Paratypes: adult ♀ (MBM 189265), dissected on three slides, adult ♀ (MBM189266), dissected on two slides, collection data same as that of holotype.

Other material.— Twenty-four adult ♀♀ (MBM189267), collection data same as that of holotype.

Etymology.— The species is named in honour of the late Professor Ruiyu Liu from the Institute of Oceanology, Chinese Academy of Sciences, for his contributions to the knowledge of the crustacean fauna in China.

Descriptions.— Total somite length, measured from anterior margin of rostrum to posterior margin of caudal rami, ranging from 1.2 to 1.25 mm (mean = 1.22 mm, n = 3), greatest width ranging from 0.57 to 0.6 mm (mean = 0.58 mm, n = 3). Cyclopiform. All somites with sensilla on dorsal surface except anal somite. Prosome four-segmented, consisting of cephalothorax and three somites bearing swimming legs 2-4, lateral posterior margin of fourth pedigerous somite processed. Urosome (fig. 2A, B) five-segmented, consisting of fifth pedigerous somite, genital somite, and three abdominal somites. First somite of urosome trapezoidal in dorsal view. Genital double-somite slightly longer than wide, with lateral expansions in anterior part, tapered posteriorly; genital area located dorsolaterally, each area bearing one small seta. Anal opening (fig. 2A) near the middle of somite on dorsal side. Caudal ramus (fig. 2A-C) about 3.5 times longer than wide, armed with six setae, two outer terminal and one dorsal smooth, one innermost and two terminal setae plumose.

Antennule (fig. 3A) 12-segmented; first segment wide and long; third segment shortest; armature formula 12, 8, 1, 2, 4, 2, 2, 2, 2, 2 + aesthetasc, 6, 6; all setae naked.

Antenna (fig. 3B): coxa small, without ornamentation; basis elongated, with setules on surface. Exopod small, one-segmented, with three unequal setae. Endopod two-segmented; first segment with outer spinules and inner setules; second segment carrying four setae: one slender seta on lateral margin; three spinose setae on terminal margin.

Mandible (fig. 3C) slender, bearing row of spinules on subterminal part of inner margin; terminal margin with row of small teeth.

Maxillule (fig. 3D): inner lobe with row of small setules on inner margin and four pinnate terminal setae and one small seta; outer lobe shorter than inner lobe, with three setae, two pinnate and one spinose.

Maxilla (fig. 3E) two-segmented, composed of syncoxa and claw; syncoxa wide, bearing long seta near articulation; claw elongated, with spinules on surface, distal part ornamented with spinules on inner margin.

Maxilliped (fig. 3F) subchelate. Syncoxa bearing one plumose seta. Basis with one spinose seta on inner margin, ratio of length to width 2.4. Endopod three-segmented; first segment with one slender plumose seta; second segment with one unipinnate seta; third segment bearing single seta and claw, with a row of fine spinules on inner margin.

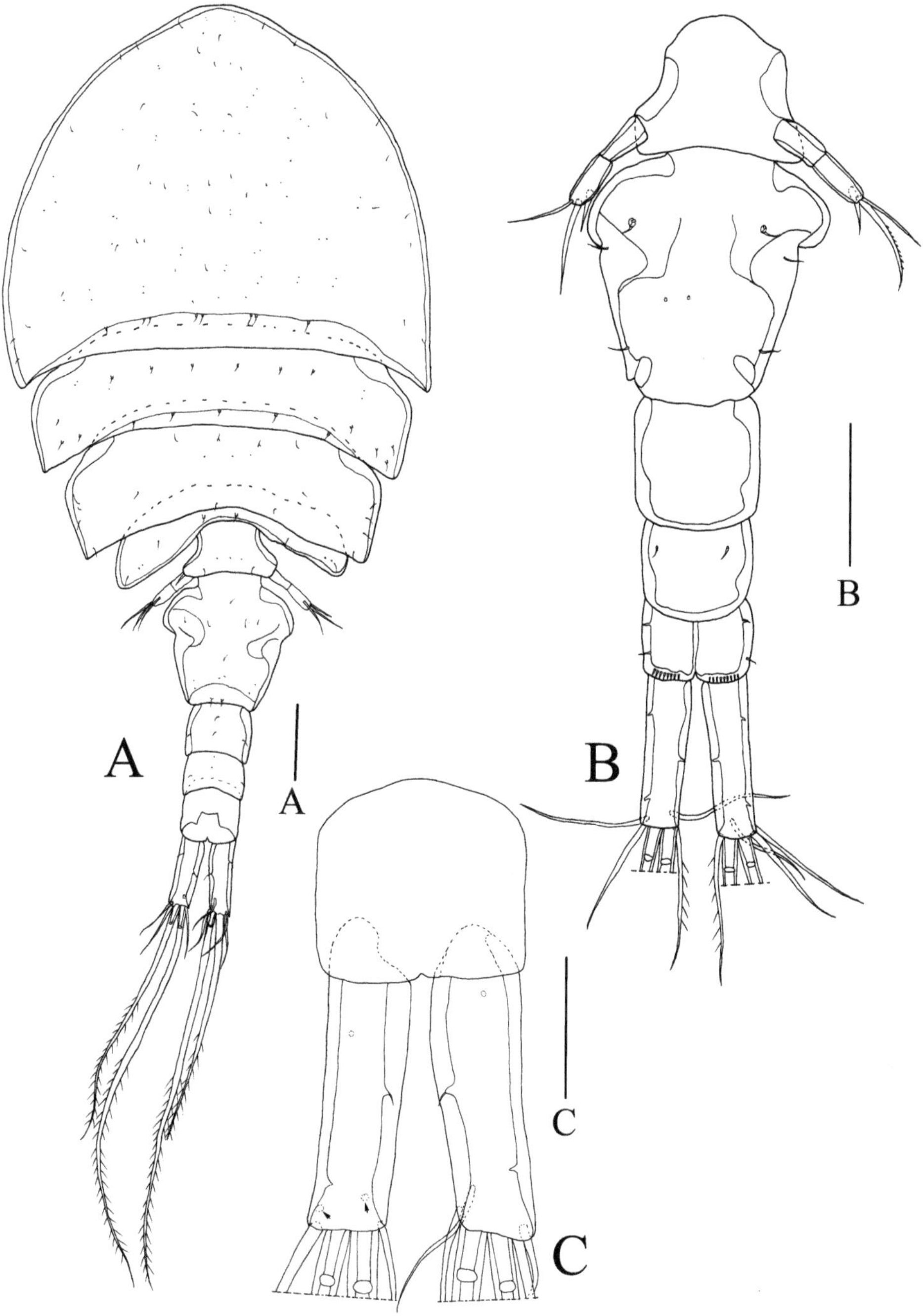

Fig. 2. *Stygiopontius liui* sp. nov., holotype. A, habitus, dorsal; paratype (MBM 189266); B, urosome, ventral; C, caudal ramus, ventral. Scale bar: A, B = 100 μm; C = 50 μm.

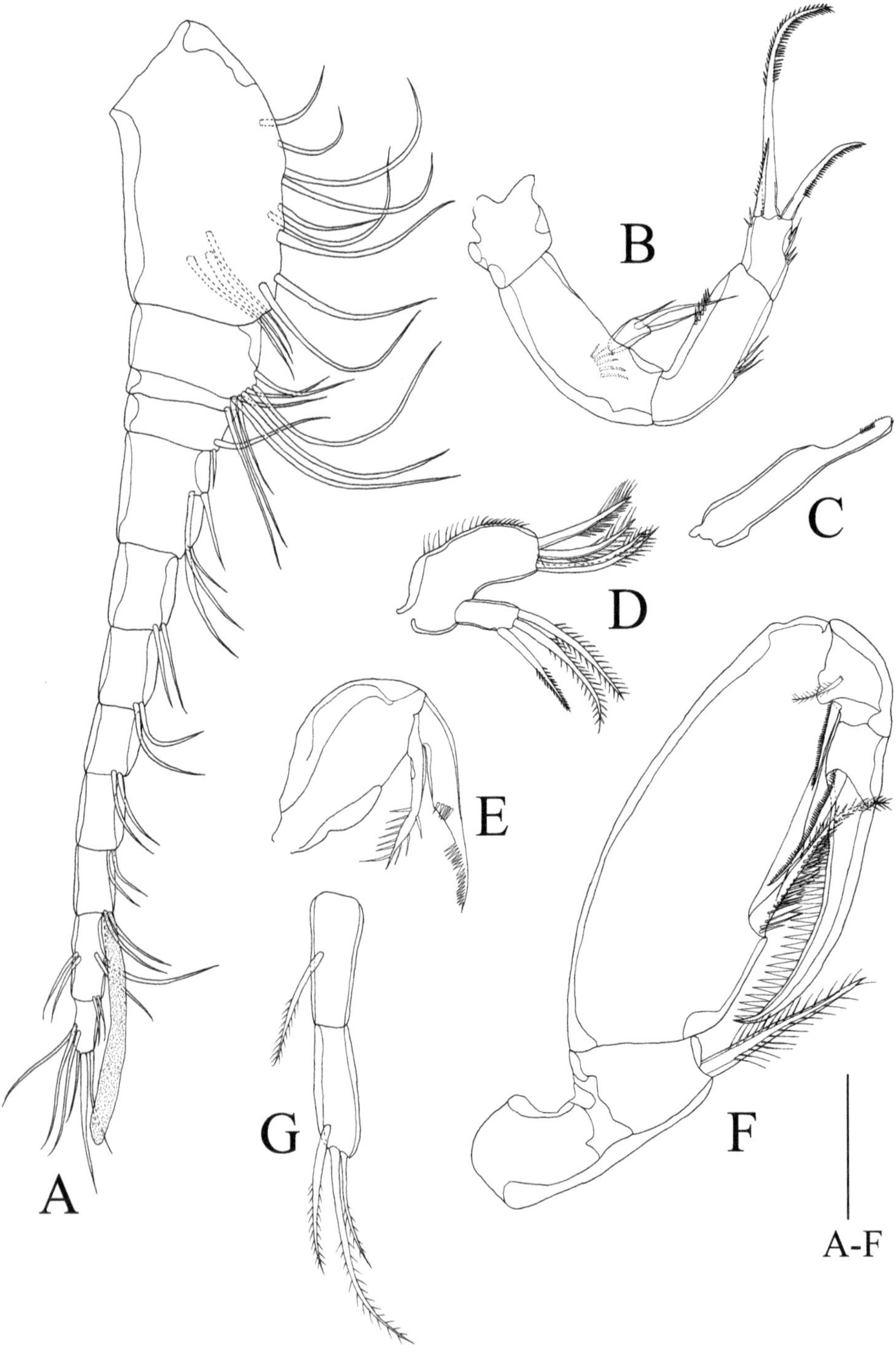

Fig. 3. *Stygiopontius liui* sp. nov., holotype. A, antennule; B, antenna; C, mandible; D, maxillule; E, maxilla; F, maxilliped; G, P5. Scale bar = 50 μm.

Swimming legs 1-4 (fig. 4A-D) with three-segmented rami, except for leg 4, with two-segmented endopod. Intercoxal sclerites without ornamention. Outer margin of endopod with row of setules, except first segment of leg 4. Formula for the armature of legs 1-4 in table I. Leg 1: inner seta of coxa plumose; basis with one plumose seta on outer margin, one stout spine on inner margin. Leg 2: inner seta of coxa and outer seta of basis slender and plumose. Leg 3: coxa without seta; basis with naked outer seta. Leg 4: coxa without seta; basis with plumose outer seta; distal segment of endopod pointed, carrying one inner seta and one serrated spine.

Leg 5 (fig. 3G) two-segmented; first segment with one plumose outer seta; second segment armed with three plumose setae of different lengths.

Leg 6 (fig. 2B) represented by one small seta.

Color of living specimens unknown.

Male: unknown.

Variability.— Most morphological features are conservative, except somite length and the length/width ratio of caudal ramus and outer edge of caudal ramus. The length/width ratio of the caudal ramus varies from 3.2 to 3.5. Outer edge of caudal ramus with one or two small processes.

REMARKS

Stygiopontius liui sp. nov. can be easily assigned to the genus *Stygiopontius* by a combination of the following characters: first segment of maxilla with inner sinuous seta; endopod of leg 1 three-segmented on both male and female; and leg 4 with two-segmented endopod, its first segment unarmed, second segment is armed with 1 terminal spine and 1 inner seta (Ivanenko & Ferrari, 2013).

The morphological distinction of dirivultid species is based on a number of subtle characters (Gollner et al., 2010). The new species belongs to a group characterized by the coxa of leg 1 with one inner seta and the third exopodal segment of leg 4 with three outer spines (Ivanenko & Ferrari, 2013). The group includes: *Stygiopontius brevispina* Humes, 1991; *S. flexus* Humes, 1987, male unknown; *S. hispidulus* Humes, 1987, male unknown; *S. lauensis* Humes, 1991; *S. mirus* Humes, 1996, female unknown; *S. pectinatus* Humes, 1987, male unknown; *S. quadripaxillifer* J. Lee, D. Kim & I. H. Kim, 2020; *S. senckenbergi* Ivanenko & Ferrari, 2013, female unknown; and *S. sentifer* Humes, 1987, male unknown (Ivanenko & Ferrari, 2013).

Within this group, the new species differs from *S. brevispina*, *S. lauensis*, *S. hispidulus*, and *S. flexus* by the following differential characters: endopodal claw of maxilliped stout, with row of fine spinules on inner margin; inner spine of leg 1

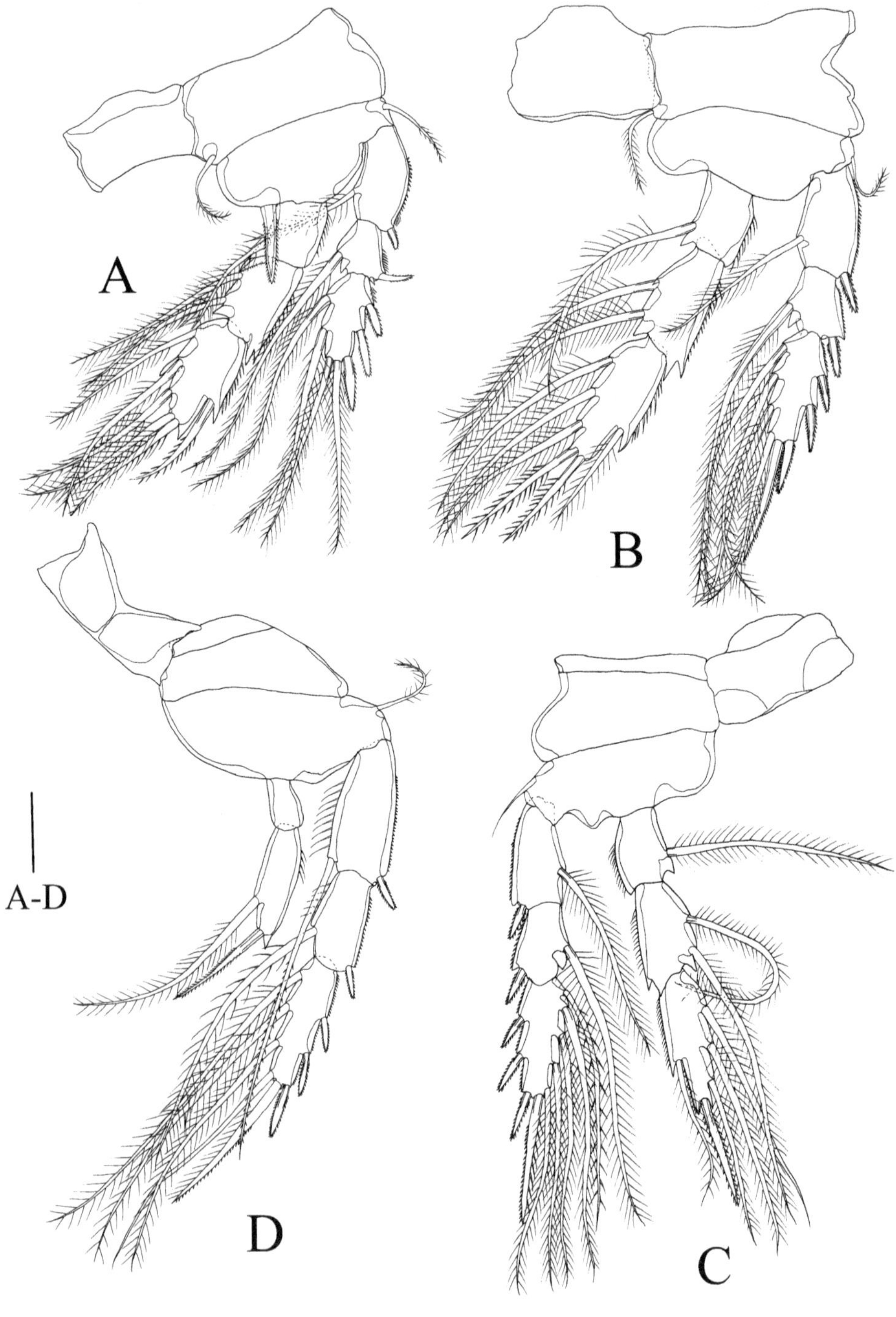

Fig. 4. *Stygiopontius liui* sp. nov., holotype. A, P1, anterior; B, P2, anterior; C, P3, anterior; D, P4, anterior. Scale bar = 50 μm.

TABLE I
Stygiopontius liui sp. nov., holotype female, armature formulae of the swimming legs 1-4

	Coxa	Basis	Exopod	Endopod
Leg 1	0-1	1-I	I-1; I-1; III,4	0-1; 0-2; 1,2,3
Leg 2	0-1	1-0	I-1; I-1; III, I,4	0-1; 0-2; 1,2,3
Leg 3	0-0	1-0	I-1; I-1; III, I,5	0-1; 0-2; 1,I,3
Leg 4	0-0	1-0	I-1; I-1; III, I,4	0-0; I-1

Roman numerals indicate spines, arabic numerals indicate setae.

basis strong, exceeded to the end of the first endopodal segment; second endopodal segment of leg 4 pointed. Additionally, *S. liui* differs from *S. flexus* by the distal endopodal segment of leg 3 with terminal spine instead of seta as in *S. flexus*. *S. liui* can be distinguished from *S. pectinatus* by the following features: terminal setae of the second segment on antennal endopod is not claw-like, the longest terminal seta is about three times as long as the second segment of the endopod; (two terminal setae on the second antenna claw-like and longest terminal seta is about 1.5 times as long as the second segment of the endopod in *S. pectinatus*); second endopodal segment of leg 4 pointed (not pointed in *S. pectinatus*). *S. liui* can be easily differentiated from *S. quadripaxillifer* by the presence of a seta on the coxae of leg 1 and 2, two inner stae on the second endopodal segment of leg 1. *S. liui* differs from *S. sentifer* in having the basis of maxilliped with a plumose seta (bearing one large thornlike, finely spinulose inner seta in *S. sentifer*).

For the species *Stygiopontius senckenbergi* and *S. mirus*, the females are unknown. However, in the genus *Stygiopontius* and related genera, certain features seem to be constant in both sexes, including the shape and armature of the second antenna, mandible, maxillula and maxilla, the formula for the armature of the exopod and endopod of leg 3 and 4 (Humes, 1987, 1991, 1996; Ivanenko et al., 2006; Ivanenko et al., 2011; Uyeno et al., 2018; Lee et al., 2020). *S. liui* differs from *S. mirus* by having the second endopodal segment of leg 4 pointed (not pointed in *S. mirus*). *S. liui* has a similar armature formulae of swimming legs 1-4 as *S. senckenbergi*. These two species differ in the following characters: exopod of antenna about 1.7 times as long as wide in *S. liui* (3 times as long as wide in *S. senckenbergi*), basis of antenna without inner setules in *S. liui* (with row of setules in *S. senckenbergi*), second endopodal segment of antenna with several spinules on inner and outer margins in *S. liui* (with row of spinules on surface in *S. senckenbergi*).

Before this study, all known species of the genus *Stygiopontius* were endemics of deep-sea hydrothermal vents (Humes, 1987, 1989, 1990, 1991, 1996, 1997; Ivanenko et al., 2006; Ivanenko et al., 2011; Ivanenko & Ferrari, 2013; Uyeno et al., 2018; Lee et al., 2020). *Stygiopontius liui* sp. nov. is the first record of a *Stygiopontius* species from a cold seep. Some crustacean species inhabiting hydrothermal vents are also found in cold seeps, e.g., *Alvinocaris longirostris* Kikuchi & Ohta, 1995 (Li, 2015) and *Shinkaia crosnieri* Baba & Williams, 1998 (Yang et al., 2016). These species were co-distributed in the hydrothermal vent of the Okinawa Trough and the cold seep of this study. In this study, *Stygiopontius liui* sp. nov. was found associated with *S. crosnieri*. Further investigations in the hydrothermal vent of the Okinawa Trough should demonstrate if *Stygiopontius liui* has a similar distribution as its host or not.

For the new species *Stygiopontius liui* sp. nov., only females were found. The phenomenon that only females or males are known is common in certain dirivultid species (including the genus *Stygiopontius*) (Tsurumi et al., 2003; Senokuchi et al., 2020). Among the 27 valid species of *Stygiopontius* at present, only 11 species were described based on both female and male specimens (Lee et al., 2020; Senokuchi et al., 2020). Species of the genus *Stygiopontius* show a different sex bias, e.g., *Stygiopontius quadrispinosus* was female biased (Tsurumi et al., 2003) and *Stygiopontius senokuchiae* was male biased (Senokuchi et al., 2020). Based on the present material of *Stygiopontius liui* sp. nov., it is not possible to tell what the sex bias of the species is. More samples and sufficient environment data will be needed to determine the sex bias and provide an explanation for this phenomenon.

ACKNOWLEDGEMENTS

This study was supported by the Senior User Project of RV KEXUE, the China Ocean Mineral Resources Research and Development Association Program (No. DY135-E2-1-02) and the National Natural Science Foundation of China (No. 41206148). Many thanks are due to Dr. Xu Zheng (Institute of Oceanology, Chinese Academy of Sciences, Qingdao, China) for providing the samples for the study, and Dr. Charles H.J.M. Fransen (Naturalis Biodiversity Center, Leiden, The Netherlands), for adjusting the language.

REFERENCES

Cuvelier, D., J. Beesau, V. N. Ivanenko, D. Zeppilli, P. M. Sarradin & J. Sarrazin, 2014. First insights into macro- and meiofaunal colonisation patterns on paired wood/slate substrata at Atlantic deep-sea hydrothermal vents. Deep Sea Research Part I: Oceanographic Research Papers, **87**: 70-81. DOI:10.1016/j.dsr.2014.02.008.

Gollner, S., V. N. Ivanenko, P. Martínez Arbizu & M. Bright, 2010. Advances in taxonomy, ecology, and biogeography of Dirivultidae (Copepoda) associated with chemosynthetic environments in the deep sea. PLoS ONE, **5**(8): e9801. DOI:10.1371/journal.pone.0009801.

GOLLNER, S., H. STUCKAS, T. C. KIHARA, S. LAURENT, S. KODAMI & P. MARTÍNEZ ARBIZU, 2016. Mitochondrial DNA analyses indicate high diversity, expansive population growth and high genetic connectivity of vent copepods (Dirivultidae) across different oceans. PLoS ONE, **11**(10): e0163776. DOI:10.1371/journal.pone.0163776.

HUI, M., J. CHENG & Z. SHA, 2018. Adaptation to the deep-sea hydrothermal vents and cold seeps: insights from the transcriptomes of *Alvinocaris longirostris* in both environments. Deep-Sea Research Part I, **135**: 23-33. DOI:10.1016/j.dsr.2018.03.014.

HUMES, A. G., 1987. Copepoda from deep-sea hydrothermal vents. Bulletin of Marine Science, **41**(3): 645-788. DOI:10.1029/JC092iC12p13185.

HUMES, A. G., 1988a. *Bythocheres prominulus*, new genus, new species (Copepoda: Siphonostomatoida) from deep-water cold seeps at the West Florida Escarpment. Proceedings of the Biological Society of Washington, **101**: 568-575.

HUMES, A. G., 1988b. Copepods from deep-sea hydrothermal vents and cold seeps. In: G. A. BOXSHALL & H. K. SCHMINKE (eds.), Biology of copepods. Proceedings of the Third International Conference on Copepoda, Hydrobiologia, **167/168**: 549-554. DOI:10.1007/BF00026351.

HUMES, A. G., 1989. New species of *Stygiopontius* (Copepoda, Siphonostomatoida) from a deep-sea hydrothermal vent at the East Pacific Rise. Zoologica Scripta, **18**(1): 103-113. DOI:10.1111/j.1463-6409.1989.tb00125.x.

HUMES, A. G., 1990. Copepods (Siphonostomatoida) from a deep-sea hydrothermal vent at the Mariana Back-Arc Basin in the Pacific, including a new genus and species. Journal of Natural History, **24**(2): 289-304. DOI:10.1080/00222939000770211.

HUMES, A. G., 1991. Siphonostomatoid copepods from a deep-sea hydrothermal zone in the Lau Basin, South Pacific. Bulletin du Muséum national d'Histoire naturelle, Section A, Zoologie, biologie et écologie animals, **13**(1): 121-134.

HUMES, A. G., 1996. Deep-sea copepod (Siphonostomatoida) from hydrothermal sites on the Mid-Atlantic Ridge at 23° and 37°N. Bulletin of Marine Science, **58**(3): 609-653.

HUMES, A. G., 1997. Siphonostomatoid copepods from deep-sea hydrothermal sites on the Mid-Atlantic Ridge west of the Azores. Cahiers de Biologie Marine, **38**(2): 63-77.

HUMES, A. G. & R. U. GOODING, 1964. A method for studying the external anatomy of copepods. Crustaceana, **6**: 238-240. DOI:10.1163/156854064x00650.

IVANENKO, V. N., 1998. Deep-sea hydrothermal vent Copepoda (Siphonostomatoida, Dirivultidae) in plankton over the Mid-Atlantic Ridge (29°N), morphology of their copepodid stage. Zoologicheskii Zhurnal, **77**(11): 1249-1256. [In Russian with English abstract.]

IVANENKO, V. N. & D. DEFAYE, 2004. A new deep-sea genus and species of the family Asterocheridae (Copepoda, Siphonostomatoida) from the continental margin off Angola (east equatorial Atlantic). Crustaceana, **77**(9): 1131-1144. DOI:10.2307/20107420.

IVANENKO, V. N., D. DEFAYE, M. SEGONZAC, A. KHRIPOUNOFF, J. SARRAZIN & F. D. FERRARI, 2011. A new species of *Exrima*, synonymy of four species of *Aphotopontius*, *Stygiopontius* and *Rhogobius*, and record of first copepodid stage of Dirivultidae (Copepoda: Siphonostomatoida) from deep-sea hydrothermal vents of the East Pacific Rise (13°N). Journal of the Marine Biological Association of the United Kingdom, **91**(7): 1547-1559. DOI:10.1017/S0025315411000178.

IVANENKO, V. N. & F. D. FERRARI, 2013. New species of *Stygiopontius* (Copepoda: Siphonostomatoida: Dirivultidae) from a deep-sea hydrothermal volcano in the New Ireland Fore-Arc system (Papua New Guinea). Journal of the Marine Biological Association of the United Kingdom, **93**(7): 1805-1812. DOI:10.1017/s0025315413000763.

IVANENKO, V. N., P. MARTÍNEZ ARBIZU & J. STECHER, 2006. Copepods of the family Dirivultidae (Siphonostomatoida) from deep-sea hydrothermal vent fields on the Mid-Atlantic Ridge at 14°N and 5°S. Zootaxa, **1277**: 1-21. DOI:10.5281/zenodo.173351.

IVANENKO, V. N., P. MARTÍNEZ ARBIZU & J. STECHER, 2007. Lecithotrophic nauplius of the family Dirivultidae (Copepoda; Siphonostomatoida) hatched on board over the Mid-Atlantic Ridge (5°S). Marine Ecology, **28**: 49-53. DOI:10.1111/j.1439-0485.2007.00142.x.

LEE, J., D. KIM & I.-H. KIM, 2020. Copepoda (Siphonostomatoida: Dirivultidae) from Hydrothermal Vent Fields on the Central Indian Ridge, Indian Ocean. Zootaxa, **4759**(3): 301-337. DOI:10.11646/zootaxa.4759.3.1.

LI, X., 2015. Report on two deep-water caridean shrimp species (Crustacea: Decapoda: Caridea: Alvinocarididae, Acanthephyridae) from the northeastern South China Sea. Zootaxa, **3911**(1): 130-138. DOI:10.11646/zootaxa.3911.1.8.

MAHATMA, R., P. MARTÍNEZ ARBIZU & V. N. IVANENKO, 2008. A new genus and species of Bychiopontiidae Humes, 1974 (Crustacea: Copepoda: Siphonostomatoida) associated with an abyssal holothurian in the Northeast Pacific nodule province. Zootaxa, **1866**: 290-302. DOI:10.11646/zootaxa.1866.1.14.

SENOKUCHI, R., H. NOMAKI, D. UYENO, H. K. WATANABE, T. KITAHASHI & M. SHIMANAGA, 2020. Sex ratio of *Stygiopontius senokuchiae* (Dirivultidae, Copepoda), an endemic copepod species at deep hydrothermal vent sites, is biased to males. Journal of Oceanography. DOI:10.1007/s10872-020-00548-4.

SHEN, Y., Q. KOU, W. CHEN, S. HE, M. YANG, X. LI & X. GAN, 2016. Comparative population structure of two dominant species, *Shinkaia crosnieri* (Munidopsidae: *Shinkaia*) and *Bathymodiolus platifrons* (Mytilidae: *Bathymodiolus*), inhabiting both deep-sea vent and cold seep inferred from mitochondrial multi-genes. Ecology and Evolution, **6**(11): 3571-3582. DOI:10.1002/ece3.2132.

TSURUMI, M., R. C. DE GRAAF & V. TUNNICLIFFE, 2003. Distributional and biological aspects of copepods at hydrothermal vents on the Juan de Fuca Ridge, north-east Pacific Ocean. Journal of the Marine Biological Association of the United Kingdom, **83**: 469-477. DOI:10.1017/s0025315403007367h.

UYENO, D., H. K. WATANABE & M. SHIMANAGA, 2018. A new dirivultid copepod (Siphonostomatoida) from hydrothermal vent field of the Izu-Bonin Arc in the North Pacific Ocean. Zootaxa, **4415**(2): 381-389. DOI:10.11646/zootaxa.4415.2.8.

WANG, B., Z. LUAN, X. ZHANG, Z. DU, S. XI, L. LI, C. LIAN, Z. LIANG & J. YAN, 2018. Surficial features and distribution patterns of the cold seep system on Formosa ridge. Marine Geology Frontiers, **34**(6): 1-7.

YANG, C. H., S. TSUCHIDA, K. FUJIKURA, Y. FUJIKURA, M. KAWATO & T. Y. CHAN, 2016. Connectivity of the squat lobsters *Shinkaia crosnieri* (Crustacea: Decapoda: Galatheidae) between cold seep and hydrothermal vent habitats. Bulletin of Marine Science, **92**(1): 17-31. DOI:10.5343/bms.2015.1031.

First received 8 April 2020.
Final version accepted 28 July 2020.

TAXONOMIC INDEX

www.ingramcontent.com/pod-product-compliance
Lightning Source LLC
Chambersburg PA
CBHW071229230926
51849CB00001B/63
9789004464339